Arups on Engineering

Edited by David Dunster

Arups on Engineering

Ernst & Sohn

Contents

Ove Arup & Partners
13 Fitzroy Street
London W1P 6PQ

Ernst & Sohn
Verlag für Architektur und technische Wissenschaften GmbH, Berlin
Ernst & Sohn is a member of the VCH Publishing Group.

ISBN 3-433-02637-8

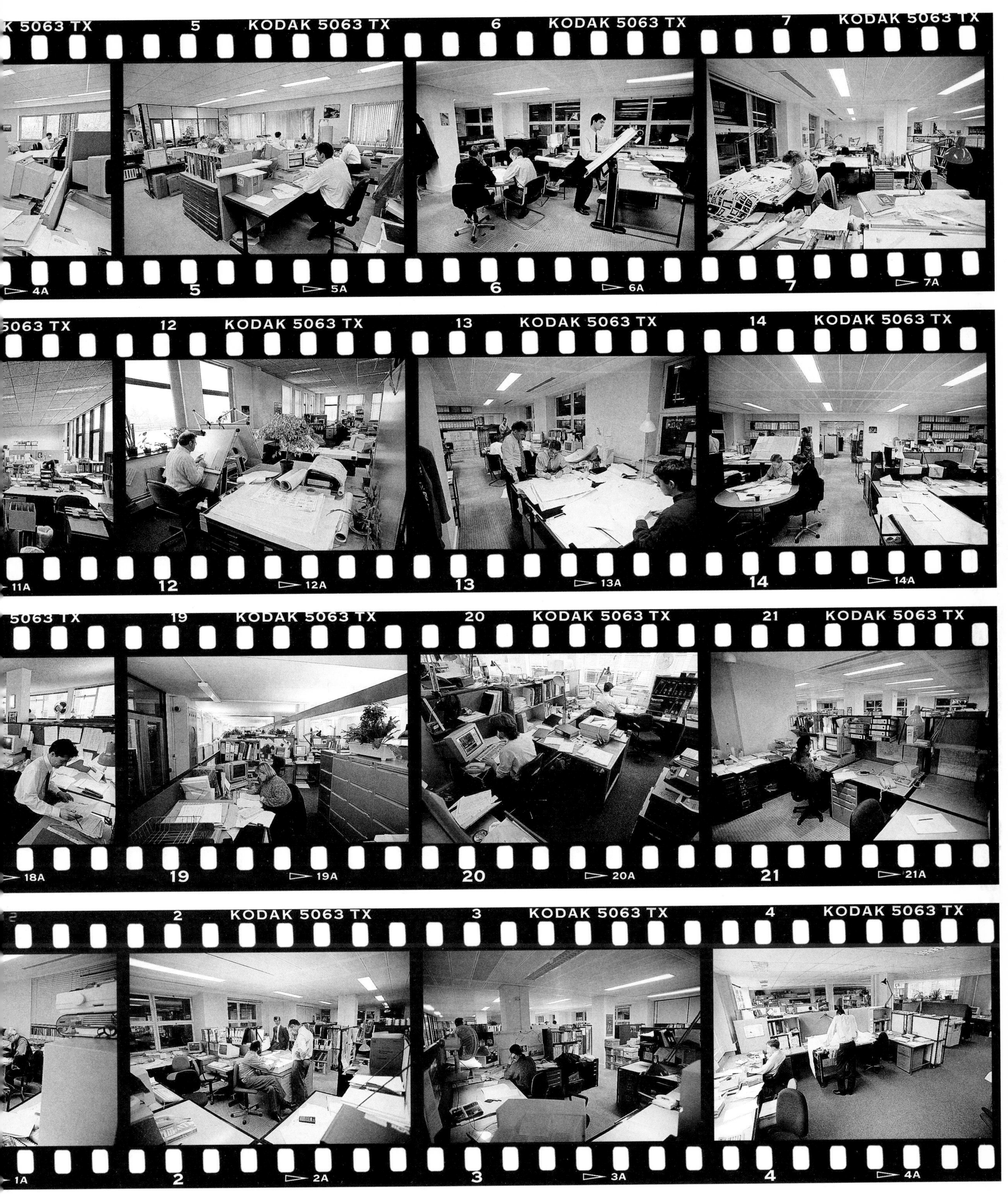
K 5063 TX
5
KODAK 5063 TX
6
KODAK 5063 TX
7
KODAK 5063 TX
4A
5
5A
6
6A
7
7A
5063 TX
12
KODAK 5063 TX
13
KODAK 5063 TX
14
KODAK 5063 TX
11A
12
12A
13
13A
14
14A
5063 TX
19
KODAK 5063 TX
20
KODAK 5063 TX
21
KODAK 5063 TX
18A
19
19A
20
20A
21
21A
2
KODAK 5063 TX
3
KODAK 5063 TX
4
KODAK 5063 TX
1A
2
2A
3
3A
4
4A

Duncan Michael

The first words

This book is about engineering design. It celebrates the 50th anniversary of the founding of Ove Arup & Partners. It includes some 30 essays written by people in Arups on their specialist topics. They are busy practitioners, mostly engineers. Their articles vary in approach – project analyses, personal emphasis – sometimes introspective, reaching past the immediate issues into a general commentary on the state of engineering. The juxtapositions illuminate the whole.

These essays are supported by substantial analyses by authors from outside the Arup organisation. David Dunster, the co-ordinating editor, opens the book with a view of our work and habits which tends to the anthropological. It has provided the theme for the unity of the book. Robert Thorne, the historian of construction, reviews the 50 years. He emphasises the inspirational quality of Ove the man and how it was possible to create the intellectual foundation for our expansion. The necessarily oppportunistic beginnings in structural engineering for buildings rapidly evolved into 'total architecture' or holistic design and construction. The activities which expressed this concept attracted an extraordinary community of talented and eminently curious people to create and then realise its many incarnations. These include multi-disciplinary engineering (which will continue to spawn many specialist practices), and the prime agent role in what we describe as Civil Engineering and Industrial Engineering. It had a very special flowering in what Americans so accurately call the A/E business. Thorne's crucial analysis points to subsequent developments but effectively culminates in – and brings out the extraordinary significance of – Ove's 'Key Speech', which articulated many of the principles that developed in the firm and was written appropriately at midpoint in these 50 years. The document itself is reproduced at the end of Thorne's essay. What now acts as an epitaph to the man is also the foundation for the firm and its future, the draft for its Constitution.

We should embrace the triple good luck that Ove Arup was an intelligent, articulate and caring man, that he had already proved himself when he set up Ove Arup & Partners at the age of 51 and attracted young talents, and that he then lived long enough to fix the direction of the firm so very precisely. This book celebrates that bequest expressed in the work and attitudes of the people who are the firm today, and given form in Ove's gently-written 'Key Speech'. Once written, that speech took on an inevitability, like Newton's Laws, to be transcended only when a finer or superior version reaches the table.

The last 50 years have seen our work dominated by various forms of 'horizontal integration' amongst designers, a view of integrated design based in teams of equals, overcoming the limitations of a class-sensitive society, and of the temporary coalitions which characterise so much in construction. This reflects one of the most positive developments around the world in recent years: the increase in highly creative and professional clients. We now see a large number of clients, developers, users, owners, and occupiers who are very knowledgeable about what could be possible, and demand a service accordingly. They make the Arup mission worthwhile.

The other notable phenomenon of the last decade in international construction has been the move to explicit and auditable systems – quality assurance, improved demands for health and safety, environmental impact, and more. It has extended the dimensions of what the engineer has to command. As such, it has proved an ambiguous change, with internal motivation being replaced by external incentive so that trust and human relationships are given no place in the paradigm. It has helped to create more consistent operations on a global basis, to exploit the new electronic technologies, but it also changes the game by prescribing the game. In an organisation which is simply the people and their shared ways of working together, it makes us very conscious of the extent to which Ove could assume and rely on the power of informal systems via intelligent well-trained people of good heart. We now have to learn how to recapture those qualities in a very different world, both inside Arups and in our fields of play.

What of the next 50 years? Our rules are already written so that we will make many changes, shifting on the basis of the past. Our service to society will recognisably have the genes of the first 50 years. Tolstoy said it all in *War and Peace*. He took 1000 pages to make the point that Russia is the subject of his love letter. The words may change, the faces will change, the markets, the projects, and the address will certainly change. We will enjoy the journey.

David Dunster

The firm and this book

As a result of continually working through the founding idea of total design, Ove Arup & Partners exemplify a new development in professionalism. As editor, I make this claim having been delighted by the responses from those in the firm to write about what interests them, and fascinated by the generous historical appreciation written by Robert Thorne. In engineering there seems to be an obsession with what engineering has achieved and how. From the exploded isometric of children's encyclopaedias to the more esoteric argument about whether history is driven by technology or by society, the public has seemed happy to accept a figure of the engineer as diligent capable and professional. But more recently, as the products of progress and indeed the very idea of progress have been questioned, engineering has had to take some responsibility for...pollution, structural unemployment, and a sense of powerlessness in the face of forces which appear out of control. The professional can no longer rely upon the halo of specialised knowledege. One ambition of this book is therefore to prise apart the professional curtains drawn around what engineers do, and offer some insight into how engineers work, and what they think about when they are working.[1]

Four main factors seem to point to this new future for the profession. The first has to do with the scientific approach to design whereby codes and tradition are continually questioned. Research and intuition both inform this approach, which while neither unique to engineering nor to Ove Arup & Partners seems nevertheless to be epitomised by the firm. The second factor derives from the collaborative nature of the work which is undertaken whether as lead consultant or not. Though engineering will always have heroes, and heroines, Arups have been singular in their interest to work with other professionals. The third factor is a self-awareness of the organisation of the firm, which I later suggest has more the characteristics of a tribe than a set of fiefdoms. This self-awareness is promoted by the structure of the firm, its co-operative basis and by the central services of which the most public output is the quarterly *Arup Journal*. As well as this, the work of the Ove Arup Foundation supports the idea that Arups recognises itself as a learning society.

The fourth dimension comes from Ove Arup's belief, articulated in the 'Key Speech' and elsewhere, that engineers should act in accordance with an ethics – not a code of professional ethics governing business behaviour but ethical standards which run continuously through life, at work and at home. From this emphasis on ethics, the gritty world of engineering so often hidden beneath a glossier world of buildings, superstructures, infrastructures, and services, no longer needs to hide its real achievements. My interpretation of Sir Ove's influence depends upon this ethical dimension, the belief that what engineers do must be informed by what they believe in.

From the outset I was also convinced that this should not be a book solely about the final products. To identify the work of the firm with specific projects would make architecture appear to dominate within the firm and this was clearly inaccurate. From the 1960s onwards, in many fields which were new for Arups, engineers were working on complex problems, taking the leading role in contracts, and yet still mindful of the moral dimension stemming from the founder of the firm they worked inside. The picture which emerges from these pages is neither a portrait nor a still-life, but more like a slice of time frozen on an endless bank of monitors, a series of individual screens which combine into a composite and volatile totality. The limiting factor on this metaphor is of course the size of the screen, the grid which frames the changing images. Working away inside the metaphor, I came to see that the 'Key Speech' acted as a document about the conditions under which the work of the firm was possible, and I began to characterise the writers as a tribe, and I will explain why later. But first it became apparent to me that a definition or two might be necessary, especially if there might be anything other than a provisional answer to the question: How do engineers perceive? Prior to this question is one defining the word 'engineer'.

What engineers do may appear inevitable, irresistible and fearsome: Inevitable not in the sense in which Martin Luther claimed he could do no other (than to found a new religion) but inevitable in the sense that common sense dictates such-and-such a particular outcome. In that same sense also impossible to resist, again not in the way that la Monroe was, but positively as the force which meets an object moves it (perhaps that is more like the same than we might choose to believe?). And then fearsome because of the sheer size of the Ravenspurn offshore platform, the bridge and link between Denmark and Sweden, the acres of poisoned land that need cleaning up, and the awesome power of computer simulation and modelling developed in Fitzroy Street and elsewhere.

An engineer solves problems by reducing problems of large scale into their smaller constituent parts. This approach gives engineers their authority with the public for whom large-scale problems appear intractable. This approach also makes difficulties for any desire for wholeness, totality or total design.

Yet there is clearly a demand for such a skill from the construction and manufacturing industries with which Arups ordinarily has dealings. The problems faced are project-led, not process oriented, and the role of the engineer is to face the holistic demand from the perspective of fragmenting the big problems into smaller parts. This tension, between the process of fragmenting and the product which must be complete, is not unfamiliar to the other arts, though the sciences and humanities can most often get away with provisional, that is still incomplete, solutions. A half-finished painting is as much use as a bridge which spans part of a river.

One way of defining what engineers are is to say what they do. This locks them into a tradition which changes slowly; they are defined always by what they have done and not by the work they might undertake. If however an engineer is defined in the context of the work done by others, then he, or more frequently now she, can be defined as not an artisan. A recent work argues that the engineer uses drawings, or representations and hence abstraction,[2] while the artisan uses experience without abstraction. While this has the attraction of a simple difference easy to understand – a plumber is not a services engineer – it fails to encompass the aspects of practice as an engineer which are close to science. Or again, technologist is a term lightly bandied about which also suggests someone operating narrowly within a particularly specialised field with no obligation beyond that field. In Arup's writings, the engineer appears to be like Renaissance Man, literate and numerate, aware of culture and science with a keen eye on the books. That part of his 'Key Speech' in which he talks of the pleasure of engineering should surely also be included, as should the idea that the whole firm is a learning organisation. Primarily though, the engineer is someone who works in teams, and brings to that group a specialised expertise and a general experience of working within large projects. Between the grand theories of society, science, and knowledge, the engineer should surely be the operator who aims to arrive at the best, most economical, and most advantageous solution when all the facts are known.

As with most definitions, this defines people and their actions, but not the context within which they work. There seems to be no *a priori* management model underlying the current form of Arups, and the managerial organisation which exists seeks an organic relationship to the processes of work in the practice. To say this fails to mark out Arups from any other kind of organisation which faces change on a daily basis. Not only other engineering firms would claim this, but so would other institutions, universities for example. Within these there has to be some kind of glue which acts to cohere and bring into creative working relationships the high levels of intelligence necessary. I want to suggest that this glue is a kind of myth.

While myths used to be considered the property of under-developed societies, particularly those studied by anthropologists, it has been argued that myth is an essential part of any society.[3] A myth surrounds the black arts of management, while the dream of transparency, which is so basic to the project of Modernism, when carried to a logical conclusion would abolish all myths in the clear enlightenment day of the future.[4] However, this would imply that thought is utterly transparent to consciousness and would therefore also deny that intuition has any role to play, that we might then lose the capacity to dream, and that little of what we now call work would be left that could not be carried out more effectively by machines.

In a very simplistic way, it could be argued that the basic characteristic of myths is that a pantheon needs to be established. Two of the Gods call them Architecture and Engineering are so mysterious that no-one is sure that they have ever been seen, though they have left behind strange objects which need to be interpreted, worshipped and propitiated. A God-like manifestation did exist though who, though no longer physically present, has left a number of gnomic texts, sufficiently ambiguously worded to warrant various equally satisfying interpretations each of which can cause endless discussion around ritual luncheon tables. These texts and the ensuing discussions serve as a compass within territory not predicted by them.

From time to time the tribe re-arranges where it lives and how it reproduces its means of existence. This can be both as the result of internal pressures, the fertility or barrenness of certain fields for example; and it can be brought on through external pressures which are generally unpredictable through the tests of risk analysis but which must be engaged with for economic and ethical reasons. While the tribe has a liberal view of the division of labour between genders, there exists an organisation beneath the overt power structures which performs prodigious feats of co-ordination and is dominated by the womenfolk. Its purpose is to permit the overt power structure to imagine that it really knows what is going on and acts as a deep structure to the language of the firm.

The wisdom of the tribe resides with the elders who meet around a table or board to consider the rites de passage from initiate through group to membership of the wise. While the elders are all individuals and are not in place because they represent certain aspects of the work of the tribe, nonetheless the elders are as one in advocating the importance of group work.

This not entirely serious analogy allows certain important points to be made about Arups. Firstly that there is a structure to the belief in what Arups do, secondly that there is an order to the organisation and thirdly that this order contains the power to reproduce itself. Of all this, the last point is perhaps the one which is most difficult to isolate from the essays which this book contains. There are two major pointers to answer the question – how does Arups maintain its pre-eminence. The first I believe is the belief in team work, the second the idea that this firm is a learning organisation.

One consistently expressed myth about the 20th century is that creativity is an individual matter. It has been pointed out elsewhere that a mastering myth of our times concerns the primacy of the individual in all matters of invention.[5] Yet many accounts given here of the process of engineering design indicate quite clearly that the dominant idea for a project emerged in discussion, and that subsequently no-one is quite sure who said it first. The conditions for team-work are also important – that no-one member feels subservient and that anything suggested will be treated with equal seriousness or equal irreverence. A further and almost more natural way in which a team can work depends upon the exhaustion of all possible alternatives so that the team feels that it has done its work because it has surveyed the field of possible solutions. Creativity therefore does not sit inside one engineer's skull; creativity, like intuition, is a term used to describe a characteristic which we can all agree upon but cannot adequately define except by anecdote or example.

In essay after essay, the codes and enshrined practices of construction are questioned. The opening gambit of the Arup engineer seems to always begin with a 'Yes...but...'. The codes and practices covering engineering have to do with the world as it exists, and cover normative engineering. Much of Arups' work is not especially innovative or operating on the edge of the codes. But equally there is a large amount of work which does, and this was perhaps the original source of the diversification of Arups into, for example, materials science. But as Steven Groák persuasively argues, even the normative or self-styled traditional techniques are failing in surprising ways, leading to a greater role for research and development and, *a fortiori*, leading to a view of all work as containing some aspect that can be thought of as R and D. Innovation exists not as something applied to the problem, a quantity which any client can buy more or less of, but becomes instead a mode of practice.

Through such ideas the concept of the team becomes more elastic, covering not just those who work directly on a job but also those who may have parallel experience in other groups and those in the specialised teams, like R and D who may have something to contribute. Conventional management logic would suggest that in organisations of the size of Arups, equal to the number of staff at a major research university in America, the principal problem is communication. Through the splendid issues of *The Arup Journal* the problem is partly addressed, but probably of more importance is the network of contacts of an informal nature which derive from the sense of loyalty to the firm and its objectives. Over this sit the Board and to one side the Ove Arup Foundation.

Sir Jack Zunz, Chair of the Foundation, explains in his essay how Arups has become a learning organisation. In this respect, Arups can claim to be unique amongst professional organisations who work within the engineering and construction industries. The component parts of this organisation have already taken the form of explicit specialisms which carry a subject forward in some depth – acoustics and geotechnics are only two examples. But emerging within the firm are specialisms which exist at a managerial, economic or policy planning level, and these emerging groupuscules cross the subject boundaries of the specialisms which previously emerged. In this cross-matrix of subjects, techniques, knowledges, and practices the firm begins to mirror the complexity of a theory of knowledge, the single most perplexing arena of academic thought that can be predicted to emerge in the next century. As a learning organisation the fundamental issues will emerge not through some *a priori* model brought in from elsewhere, but from inside the organisation. Rather than a tree growing from a seed, the organisation may be more like a tuber which reproduces itself through the rhizome.

Most vanity publishing trumpets the unique and special characteristics of the work of the practice which seeks to be different by treading a well-known publishing path. I hope that this book faces that problem in a slightly different fashion. By their nature, the essays talk about the successes and not the failures, but they do talk about the problems as well as the solutions. The Webster's *Collegiate Dictionary* defines an engineer as a person who makes engines, and prefaces that definition with '(Rare)'. Whether the metaphor for the firm is animal, vegetable or mineral, there is a remarkable record of an endeavour in engineering in these pages. And the most remarkable potential comes from the organisation, an engine designed by a rare engineer.

References

1. 'Shaping the 21st Century', The Ove Arup Centenary Seminar, 1995, is the proceedings of a one-day seminar during which architects and engineers looked at how they might better collaborate on matters relating to construction in the next century. As a result of chairing the organising committee responsible for this event, with Patrick Morreau of Ove Arup & Partners and Mark Whitby of Whitby & Bird, I gained a considerably different perspective on the attitudes of engineers, and I believe that the proceedings of this important event deserve wider dissemination.

2. Eugene S. Ferguson, *Engineering and the Mind's Eye*, (Cambridge Mass, 1992) pp.3-9 continues a tradition of medieval adulation for the artisan that was earlier advanced by George Sturt in 'The Wheelwright's Shop'.

3. Most notably in the works of Claude Lévi-Strauss, especially *The Savage Mind*, (London, 1972).

4. See Jürgen Habermas, *The Philosophical Discourse of Modernity*, (Cambridge, 1987), especially chapter 1, 'Modernity Consciousness of time and its need for Self-Reassurance'.

5. See for example, Thomas P. Hughes, *American Genesis*; Alan Trachtenberg, *The Incorporation of America*, (New York, 1982); Alfred Chandler, *The Visible Hand*, (Cambridge, Mass, 1977).

Design attitudes

Engineering is pragmatic, precise, and rational, but also fundamentally intuitive. Early in design, intuition selects solutions for analysis. During resolution, it tells us what models to use, and when enough analysis has been done. Computers aid the process and free intuition to focus on engineering's boundaries and interfaces.

Richard Hough

Intuition in engineering design

Designers' intuitions led to very different engineering concepts for structure and services distribution at the Centre Pompidou, Paris (above) and the Neue Staatsgalerie, Stuttgart (opposite).

Intuition

Intuition is the immediate apprehension by the mind, without reasoning. At the beginning of a design, it is how our mind first starts to select and reject from the infinite range of possibilities. It helps us to pass freely through the barriers we build with reason, habit, and language, and so to notice parallels, relationships, and analogies.

Intuition is also important during the resolution of the design. It is not a good defence when claims of errors or omissions are brought against us. And yet we could never bring our designs to completion without it. It tells us when enough analysis has been done, when enough detail has been examined, when enough checks have been carried out. We draw a line and trust our inner judgement.

All of this is incongruent with the commonly held view of engineers: that their methods are deterministic, their judgements absolute, and their product perfect. The idea that science and technology might be omniscient has held us in its thrall since the Age of Reason. Only recently has the pendulum started to swing. Ove Arup contributed to the search for a new balance by showing us how to find, or at least recognise, the human face of technology in our design. To this end, one of the binding themes through the work of the practice has been an acknowledgement of, and a cultivation of, the role of intuition.

Design conception

As engineers, we contribute to the design of things that make for a better life. Big things like roads and bridges, medium-sized things like buildings, and small things like manufactured products. At its best, our contribution not only works efficiently, but is fully integrated into the whole. It supports the overall intention so persuasively that the owner and the user not only benefit from it practically, but apprehend and enjoy it at more sublime levels as well. The Sydney Opera House roof doesn't just keep the rain out.

Our perceptions about the better life and how to get it, are subtle and complex, so the scope for the designer's intuition is enormous. Amongst the big things, designing a new route for a railway requires the modelling of whole ecosystems for environmental impact assessment. Models keep improving, but their realism is still beggared by the subtlety of the system itself. Some relationships have to be simplified, some input truncated. Environmental engineering has become the lively arena in which the intuition of civil engineers mixes with the pragmatism of politicians and decisions get made.

At the smaller scale, our contribution to impact and safety studies for the automobile industry is part of a jigsaw that includes things like styling, comfort, durability, body weight, and fuel efficiency. As with most engineered products, human perceptions vary as to the relative importance of the criteria, as well as how to meet any one of them. To find a well-integrated solution for such a product, on the basis of limited and subjective user feedback, requires engineering intuition.

In the engineering of architecture, technical solutions that meet human aspirations are also important to find, given the power that buildings have in our lives. What use of technology will be appropriate for, and supportive of, a desired experience of a building or a space within a building? Our perception of, and response to, the way a space is structured, and the materials it is made of, spring from deep sources. These sources were tapped by the designers of Gothic cathedrals, of Mayan temples, of Victorian railway stations. Our palette of materials is now richer, and our history longer, so the territory our intuition can explore is that much broader.

The sculpted stone-clad concrete structure of the Staatsgalerie in Stuttgart and the see-through steel frame at Centre Pompidou each supports a very different experience of the idea of a museum. The double-wall, double-roof air-distribution strategy at the former and the open lattice of ductwork at the latter each reinforces different architectural intentions. The structure, materials, and air movement strategies employed at the Menil Gallery in Houston are all carefully chosen to support a spatial experience quite different from the other two.

With an intuitive feel for materials, an engineer can make choices early, knowing the demands of the task, like strength, stiffness, durability, combustibility, toughness, formability, density, and economy, and the characteristics of the available materials. While these choices are testable, an intuitive designer who avoids backing into corners, saves an enormous amount of time and effort. A 'feel' for the brittleness of stone and glass underwrites the system, the geometry, and the detailing of the glasshouses at the Parc de la Villette Science Museum, Paris, and the stone arches of the Pavilion of the Future at Seville's Expo '92. In seismic regions, a 'feel' for the relative ductilities of both materials and structural systems allows a safe hierarchy of energy absorption to be built into a design before the process of structural analysis begins. In forging, pressing, casting, and extruding processes, an intuitive sense of a material's formability is still the first source of guidance in designing new product shapes, despite the growth of large-deformation non-linear stress analysis.

What are the sources of our empathy with engineering materials? By observing the world from early childhood, we make sense of force and movement, light and energy. An engineer who has been open to these lessons has a good intuition, and closes quickly and easily on the best physical solution to meet a set of constraints. When it clicks into position, he knows it to be right at many levels of his being, as do his colleagues, and the final users, who all share a common experience of the physical world. Ove Arup's early concrete work, of thin shells, sculpted beams, elegant footbridges, and precast assemblages, all had this 'knowingness' that outlasts fashion and continues to give enjoyment. An openness of eye, mind, and heart also allows us to leapfrog over pessimists, skirt obstacles, avoid distracting detail, and buoy our colleagues. The process is contagious: the group builds on itself. A kind of collective intuition comes into play. Boundaries disappear and integrated multi-disciplinary design results.

Imagination is an attribute of intuition: the ability to visualise solutions in our mind's eye. It is important for conceiving and developing three-dimensional systems, which of

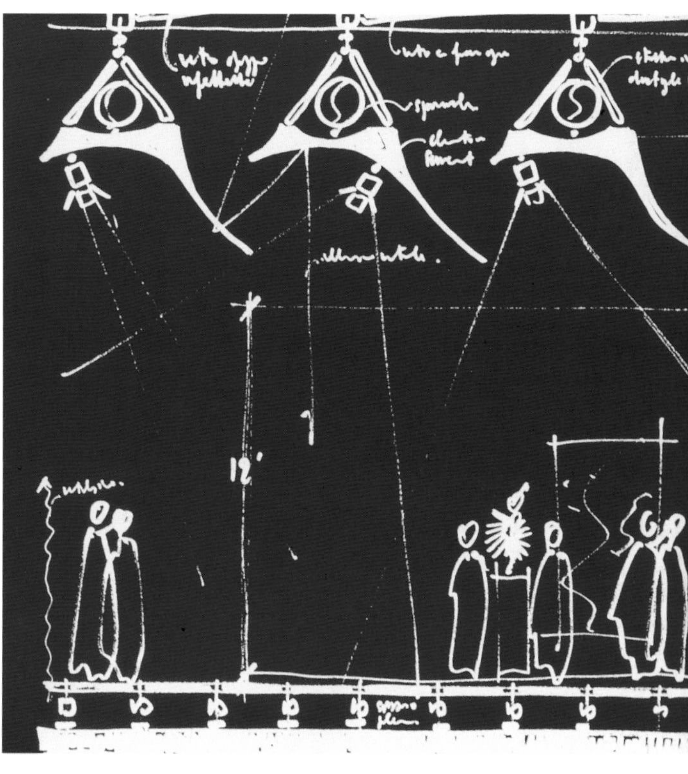

(Right and opposite) The engineering of the Menil Gallery, Houston supports a particular spatial experience.

course means all systems. Stories abound of engineers getting into or out of trouble by overlooking or noticing some force or movement out of the two-dimensional plane of first imagining.

Since intuition draws on the breadth of human experience, it reflects cultures as well. Among cultures based in the Mediterranean, the arts have a more tactile tradition, that nurtured the engineering of Nervi, Torroja and Candela. More recently, projects like Bari Football Stadium, the Franciscan church at Padre Pio, and Reina Sofia have all drawn on the sculptural uses of engineering. In France, love of design has produced whimsy and flamboyance as well as the expression on a very large scale of the most abstract of spatial ideas. Engineering intuition has been able to make a big contribution.

In the English-speaking world, the surge of engineering activity during the Industrial Revolution spawned a tradition of inventiveness that is more intellectual than intuitive. Yet intuition itself is well rooted in the culture, so that it gets due respect in engineering. The United States is a particular case. Fear of litigation and the rigour of commercial competition mean that speculation and innovation carry high risks. And yet the level of effort, and the huge turnover, ensure that remarkable engineering can still be achieved. Intuition is richly rewarded when it provides profitable results.

The privilege of working in a multi-cultural design environment lies not just in extending the boundaries of personal intuition, but helping to grow something bigger, a kind of collective consciousness that contains the intuition of many cultures.

Design resolution

Once intuition has helped establish the basis of an engineering design, it remains a key factor during development and resolution of the design. How is that so when engineering, certainly post-concept, is supposed to be all precision and determinism? Engineering is to science as politics is to philosophy: the art of the possible. Time and cost constraints limit the practical extent of analysis and resolution of detail. That doesn't mean unsafe structures or unreliable machines: it means that intuition is important in deciding when enough analysis has been done.

Sometimes this intuition is codified. Building codes and regulations represent the sum of a society's construction experience and understanding. They are continually evolving as new information becomes available. After the Los Angeles earthquake of 1994, hundreds of modern steel-framed buildings were found to have serious weld cracking. The collective, codified intuition of a generation of Californian engineers and builders had failed to assimilate the importance of cyclic plastic straining in joints undergoing bending reversal. Intuition includes looking over the fence periodically to make deductions from parallel experiences. There was no shortage of fences to look over in that case.

Sometimes engineers work beyond the scope of available codes, especially to achieve something new. The concept of 'effective length' for understanding the behaviour of compression members like columns and arches is deeply entrenched in most structural engineering codes. Some structures, like the tie-stayed arches of the Chur or Lille station roofs or the Louvre courtyard roof, are not amenable to modelling that way. Rather, all the real-life imperfections likely to distort them need modelling explicitly. Intuition is needed to encompass

(Above)
Louvre courtyard roof.

the effects most likely to be significant, and to choose the critical loadcases from an infinite number.

Similar modelling techniques were used for the slender stone arches of the Pavilion of the Future at Seville's Expo '92. The precision of modern Spanish stone-masonry, and the best in modern computing, still did not supplant the role of intuition. The test came when cracks, from the erection process, were discovered at some of the glued and dowelled stone-to-stone joints. That the subsequent re-analysis with cracked joints instead of glued joints still demonstrated the stability and integrity of the system was a tribute to the intuition of the engineer who conceived the system.

There is rarely time or money available to carry out comprehensive prototype testing or analytical modelling of unusual structural components like metal castings, forgings, or heavy weldments. The risk of crack initiation and growth depends on a very wide range of conditions, and it is ultimately the engineer's intuition that chooses a balance of computer models, prototype tests, material quality tests, and workmanship tests. Often, the schedule requires that modelling proceed in parallel with production, which puts greater reliance on the initial intuitive design.

Ground works

Intuition is nowhere better tested than in choosing a geotechnical site investigation regime, prior to foundation and substructure design. Clients are circumspect about the costs of exploration, so boreholes and trial pits sometimes stop just short of, or pass by, important buried features. From the initial review of historical photographs and maps, to the final sample on site, geotechnical engineers are invited to be intuitive wizards. In the construction industry's scramble to shed risk, geotechnics remains a major intractable area. It is the only area allowed by some clients to carry a contingency sum in the costplan, and then only if the builder refuses to contract for the risk.

In seismic zones, intuition becomes important for unusual structures that approach the bounds of code applicability. No code covers every combination of asymmetry, redundancy, ductility, and expected ground motion. The combination of dynamics and non-linear behaviour is particularly challenging, and intuition will point to different levels of modelling effort in different cases. Engineering becomes more of an art.

Earth, wind, fire, and water all exhibit engineering complexities that limit the accuracy of modelling, even where time and money are available. Fire engineering in particular grapples with a large number of recalcitrant variables. That is an area where a traditionally heavy reliance on intuition is finally coming into a healthier balance with reason.

A consequence of using intuition at all is that occasionally there will be failures. This is understood in most avenues of life. We continue to use referees to adjudicate at sports, pilots to fly aeroplanes, and surgeons to carry out medical operations, while knowing their judgement to be fallible. In all pursuits, people use their intuition to engage all their faculties with their skill, experience, and imagination, to make choices under pressure of limited resources, particularly time and money. That is very true in engineering, and yet we have allowed, even encouraged, a false image of certainty to grow up around us.

The language of risk management offers the way forward.

The new generation of seismic engineering codes in America will give building owners a choice of level of risk, and the risk will include the level of engineering analysis and hence the level of confidence in its predictive value. Certainty will be expensive, while confidence will remain affordable.

Intuition and computers

More often than not, computers are still a distraction rather than an aid to intuition, but that also will change. They are becoming intelligent in the sense that they will occasionally reveal an aspect of the basic behaviour of an engineering system that most intuition would overlook. There are instances of unsuspected dynamic mode shapes, or trends in non-linear models, revealed by routine analysis, that surprised even perceptive engineers. Computational fluid dynamics has revealed air circulation patterns in large volume spaces that periodically change their basic nature and defy simplistic predictions. The responsibility for boundary conditions and constitutive models is still fairly and squarely on the programme however, and a surprise to the intuition is still preceded by 100 false alarms.

Nevertheless, the direction is clear. Computer modelling works increasingly in parallel with our intuition rather than subsequent to it. The feedback loop is shorter and more frequent. A consequence is that the subject of intuition movesupstream. Ray-tracing has made room acoustics more accurately predictable, so intuition is now needed to understand and predict the criteria that will satisfy different groups of users. Now that thermal comfort is more accurately predictable through fluid dynamics, intuition grapples with the integration of other environmental parameters affecting psycho-physiological comfort, and their total impact on overall well-being. We can simulate daylighting, but we still need to relate the simulation to real experiences. And while the tools become increasingly sophisticated, the ends to which they are applied often become increasingly simple. To have confidence that a naturally-lit, naturally-ventilated space, with a light structure and simple cladding, will be comfortable, durable, and safe in a given physical setting, may require the very best in predictive software, supported by the shrewdest and most experienced of human intuition.

Whatever the state of the technology, intuition will always be our guide in applying it: in first framing the question, in finding the shape of the answer, and then in the technical details of its resolution. That will remain true as long as our engineering supports a human end and as our long as we remain interested in seeking the human face of technology.

Both concept and detailing of the glass-houses of the Parc de la Villette Science Museum (above) and the stone arches of Seville's Expo '92 (right) spring from an intuitive appreciation of the engineering properties of glass and stone.

(Above left) Seoul railway station roof proposal. Optimisation of slender compression systems by non-linear analysis would quickly become intractable if intuition were not used to narrow down the likely critical conditions.

Like a red-cliffed mesa in the Arizona desert, with a water pool oasis, saddlebag sides, and a cloud floating on top, the new Phoenix Central Library (this and following page) is as rich in images as it was in oppportunities for designer's intuition.

Structural design is based on knowledge and on science. Manning illustrates this with the design of a prototype car, problems of detailing, the emerging airport type, and design influenced by site constraints. Unusual projects are design led and emphasise the scientific principle involved.

Martin Manning

Engineering design and first principles

Two polarities characterise structural engineering design; one is based only on historical knowledge and the other principally on science.

In engineering based on historical knowledge, the forms of the structures are known, the materials used are conventional, details are well understood, the construction sequence is readily agreed, and the calculations are laid down in codes of practice. The contribution engineering designers make to such structures is principally production engineering.

In science-based engineering design much more must be determined by the designer. The structural form is open to alternatives and development, although clearly it must relate to the function of the building. The materials can be uniquely specified, the details have to be developed, alternative construction sequences investigated. New methods of calculation have to be evolved and appropriate models analysed. Codes of practice may be inappropriate. In such projects the engineering designer has to decide – What should we do? – Why should we do it? When these questions are answered there still remain the problems of how the structure will work and carry the necessary loads successfully. Projects such as this are design led and the designer must understand both the physical planning issues and the other technologies involved in the production and design of the project. The size, scale, shape, and finish of the structural elements will matter. Clearly, cost and time will matter to all. The need to be able to construct the building in as flexible a way as possible might also be important. In science-based design the work for the engineer requires a conscious review of previous habits of thought and a fresh start with each job. In the following projects the unusual was achieved because of a preparedness to work from first principles. These will be discussed with respect to form, materials, typical structures, and construction sequence.

Form

In the late 1970s senior management at Fiat decided that their car designs needed review. Like most mass-production cars of the time the body was made of pressed steel sheet; the cladding was the structure, and the body and the styling were inseparable. The whole was more determined by ease of production than by issues of experimental or efficient engineering design. This resulted in cars which were heavy and thus energy consumptive; liable to corrode and thus of limited life. They were inefficient and complex structures and their behaviour under impact was difficult to understand. Styling changes were costly because huge investments were required to change the steel presses.

The response was to propose a scheme with a much lighter tubular steel frame, clad in plastic panels. There was less steel so it was easier to protect against corrosion; the structure was simpler and so it was easier to make it more efficient and understand precisely how it would behave in crashes. It was lighter so that its energy consumption would be improved, and plastic panels allowed styling to change more easily.

Other changes were proposed. The car could be divided into three separate structural parts, each with a different function. These could be assembled in different places and only brought together at the last stage. The front part of the

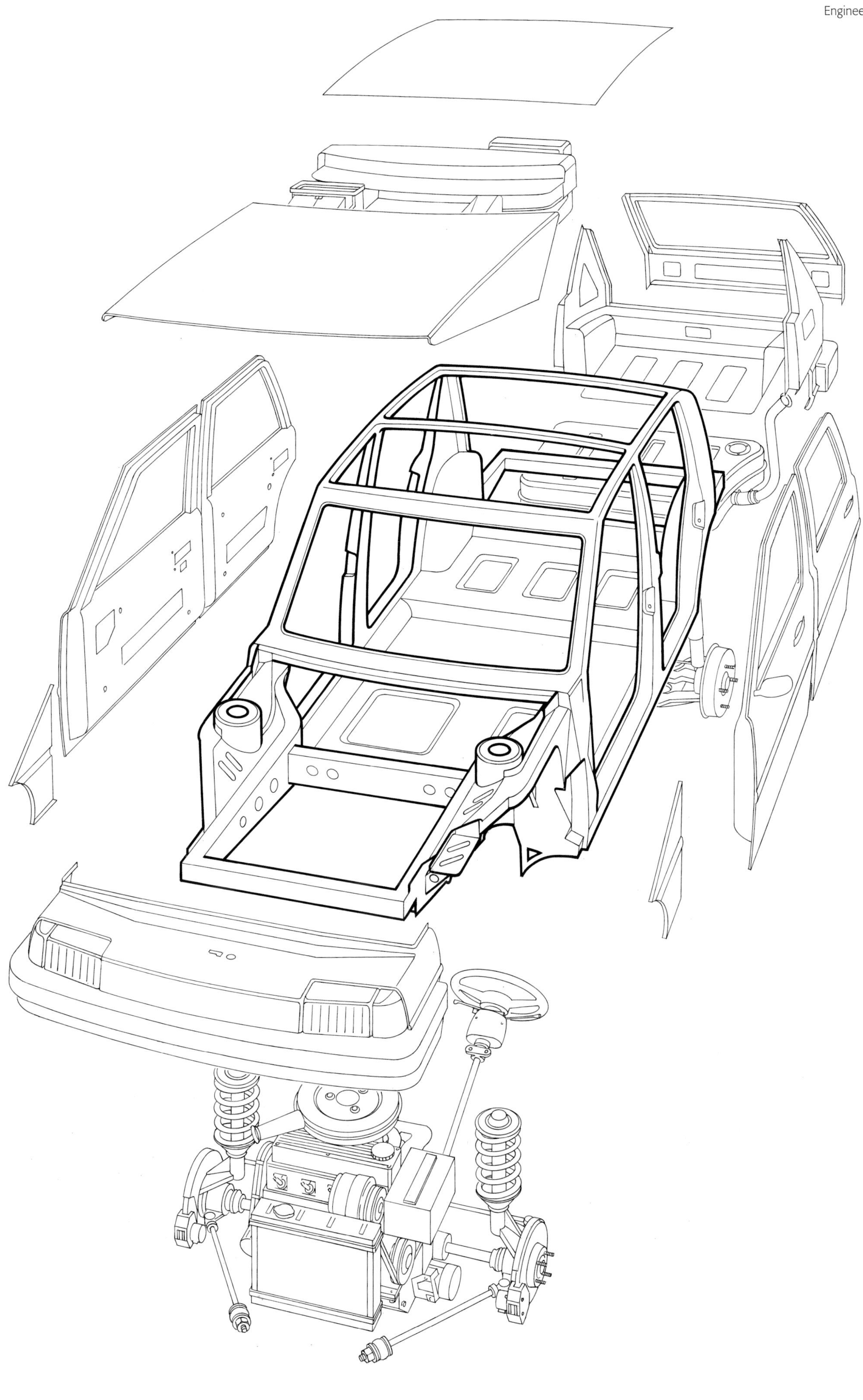

Fiat: A tubular steel frame designed as structure and plastic panels designed as cladding.

car supported the engine, the front suspension and the steering, and resisted head-on collision. The rear part supported the rear suspension and resisted rear-end collision. The centre part of the car could be purpose-designed to house the driver and passengers, could be capable of change, and be a properly ventilated and acoustically treated environment. The perceived benefit of assembling these different parts in different places was that work forces could be kept reasonably small and therefore easier to manage. Eventually Fiat solved that problem with robotisation. But the principal proposal, that is, of articulating the functions in how the car should be made, was still valid. The problem, we discovered, lay in discussing with Fiat's engineers what the performance criteria should be, especially true for the more conventional characteristics of car design. Designing for head-on collision was, though scientifically difficult, curiously easier to progress. There were no empirical laws. Everyone agreed that it should be done from first principles. Much more difficult both to discuss and agree were the standards for which the car should be designed for normal circumstances of driving. Traditionally, body shells had been tested for torsional stiffness. Cars with a torsional stiffness-to-wheelbase ratio greater than a certain value were considered 'good' cars but were heavy and increasingly expensive to produce. Cars with a ratio less than that value were considered to be of lower quality though cheaper to build. The aim was to develop a car design which had the 'right' torsional stiffness, but as measured values always differed from those calculated, making innovative design proposals proved difficult.

We found the sophistication of analytical work and computing was remarkably high, much higher than that commonly found within the building industry. But paradoxically, the level of understanding of how the structure behaved as it did, or why, was lower. We were unable to discover whether the torsional stiffness requirement was a function of noise, fatigue performance, handling characteristics, or the fact that, with more flexibly framed cars, doors opened unexpectedly going around corners. However at the end of this project a prototype was built and tested. Even now, years later, parts of that prototype study appear in car designs although our precise structural design hasn't yet. Here it was evident that the calculation method is a product of the design. It doesn't determine it.

Material

Another car company, this time Renault, wanted to have a new parts distribution warehouse in Swindon. Part of the brief to the architect sought a building solution which demonstrated the company's commitment to high quality and innovative engineering and design. The design developed along the lines of an exposed steel roof structure composed of 24 x 24m bays with the beams supported along their length by cables hung from the tops of masts. One particular detail which occurred up to 16 times on any one mast (and there were 59 masts) was the junction between the cable or tension rod where it had to be attached to the tubular mast.

Because of some of the characteristics of the design, a cold-formed alloy steel was chosen for the tie rods. This was not readily weldable and so some form of mechanical connection was required between the tie rods and the masts. It had to be simple to erect, and require low maintenance, particularly with regard to corrosion protection. The solution also had to be relatively cheap.

We decided to pursue the idea of using castings. Castings provide the designer with an opportunity to integrate many design issues into a single piece. We also decided to try and use a material which at that time had not been used as a purpose-designed component for a building – spheroidal graphitic cast iron. This material was designed to be cast, and has tensile capabilities similar to those of steel though it is much cheaper. The material had been developed originally for the automotive industry and particularly for the manufacture of engine blocks. Its performance in buildings, however, as a relatively highly stressed, statically determinate structural element was untested. We therefore returned to the basic physical properties of the material, not the characteristics measured by a Charpy Test, but the fundamental ductility characteristics measured by COD testing. Allowable stresses are related to how the material is made and cast but also how much of what sort of testing for defects can be afforded, in terms of both time and money.

The practical problem for the engineering designer with such a proposal is that the technical process appears to be incredibly scientific and academic. However, in terms of the construction process, tests have to be done that are unfamiliar to the steelwork contractor accustomed to normal building codes, while the founder accustomed to working in another sort of industry concludes that building engineers are conservative and difficult. Both architect and client worry increasingly about cost and programme, and therefore just at this stage when the engineering designer tries to justify a new form of structure there comes a powerful sense of isolation. The ideas are his, but the benefits claimed by him for his solution, and required by all, must be nervously awaited until he is satisfied. Such things require nerves and confidence.

Typical Structures

Most of us spend our careers designing buildings with the structural concept of Le Corbusier's Domino House of 1913 – which leaves the elevation and the arrangement of walls free from the necessity to carry anything other than self-weight. True, the buildings have more storeys, more bays, but they are essentially that same early 20th century structure. Occasionally, however, a new building type emerges for which the structure and details have to be re-thought from first principles. In that case we must understand the function of the building, how the brief will be developed, what will be the design process and the overall construction context. In the

Renault: One type of mast (590A) and one type of grid line beam (1050H) assembled into 42no. 24 x 24m bays.

Renault: SG iron castings, foam-filled and silicon-sealed for corrosive protection, anchoring the tension members to the tubular columns.

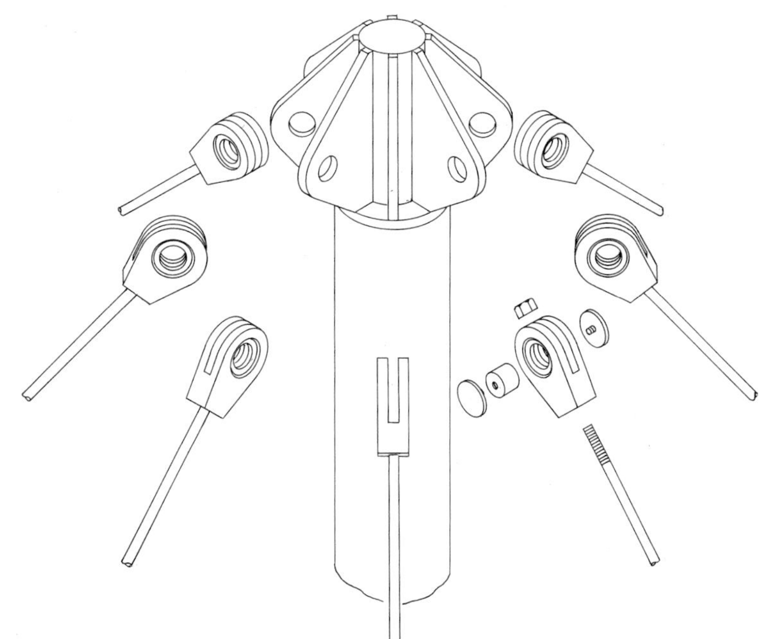

latter part of the 1980s and the early 1990s, airport terminal buildings have inspired such a rethink.

The buildings through which passengers pass to and from planes fall into two types. Firstly, there are the central processing buildings in which passengers are separated from, and reunited with, their luggage; secondly, there are the piers against which the planes are parked. Movement between the two is by walking, travelators, or tracked transit vehicles, depending on the size, complexity, and design of the airport.

Consider the central processing buildings; their common features are a departure area, an arrival area and then the mechanical plant areas. In departure, passengers check-in, pass through customs and security, and buy their duty-frees before setting off for the pier. These are areas characterised by long spans and large plan area, and hence great height and much natural light. They are arranged to be calm, clear, and convenient. They are ventilated from a relatively low level and at regular but infrequent centres. New fire engineering policies have to be adopted for smoke logging and escape.

Arrival areas can be shorter-span and possibly lower in height than departures. People move through them more purposefully and spend less time waiting. There is likely to be a need in them for office space for customs, immigration and airlines, but the brief for this will only become finalised probably late on in the design process and possibly even late on in the construction process.

Lastly, there are the mechanical plant areas which house not only the main items of environmental plant, but also the baggage handling machinery. These can be relatively short-span and need a robust slab above them, as much of the mechanical engineering will be hung from the slab.

The requirements of these three types of area have led to a form of airport structure which has been adopted on a number of projects. What the floors are made of, and the roof design, continue to be things that each project team chooses as a function of architectural desire and local construction technology. Engineering and design solutions are typified by at least interdependent factors: a light roof with spans of 30-40m, 10-15m above the departures level slab; a departures level slab supporting crowd and storage loads for duty-free shops with spans of 10-12m, 8-10m above the arrivals level slab; an arrivals level slab supporting concrete loads and heavy machinery loads above the baggage hall with spans of 10-12m, 8-10m above a base slab; areas of infill mezzanines in the two lower double-storey height spaces; buildings with a large footprint about 180m wide by up to 400m long, capable of being built in phases; and finally unbraced frame structures because of the difficulty of locating walls or bracing in the baggage-handling areas.

The usual problems raised for the engineering designer are stability, differential thermal movement, construction access and the need for economy. However, the structural system must also be able to accept changes in the size of the building and the amount and position of the infill mezzanines. It is this need for a robust design concept, which can readily accommodate changes in the brief during the design phase, which has led to the idea for a two-storey height primary frame supporting a secondary light envelope and infill mezzanines with tertiary mechanical engineering support systems. This provides the economy from refinement and the necessary flexibility. While this may describe the typical problems associated with the new generation of airport buildings, each new design must appear to be unique and identifiable.

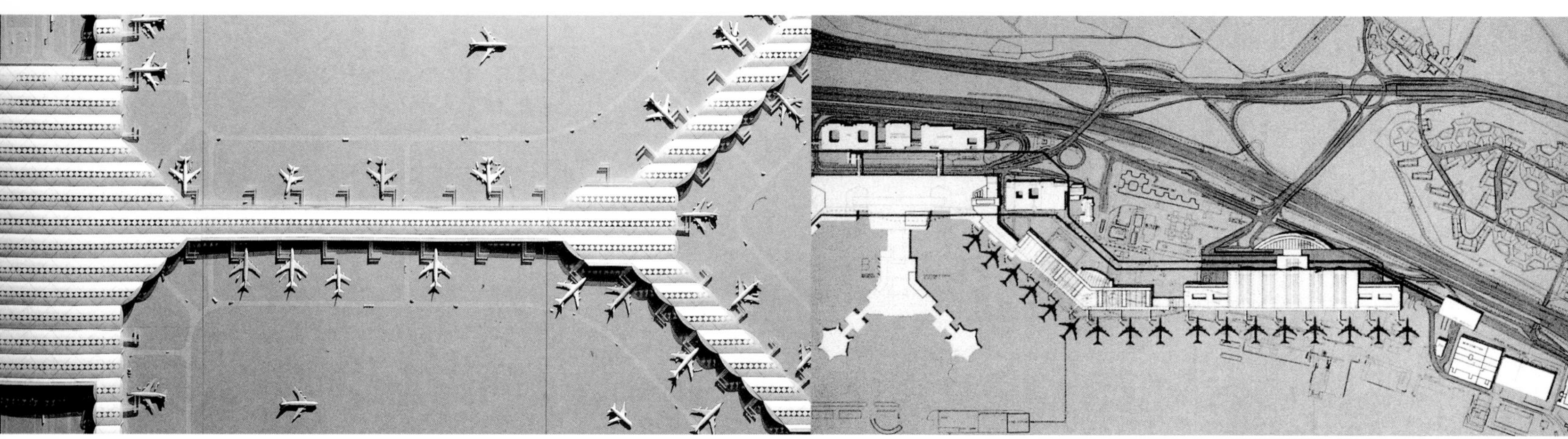

Construction sequence

The issues raised by a new generation of building and engineering types was also raised in the late 1980s when a new type of office building was constructed in London. Examples of it all had a number of similar commercial characteristics – large sites with a number of separate buildings, large floor plates, and short construction programmes. Perhaps inevitably, the solutions which were built also had a number of identical technical characteristics – no basements, braced steel frames, panellised cladding, prefabricated toilet modules. Solutions differed as to whether the air-handling plant was located on the floors or whether the central plant was at ground level or on the roof. The planning regulations gave some benefit to shallow basements.

Another common strategy was to maximise the separation of the different elements of the building, both physically and contractually. This encouraged the growth of construction management as the overall contractual context. At one particular project, Minster Court, we set out to maximise the use of the site by reconsidering what was, by then, the conventional wisdom. The large Central London site, 120m x 80m, was divided into three buildings on plan but more importantly two further buildings in section, making six different working areas in total. An 8m deep basement was constructed over the whole site from ground level down, at the same time as multi-storey steel frames were erected upwards. This allowed an early start for the steel frame and cladding, but an even earlier start for central plant installation in the basement. The ground level slab was designed as a heavy duty traffic deck for site access.

The behaviour of the foundations needed very careful and sophisticated analysis with such a phased loading, especially considering the proximity of the movement-sensitive Circle Line under Great Tower Street and the construction of the Docklands Light Railway tunnels during the contract. In the end the construction was easy, but only because of unusually refined and integrated structural/geotechnical design and analysis. The temporary works interfaces between the various packages also required a much greater than normal involvement of the consulting engineer in the management of the site.

Conclusions

Innovative engineering design is necessarily based on knowledge of the conventional. How else can the designer know if his supposed innovation is worthwhile? But innovation is also based on a real understanding of why certain structural configurations work. The analysis cannot just be of a type which checks whether one is sufficiently close to the norm. It must also be of a type which has identified roots in physics. All members of the team want the benefit of innovation but only the engineer can really be responsible for achieving it. To propose change and campaign for it requires courage, and that is easier when we are being encouraged within a well-led team. All of the above projects have benefited from such leadership and from teamwork. But in the end that is as necessary for any project as physics is for the engineer. The task is to build buildings, not just design structures. A structural design has to be built and as such recognise all of the consequences of the process. Holistic design understands that.

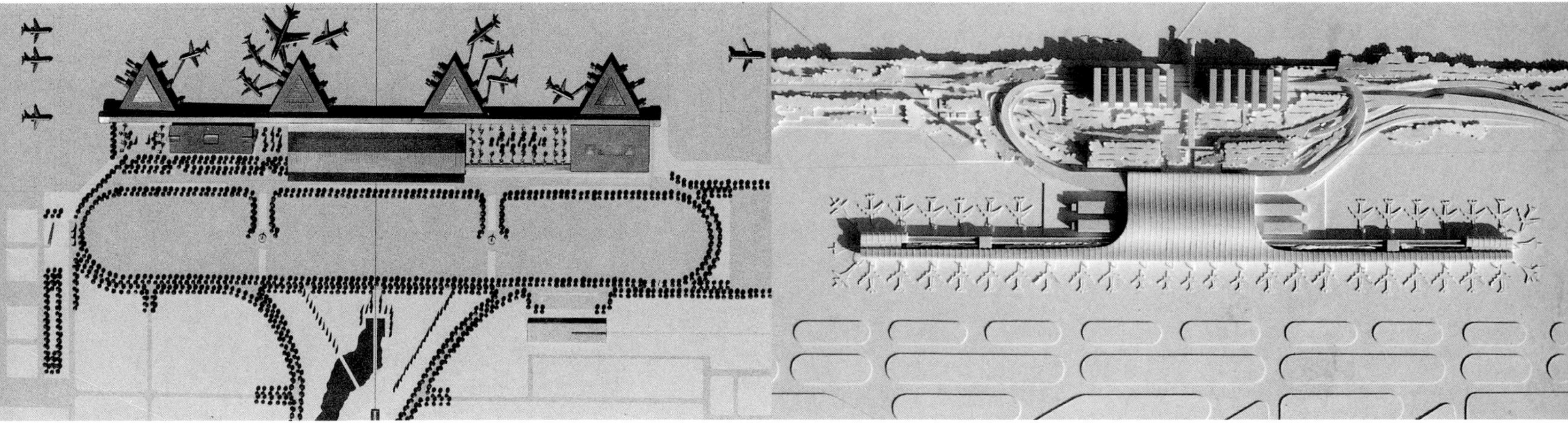

(Above, from left to right) Chek Lap Kok Airport, Hong Kong; Frankfurt Airport, Germany; Barcelona Airport, Spain; Kansai Airport, Japan.

The involvement of engineers in building has profoundly affected the way architects design, with technological change acting as a catalyst. The author looks at the relationships between form and function and force, and argues that the engineer has an increased role as both catalyst and as enabler.

Jane Wernick

Technology as catalyst

Over the last 200 years significant changes in the way our society has been organised, combined with major technological advances, have led to our conception of the engineer whose skill is underpinned both by practical experience and the results of scientific investigation. During this period we also saw the emergence of the 'general builder' and the 'main contractor', who in turn began to organise their work by taking advantage of the productive power of the factory-based manufacturingindustry. This led to and enabled the large-scale building programmes associated with the major urbanisation that started in the 19th century and still continues.

During the mid-to-late 19th century, many theorists and historians began to explore the relationship between science and technology and architectural design. For example, Gottfried Semper in *The Four Elements of Architecture and Other Writings* examined the development of an architectural syntax and proposed a flexible method of understanding structure's contribution to architecture. He examined the relationships between expression of the material of the structure, and definition of the spatial

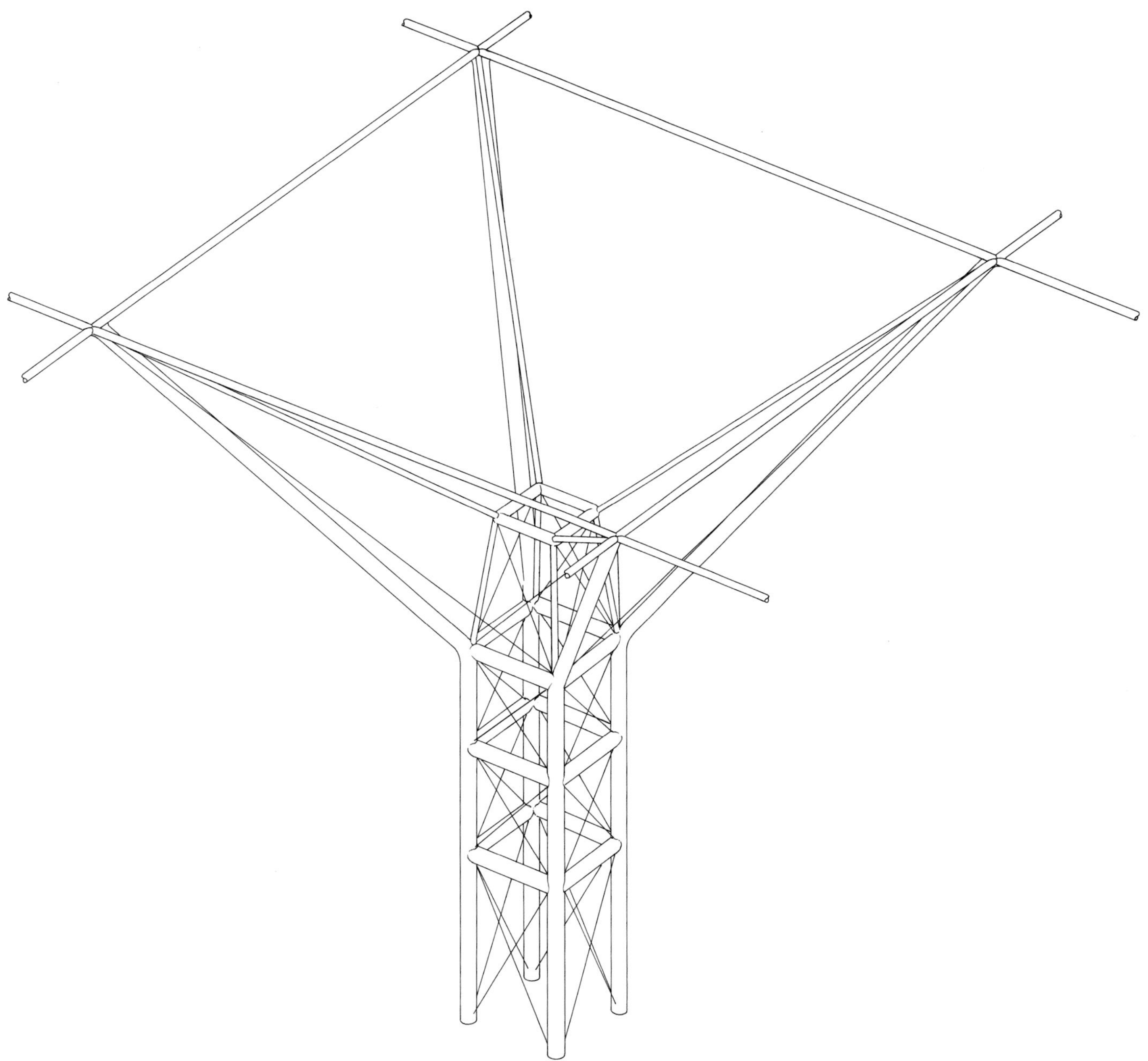

form. He also allowed the separation of cladding from structure. Another historian, Eugéne Emmannuel Viollet-le-Duc, in his *Lectures on Architecture*, sought to articulate the structural rationalist view. He encouraged architects to search for the appropriate structural forms and to use new materials.

In the UK, the extraordinary and heroic project of that period – the Crystal Palace – remains an important model of how it was possible to achieve good design through the new understandings. That design was applauded at the time by the general public, the client, architects, engineers, and builders. Remarkably, nearly 150 years later, we still hold the project in high regard.

From today's perspective, amongst other changes it is possible to argue that one significant but gradual change in the building process has been the increasingly explicit role of technology and the willingness of architects to exploit the possibilities made available. Throughout this period of change technology has largely been a catalyst – not so much a determinant, nor simply a consequence of architectural form.

It can also be argued that the role of the engineer in that transformation has been – at least in part – that of interpreter and proposer of those technological possibilities. Initially, it was the structural engineer who developed this creative relationship with the architect; Ove Arup and Felix Samuely, being the two engineers who brought about the greatest change in the role of the consulting engineer in this country. Many other engineering disciplines have since increased their contribution to the design process, the crucial feature being their early involvement in the process. Today this occurs often at the very outset, at the design concept stage.

There have been some significant ideas and technical advances which have continued to give vitality to this process. Within structural engineering, particular mention should be made of five major aspects:

- *Manufacturing processes and their relationship to on-site erection and fabrication.* These have been enhanced by the advent of computer-controlled processes which link design to manu-

(Above) At Stansted Passenger terminal, UK, technology acted as a catalyst of form. The tree-like structures carrying the double-curved, single layer tubular steel grid also house the air supply, the uplighters, and passenger information. This one key element repeated at 36 m centres was refined to optimise the weight of the steel, stability against overturning, torsional buckling, level of prestress in the rods, and clarity of form.

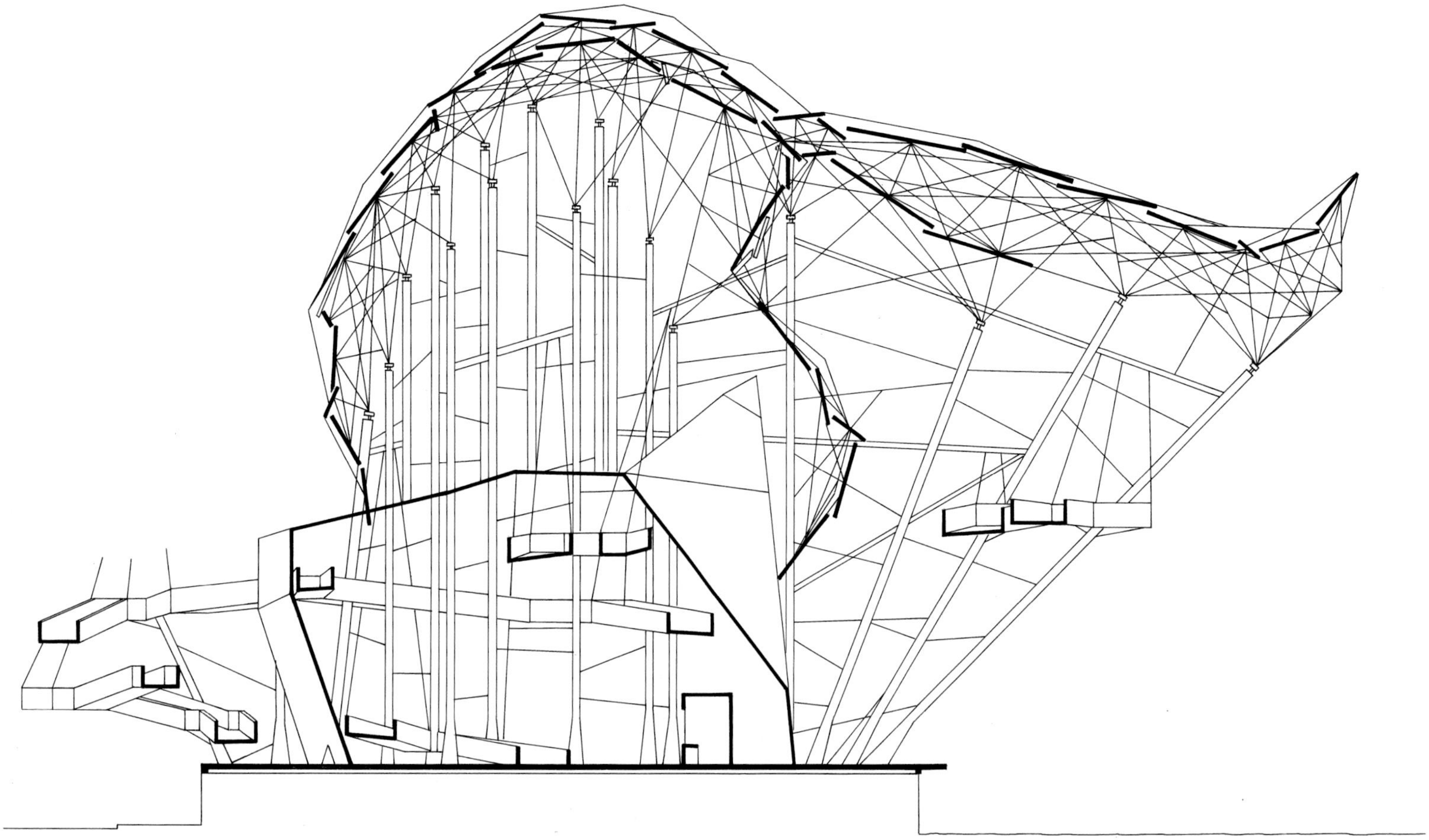

facture; by increasingly sophisticated methods of quality control and methods of measurement, and greater standardisation of all types of component parts, together with the use of modular construction.

- *Understanding of materials used for structural purposes, allowing us to be much more certain of their reliability and consistency, their strength, and their cost.* Among the materials we have seen emerge and significantly improve over this period are: reinforced (and then prestressed) concrete; steels; laminated timber and, later, stress-graded timber; fabrics, nets and membranes; plastics and polymeric materials; glass (as a structural material); various composites (eg carbon-fibre).
- *Analytical methods, particularly enhanced with the advent of the computer and finite-element methods.* In particular, this has allowed the analysis of irregular three-dimensional forms, and of non-linear structures such as membrane structures, cable trusses, and suspension structures. The analysis of structures under dynamic loadings such as earthquakes has also become possible.
- *Simulation and testing methods, enhanced by computing and also with much more confidence in using various forms of physical analogue modelling.* Standards and codes of practice have a role as one means of capturing these into the standard design process.
- *Increased global communications.* These have also led to sharing of knowledge and the establishment of some international standards, eg the emerging Eurocodes.

Together, these have not only advanced the contribution that the engineer can make generally. They have also given a credibility and confidence to which the architect can relate. Architects have often contributed to this process by the problems they have posed, the ideas they have conjured out of the changing social situations to which they respond. Engineers have demonstrated that they are able to make use of these developments and have, in the process, changed the way in which buildings are designed. Konrad Wachsmann, in his book *The Turning Point of Building*, describes how the design of a building is no longer dependent on the skills and talents of one master, but has become the concern of teams where 'problems are attacked simultaneously from all sides and brought into direct relationship with each other'.

The experts who comprise the design team bring to the process knowledge of their own specialities, and, when the team is most successful, an awareness and sympathy for the aims and aspirations of the architect and the other specialists.

The attitude that architects take towards the contribution of structure to the form of the building can vary enormously. The way in which the structure can be used can be considered to fall roughly into four categories. It is interesting to discover how quite different architectural styles can emerge from any one of these categories. This indicates that the application of technological expertise need not constrain the architecture to develop in a certain direction, but can really be used as an enabler:

- *'Kit of parts'*, the idea that the building is developed from a predetermined repertory of industrially-produced components. The Crystal Palace was the progenitor, but it has continued as a major preoccupation. More recently, it has been further associated with internal flexibility of space – eg Pompidou Centre, and Stansted Airport. The proposed Atlanta Pavilion, which uses straight pieces of glulam timber and spherical nodes to produce a totally irregular space frame, is another example. The use of the standardised method of putting the straight pieces together produces here an architecture that is quite unlike the systematic and ordered form of the Stansted building. The IBM travelling exhibition was also a good example of a structure made of accurately manufactured parts – this time of timber joined by standardised aluminium case nodes.
- *'Form follows function'* – the motto of Louis Sullivan, whose protegé Frank Lloyd Wright gave it the clearest statement in his remarkable structure 'Falling Water'. This approach could also be said to apply

For the 1996 Olympics in Atlanta, a Visitors Center sitting on ground over a subway station and surrounded by high-rise consists of the Center sitting below a complex shading canopy. The canopy will be made of timber because of loading restrictions and a donation of timber to the project. This suggestion was made as one way to support a doubly-curved iregular surface up to 80 feet above ground, which can be considered as a timber spaceframe supported on raking timber columns. The connection between the spaceframe and the columns was designed as a special system of spherical steel nodes that could be bolted into the timber by means of special inserts at almost any angle. All timber elements are made from glue-laminated southern pine.

(Below) Axonometric of roof structure.
(Left) Section through Visitors Center.

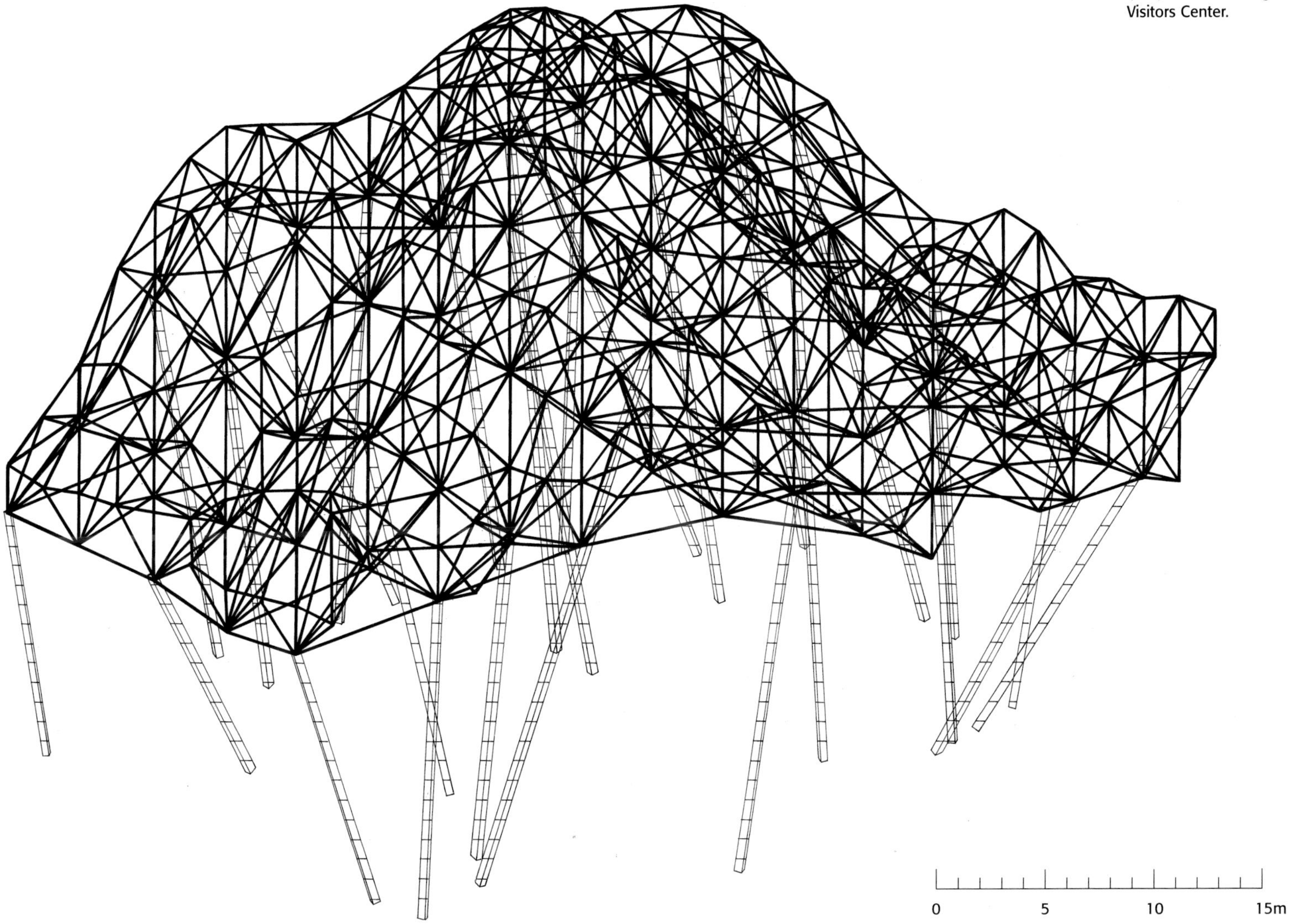

0 5 10 15m

to the proposal for the Cardiff Opera House. The building comprises a linear part which is partially raised off the ground and is wrapped around a central space. Generally that linear part houses offices, dressing rooms, technical spaces, parking. Whenever an activity which highlights the difference between an Opera House and any other building occurs, such as the auditorium, the rehearsal rooms, and the flytower, that particular activity is given a specially formed box or 'jewel' than can clearly be seen from outside the building.

- *'Form follows force'* – the characteristic of many works of Antonio Gaudì and in particular La Sagrada Familia. Here the structure is deliberately placed along the lines of force, the form of the structure being found by using a loaded hanging model of the church which was then inverted. A later example is Le Corbusier's *Five Points of Architecture*, which separates the cladding structure from the load-bearing structure, and which was subsequently developed in many designs of the 1950s, where structure is physically differentiated from the rest of the building in a type of structural expressionism. Membrane structures, and cable structures where the force stays within a small element such as a filament of glass or a steel cable which moves as the applied loads change, are also clear examples of 'form following force', as is the 'Image of the Future' bridge which was conceived by the architects David Marks and Julia Barfield with the author of this paper for an ideas competition. The bridge clearly expresses which elements are in tension.
- *'Form controls force'* – in a sense a reaction against 'form follows force', where there is a visible expression of structure but its form does not explicitly express the most direct stress paths. Examples include the works of Pier Luigi Nervi and more recently Santiago Calatrava. This could also be said to apply to the TGV station roof at Lille where the arches are made of thick-walled tubes which are stabilised by tension rods in order to keep them slender.

All of these categories are properly

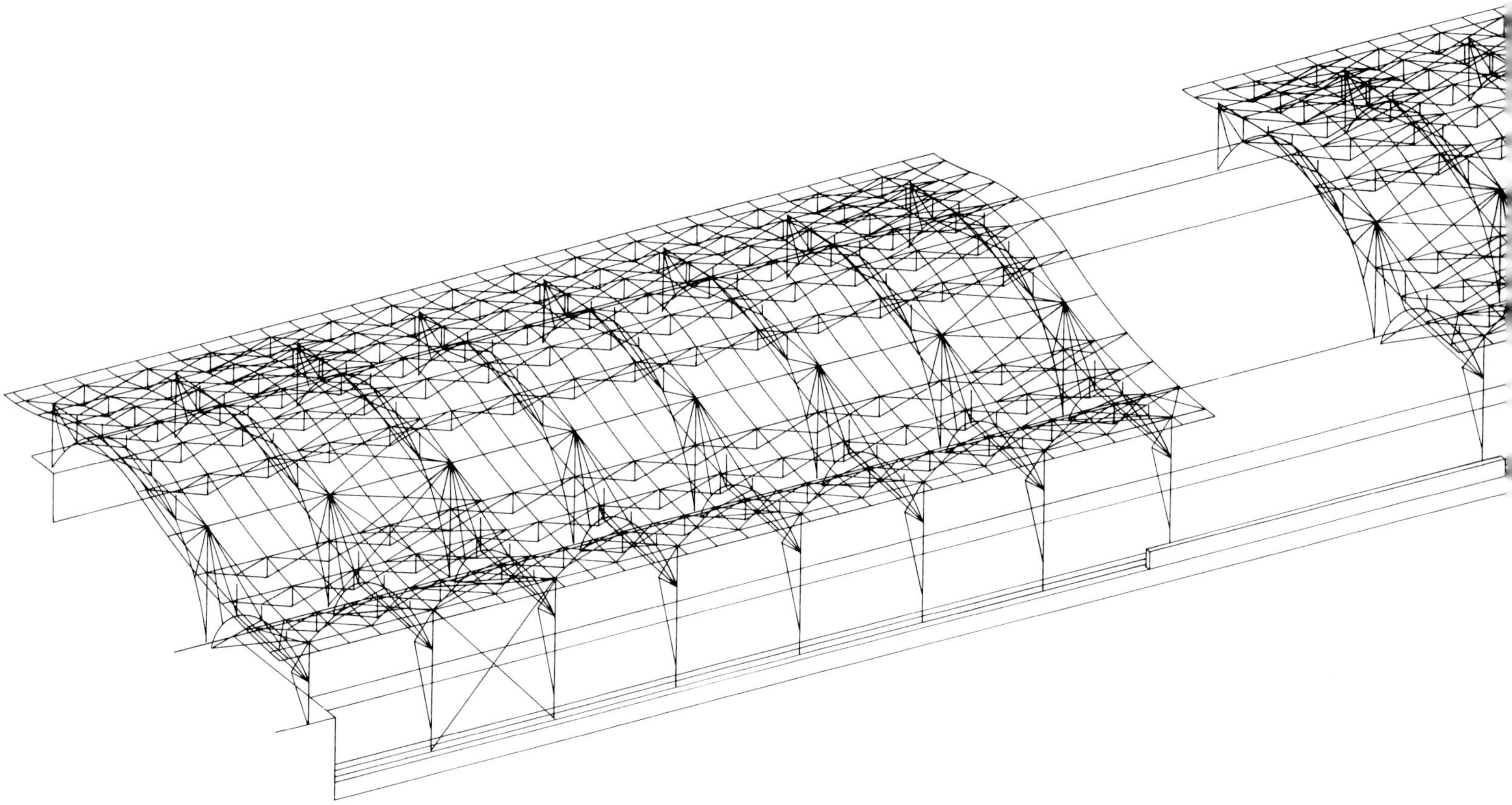

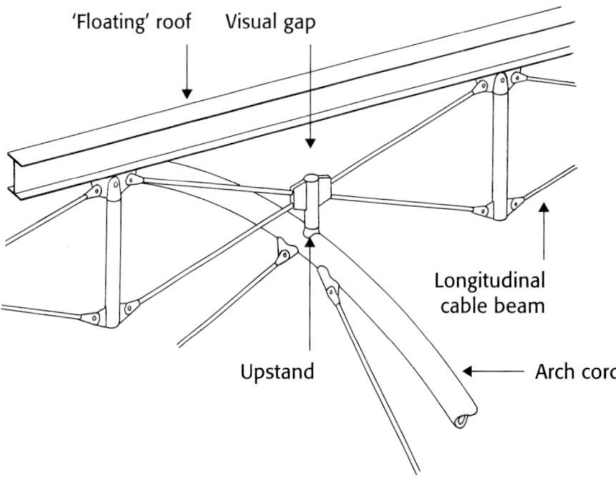

In the TGV station at Lille, the roof was designed to appear to float supported by the slenderest possible members. This was achieved by placing rods in the plane of the arches to control their in-plane buckling; by providing a moment connection to reduce the likelihood of the arch buckling; and by using small diameter thick-walled hollow circular section rods.

Secondly, the vertical connection between the arch and the main roof beams is made by rods that are arranged perpendicularly to the arches, under the roof beams, thereby giving a clear visual separation between roof structure and arch.

(Centre) Axonometric of roof structure.
(Bottom) Detail of structural connection.

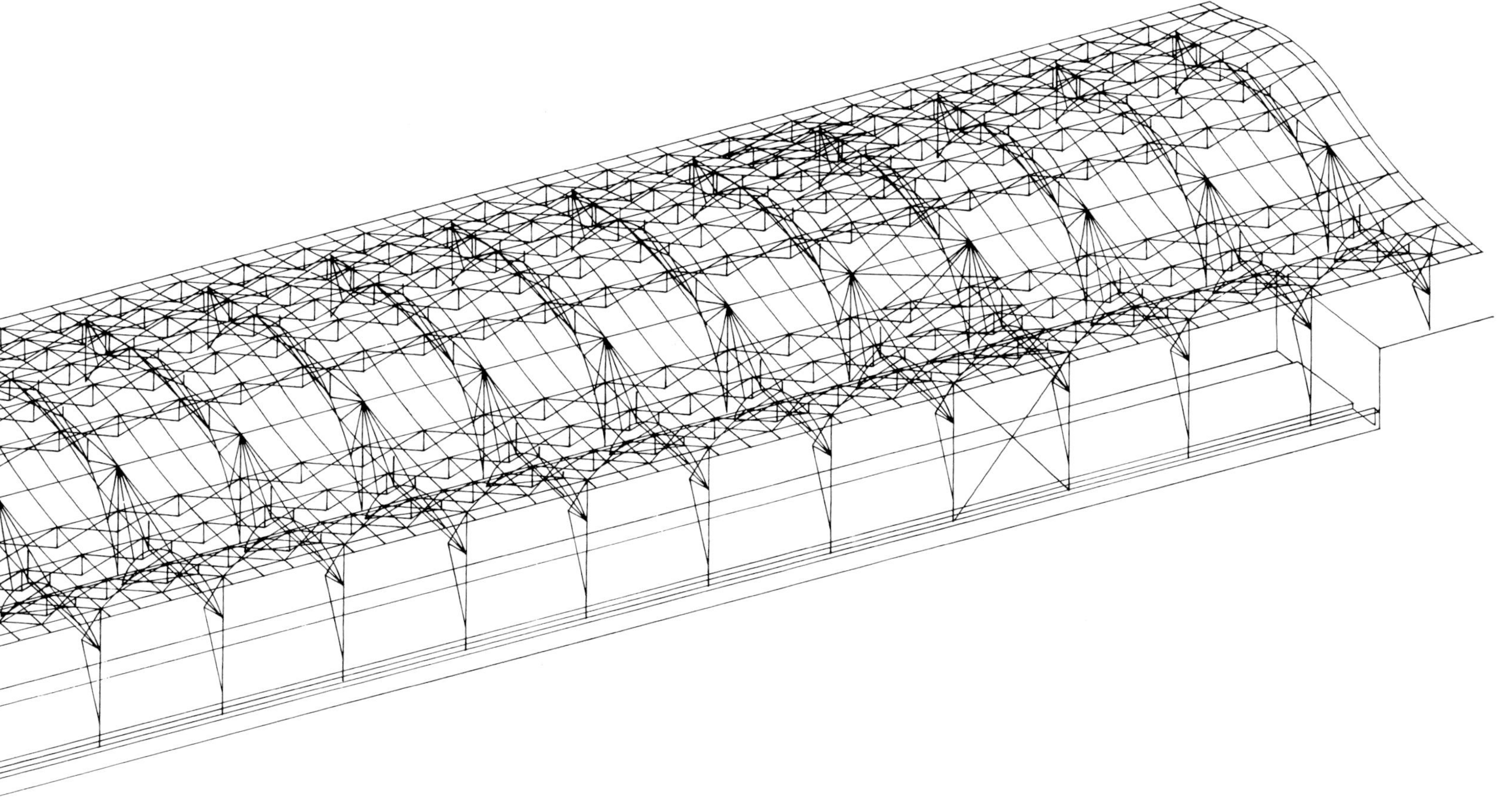

understood only when one examines the design development at a detailed level. Often a building will reflect more than one approach. The skill that the structural engineer needs is that of distilling the aims of the architect and providing the knowledge and technical ability to enable him or her to develop a design which satisfies those aims, and is also buildable and satisfies the client's requirements of function, cost, and programme. It could be argued that the potential for the engineer to contribute to the design process grows ever richer as the palette of materials, and methods of analysis and of construction, develop. But it is the conversations that take place between the members of the design team at the start of the project that provide the greatest opportunity for advancement.

It is possible to speculate how all of these changes will affect the next generation of buildings. It is now impossible to imagine that any one person would have all the skills and knowledge necessary to enable him or her to work on their own. In fact, the emergence of the design team can be seen as a positive means of enriching design, and of resolving the hugely complex design problems. These problems are enmeshed in society, in a way which may not have been envisaged – some of them brought about by technological advances themselves. Does this mean that the role of the architect is diminished? Or does the architect act as the co-ordinator, and director of the team, which achieves the final synthesis of all the requirements and possible solutions to produce the finished design? In order to achieve this synthesis, each member of the team must totally understand the nature and implications of the solutions. It is this necessity that can lead to the most interesting and challenging part of the engineer's work. Because the engineer now has a chance to imagine, investigate, and propose solutions of fundamental importance to the design, the challenge must be to chose appropriate solutions and to present them and explain them convincingly and thoroughly. If successful, the engineer's role is integral to the design process and becomes that of catalyst, and also of enabler.

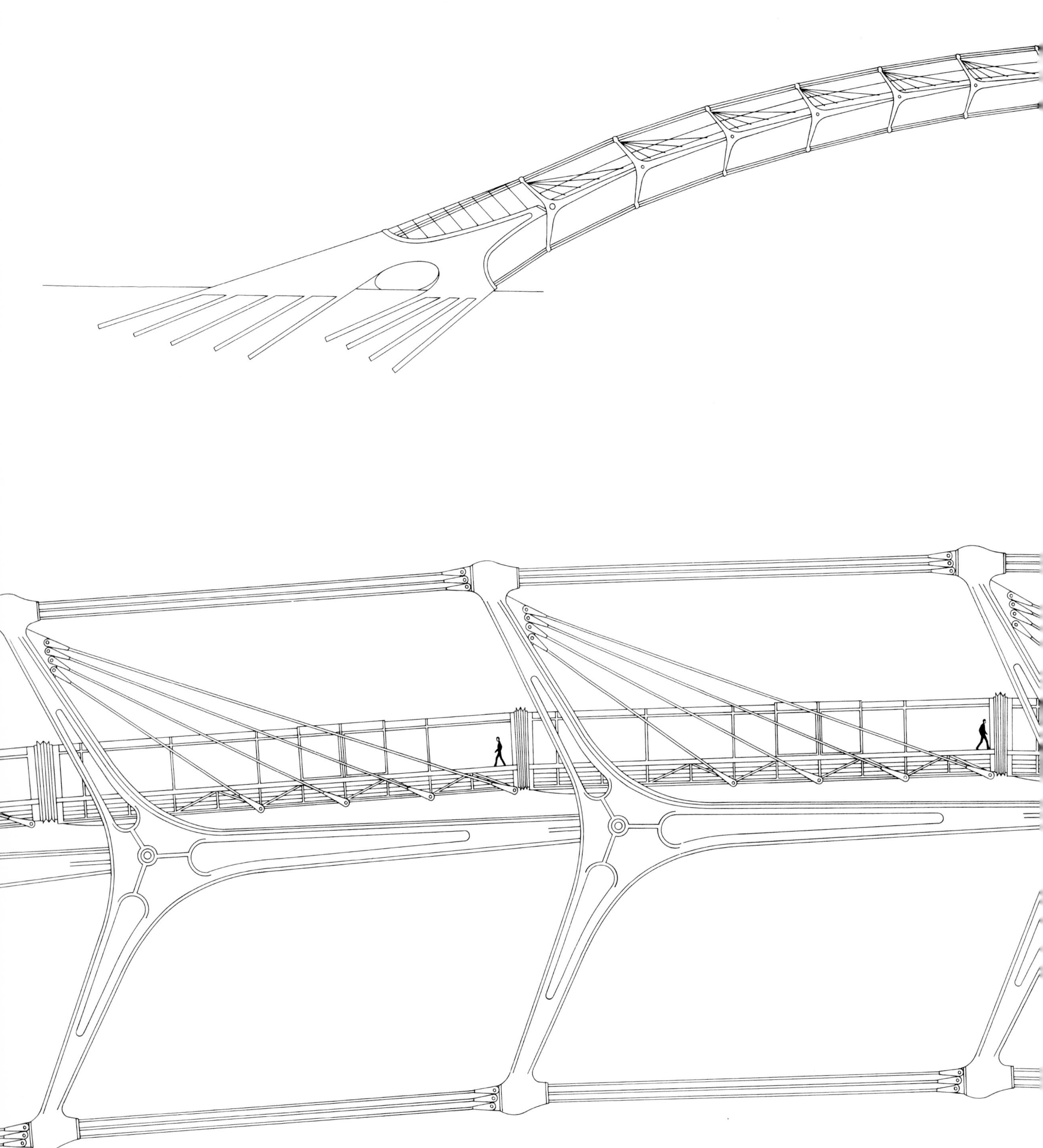

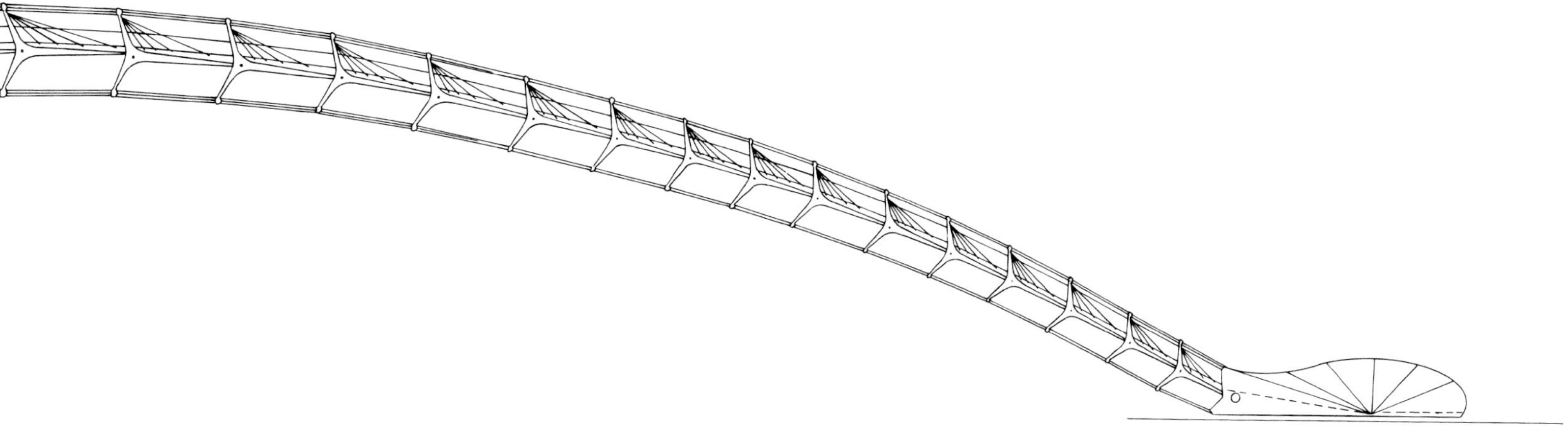

In *On Growth and Form* D'Arcy Thompson made an analogy between a bridge and the body of a four-legged animal. The body can be considered as a bridge between the front and rear legs, with the vertebrae carrying compression and the tendons controlling deformations under various loads. This imaginary pedestrian bridge was designed to deflect and change shape as people walked over it so that this dynamic behaviour could be experienced. While fixed at one end, the bridge would be supported vertically at the other, allowing it to stretch and extend as loading changed.

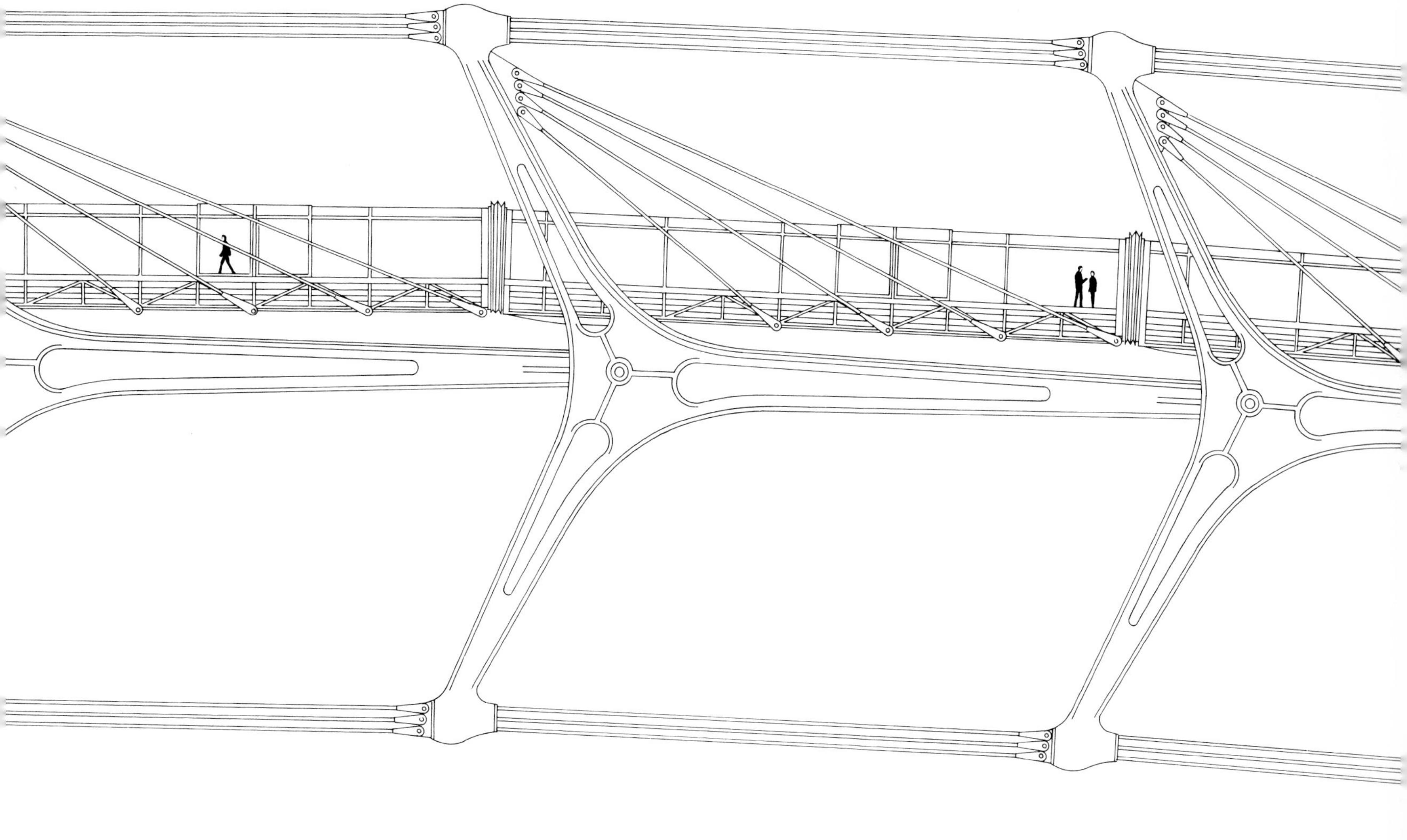

Calling upon his memories of the early days, Jack Zunz encompasses working in the firm with a unique commitment to answering the question – How can Arups continue into the future? His response depends upon the concept of 'the learning society', committed to the idea of client service through professional and lifelong learning. Both specialised knowledge and a general approach to thinking are argued to be essential aspects of this future.

Jack Zunz

Arups – a learning society

'In the beginning there was Ove' and Ove was a very special kind of person.

Not only was he a man of exceptional technical and creative ability, but also one of quite extraordinarily wide-ranging cultural interests, which went beyond everyday engineering analysis and synthesis and which stretched out into the community.

These broadly-based cultural interests manifested themselves in various ways. firstly, our location – we started in Soho but since 1949 we have been based in Fitzrovia, where we have had the majority of our offices ever since. In 1946, Victoria Street was the professional ghetto of consulting engineers where most had been based since the railway years of the 19th century. Apocryphal sources have it that Ove felt that there weren't any good restaurants in the Victoria Street area – Soho and its extension north of Oxford Street were known for their generous sprinkling of well-known eating establishments and he didn't want to be too remote from having easy access. Fitzrovia also had a reputation for cheap accommodation and living dating back to the 19th century and, as a consequence, attracted the kind of people which turned it into a breeding ground for new ideas in the arts, in literature, and in politics (to which can now be added engineering!). Our offices in No.8 Fitzroy Street had a blue plaque commemorating Octavia Hill (1838–1912) who played a leading role in housing reform. Later in the 19th century, James McNeil Whistler lived and worked there before moving to Hampstead, while Walter Sickert later took over Whistler's studio. It is doubtful whether the legacy of these great artists had a direct influence on Arups' culture, but what is probably more to the point is that many architects with whom we collaborated in those early years were based in Bloomsbury or in and around Gloucester Place.

In 1946 Fitzrovia was not only home for sections of the rag-trade where the sound of sewing machines preceded today's roar of traffic, but also the ever-present artistic fringe was able to express itself more freely than in areas of greater convention. The location of one's workplace influences its culture and all this had a deep influence on the development of our firm. We were exposed more readily to the broader context in which we work and practise our professions.

A direct consequence of our location was the lack of formality. The 1940s and 1950s were still the days of furled umbrellas and bowler hats. Someone once said a bowler-hatted man in our reception space must be either an accountant or a quantity surveyor! The lack of pomp and formality was uncommon and helped to create a fertile atmosphere where creative ideas could flourish and initiative was free to express itself.

In describing the origins of the culture in which education and training developed in Arups, our location in Fitzrovia is important, and our comparative informality had its place, but they were all part of the fountainhead which was Ove. Although he grumbled about the time people, particularly his partners, spent attending lectures, courses, and conferences, he quite inadvertently kindled their curiosity and provided the stimulus to broaden and widen their horizons. He provided the basis, the foundation for the structure which was subsequently erected and which now includes one of the most sought-after graduate training programmes, in-house seminars,

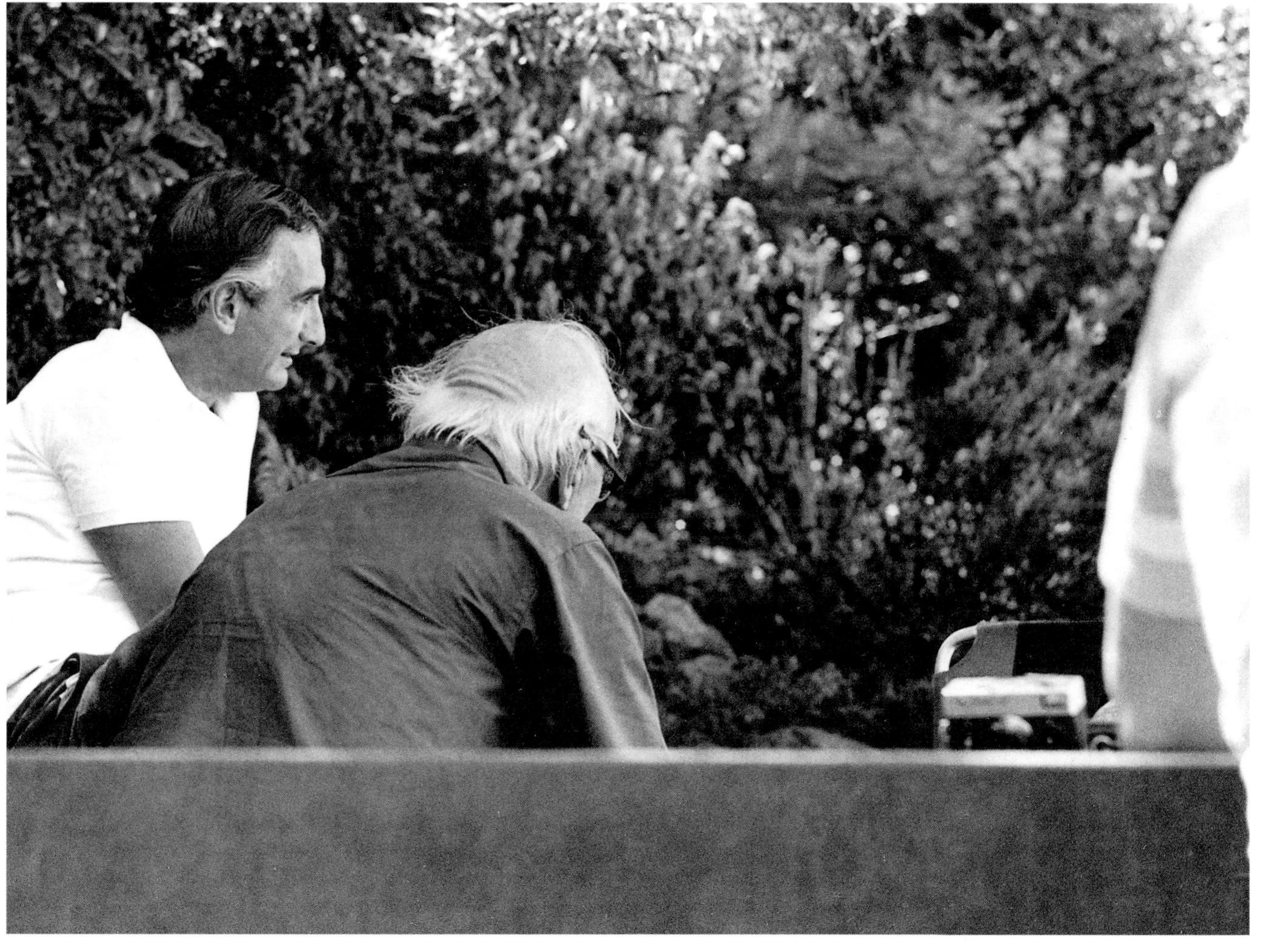

courses and conferences, a positive policy towards attendance at external courses and conferences, and study for further qualifications. Our off-beat origins have more or less fortuitously lead us into what is today called a 'learning society'.

Many will remember one of the greatest football managers of recent years, if not of all time – Bill Shankly of Liverpool. His knowledge of football, its tactics, and its players was legendary, but so was his integrity and his passion for what was central to his life. Shortly after his retirement, and not long before he died, he was interviewed on television. The interviewer asked what football had meant to him. He said that for him 'it wasn't a matter of life and death – it was more important than that!' In his own earthy way, said with that gritty Scottish accent, Bill Shankly expressed how he felt about his life's passion. He may not have been the first to use these words, but they have become immortalised not only in the world of football, but also as a way of describing with as much emphasis as possible any matter of importance and deeply-held conviction. I believe education, and particularly lifelong learning, today is a matter of life or death and, if hyperbole helps to press home the point, I will gladly borrow Bill Shankly's words and say that it is even 'more important than that'.

Arups in the late 1940s, and 1950s, was a very different place from what it is today. The post-World War 2 era is difficult to understand in today's context. It is not easy to comprehend the euphoria, the feeling of optimism, the immense sense of elation which pervaded and engulfed us. There was a universal feeling that something very evil had been overcome. Despite the dawn of the Cold War, there was a consensus of optimism, fuelled by a naive belief that the benefits of technology were only just beginning to pave the way for a Utopian future for all mankind – naive in the context of hindsight. There were perceived to be no limits to growth – or so most people, particularly young idealistic enthusiasts, thought at that time.

50 years ago education was considered to be not so much a right as a privilege. Secondary education was thought to be a bonus – most young people left formal education at 14 or 15. Certainly tertiary education was for the few and was thought be a passport to reasonable living standards. One was then expected to obtain some of that magical 'experience' of one's chosen profession or vocation without which one could not hope to aspire to positions of responsibility, but the phrase 'lifelong learning' – let alone 'continuing professional development' – was certainly not the coinage of everyday language.

At the same time, society was more ordered. There were more accepted codes of conduct, influenced strongly by the family, by religion, and by social convention, than appears to be the case today. These so-called accepted values, coupled with the prevailing optimism about the future of mankind, gave generally a feeling of security, maybe a false one, but nevertheless a kind of reassurance of one's place in the world. But it obviously wasn't a perfect world – far from it. A great deal of the religious and family coding was more a matter of habit than conviction, instinct rather than logic: a reassuring façade perhaps behind which lay much artificiality, hypocrisy and sham. Life wasn't necessarily better – it was totally different! Today life is no less thrilling, no less challenging, but it is a life where education is not just a preparation for life, but its generating current.

50 years ago the world's population was less than half of what it is today – and it will probably double again by the middle of the next century. The problems generated by these burgeoning millions are acute and critical. The issues of pollution, with all its threatening manifestations, were thought to be generally limited to tactical problems of having to deal with the occasional smog – the strategic ramifications of general pollution of our planet and of the atmosphere were not yet on the daily agenda. Global warming or holes in the ozone layer may have been mentioned in the upper reaches of academia, but had not yet become the coinage of public discussion or debate. One didn't discuss the 'quality of life' – nor was there any question of the unthinkable – the issue of human survival on this planet. Acquisitive habits generated by our free market societies were in their infancy, as were modern means of communications, both physical and electronic. The concept of the 'global village' had not yet been articulated. Traditional moral and ethical values were rarely questioned – these conventions provided apparent, if false, secure boundaries.

There is no nostalgia intended in recalling the early

beginnings because, in so many respects, today's world offers immense benefits and opportunities despite the problems which have been identified in the recent past. It is a typical example of the Yin/Yang of life – the positive and the negative. We can live and deal constructively with the problems which confront us, provided we understand that, in this contemporary world, education and a proper understanding of the world around us are perhaps the only guiding stars in an otherwise featureless landscape, the only reliable signposts in a bewildering confusion of false trails and dead ends.

All this means that secondary or even tertiary education is no longer uniquely that passport to a useful and productive life that it once was.

Greater liberty and freedom mean more responsibility, and self-control, and self-discipline which, in turn, mean that a full and proper understanding of the world around us is immensely important; the concept of lifelong learning is an essential ingredient of a civilised society, like food, clothing, and shelter.

An implicit understanding that learning is an ongoing experience is nothing new. Trite aphorisms like 'you learn something new every day' abound. Even Winston Churchill admitted that 'Personally I am always ready to learn, although I do not always like being taught'. But the explicit articulation that lifelong learning is now a necessary adjunct to a civilised and economically successful society is a relatively recent phenomenon and has arisen as much out of need as from desire.

The explosion of knowledge in most disciplines has created the necessity for lifelong learning. Our professional institutions have begun to formalise lifelong learning requirements as a condition for continuing membership. This formal requirement for 'Continuing Professional Development', as it is now known, is on the increase and is likely soon to encompass most professional and vocational walks of life.

We have been ahead of public perception that education and training form a continuum which goes on throughout our professional lives. The underlying theme to our strategy has been to train and educate:

- for quality and excellence in our work
- for personal self esteem and satisfaction
- for the benefit of the public good.

To explore, to innovate, and to inform have always been an essential part of Arups' intellectual baggage, so that general acceptance and encouragement for lifelong learning in one form or another, the concept of education from the cradle to the grave, is now with us, is very welcome and supported by us with all the means at our disposal.

But there is another aspect to education which is particularly relevant to engineers. General Sir John Monash was the Commander-in-Chief of the Anzac – the Australian and New Zealand – Forces in Europe in World War 1. Historians, indeed Lloyd George himself, acclaimed Monash as one of the ablest, if not the most creative general in that awful conflict. Few generals in that ghastly waste of human life emerged with their reputations enhanced. But Monash was one of them.

He was not a professional soldier. He was a distinguished civil engineer who came into military service through part-time participation while a civilian in Victoria. He had, incidentally, introduced the use of reinforced concrete into South Australia and into Victoria. After the war, when he had been acclaimed as a brilliant military strategist and hence been propelled into the upper reaches of public life, one of the positions which he held was that of Vice-Chancellor of Melbourne University. The University which now bears his name had not yet been founded. He held strong views on the education of young people and he articulated these well and forcefully. At the time the professions in Australia were dominated by narrow, poorly-educated men without much cultural background – a situation which has not dramatically altered in many countries in subsequent years. Monash said:

'The greatest mistake in life is to specialise education too early. Nothing is more absurd than for a boy of 14 years to decide what calling he shall pursue. The first essential is to be an educated man. When such foundations have been laid it is time enough to select a walk in life. He should have a knowledge of the laws of nature, of the history of civilisation and of art, music and literature. To whatever extent we lack these things, to that extent is our vision and outlook limited and cramped.'[1]

This was said in 1923 and we can well reflect how even more relevant this need is today than it was more than 70 years ago. Science and technology are today under public scrutiny and criticism and rightly so. Gone are the days when the wonders of science and technology, which at one time promised to take us to Utopian heights undreamt of in the pre-industrial era, are accepted without question and debate. Science and technology are perceived to be a threat to humanity, and civilisation will only survive by an intelligent interprepation and application of what science and technology have to offer. It is therefore ever more necessary for engineers and scientists to have a comprehensive understanding of the world in which they live and which they serve.

In his 1959 Rede Lecture C P Snow first articulated his thesis of 'The Two Cultures' in what he admitted was a necessary simplification. He stated that in moving between the two groups, the scientific and the literary – groups of comparable intelligence, of similar racial characteristics, social origin and income – he found that they had almost ceased to communicate and had little in common in moral, intellectual, and psychological matters. He subsequently summarised his thesis as follows:

'In our society we have lost even the pretence of a common culture. Persons educated with the greatest intensity we know can no longer communicate with each other on the plane of their major intellectual concern. This is serious for our creative intellectual and, above all, our normal life. It is leading us to interpret the past wrongly, to misjudge the present, and to deny our hopes of the future. It is making it difficult or impossible for us to take good action.'[2]

This polemic was underpinned by Snow's deep concern for social justice, for Britain's place in the world, for the plight of poorer countries and their need for appropriate assistance, and for the future of the planet itself.

While C P Snow's division of our society into 'Two Cultures' may be a fact of life and a consequence of our social and educational legacy, I have never been able to accept that it is necessary, and it is downright irresponsible at this time in our history not to take appropriate action. Why a sound scientific and technical education should be mutually exclusive with a broader understanding of the society in which we live, its history, its cultural roots and so on, I shall never know and will never accept. Society cannot and will not be well served by its engineers if they practise a kind of 'apartheid' from their fellow citizens and the society which they are meant to serve.

Ove Arup straddled the two cultures and, since its early beginnings, our leaders have sought to give the firm that extra dimension with an interest in the humanities, a dimension which has been embraced and continues to flourish as we have grown and broadened our technical and geographical interests.

These two strands of learning are central to education and training in Arups. On the one hand lifelong learning is now an imperative if we are to grapple successfully with the world in which we live and which we serve, while on the other, a broad knowledge of our society and its culture is essential for all, but particularly for us as engineers if we are to direct science and technology for the benefit of mankind.

What then are the key features of Arups as a learning society? It is a mixture of the formal and the informal with some emphasis on the latter. The formal includes:

- A graduate training programme accredited by the relevant engineering institutions which is highly regarded in the industry. Indeed, the firm has been described as 'a University for Engineers.'
- Sponsorship for those who wish to study for higher degrees technician training scheme, again approved by the relevant professional institutions
- Short courses for all levels of staff to improve technical, personal and interpersonal skills; these include skills in information technology at one extreme to presentational and language proficiency at the other.

The informal varies from master classes to on-the-job training, from lectures by in-house or external speakers on subjects as varied as racing car design and architects on their own and others' favourite buildings, to operating modules of special, generally in-house developed learning programmes. There are skills networks and external conferences and courses which we attend and in which we participate. We rely on cultural rather than organisational pressures for people to participate – by example and exhortation rather than prescription. A true learning society is not simply reactive to what are nowadays termed business or industry's 'needs'. It invests sufficiently in people so that they can live a fuller life and develop beyond their expectations, work at the frontiers of their chosen profession, and from there fulfil the needs of the future.

Education and training will enable us to aspire and attain the standards of excellence we set ourselves – in our work, our dealings with our colleagues and with the world at large.

In 1920 H G Wells in his book *Outline of History* said that:

'Human history becomes more and more a
race between education and catastrophe'[3]

Prescient words written 70 years ago, yet how apposite they are today. Arups must continue to lead a learning society. But I will let Ove have the last word:

Wisdom's not a thing you find
It is more like a state of mind
and those who have it do not show it
while those who show it do not know it
Indeed a most elusive thing!
To get what's needed of the stuff
A life span isn't long enough.[4]

References

1. Geoffrey Serle, *John Monash: a biography*, (Carlton, Victoria, Melbourne University Press in association with Monash University, 1982).

2. C P Snow, *The Two Cultures and the scientific revolution*, (Cambridge, 1959).

3. H G Wells, *The Outline of History; being a plain history of life and Mankind*, (London, 1932).

4. Ove Arup, *Doodles & Doggerel*, (Ove Arup & Partners, London, 1989).

The ideal of total design advanced by Ove Arup demands a change of attitude amongst design professionals beyond improved integration of the various components. Only through serious co-ordination can there arise the effortless blend of elements into a whole which marks out the truly satisfying end product. For this to occur, there should as well be no rigid boundaries between the intent of the parties concerned. Professional definition thus becomes a phenomenon of institutional difference which must be left behind in the building team.

Bob Emmerson

Integrated design

Brynmawr – an early example of integration.

I was first attracted to Ove Arup & Partners by Sir Ove's approach to total design or perhaps what might be better called holistic design. Ove took the view that the professionals involved in the design process should work closely together, understanding each other's problems to synthesise a design which rose above what might be produced by an individual designer taking contributions from his fellow professionals. He felt that the whole should be more than the sum of the parts. A structural engineer might produce a brilliant structure, which was economic by itself but which totally ignored the architectural constraints and not least the requirements of the services engineer. Similarly, but more rarely in those days, the services engineer might hypothesise the ideal environmental response to a problem which ignored both the special requirements of the architect and the framing requirements of the structural engineer.

He felt that these problems could not primarily be solved by structural changes within an organisation or changes to its methods of working, although they might be facilitated by them, but rather by changing the attitude of the various professionals working within the design process. He encouraged his engineers to think beyond the boundaries of their own design profession. He encouraged his structural engineers to try to understand the aesthetic aspirations of individual architects, to look at the planning and spatial problems posed by the brief, and to understand some of the problems posed to the services engineers. Not least, since for many years he had worked for or as a contractor, he encouraged us to think how a building would be put together and assembled in the most convenient manner.

An early example resulting from this approach is Brynmawr rubber factory where the columns are integrated with the ventilation system and spray out to support the roof shells, designed by Ronald Jenkins, which enclose and define the space and at the same time provide natural light through circular roof lights.

Many designs have one or other dominant defining features which give them a unique character; this may be the treatment of their façades, the way spaces are defined and interrelated, the expression of the primary structure in the architecture, or the blending of the

University of Birmingham, Department of Mining and Metallurgy. Large precast concrete floor unit being lifted into position (extreme left)

The colonnade. The column cluster is capped by extract hoods from fume cupboards (left)

Perspective section showing elements of construction (bottom)

environmental solution into the design. This is not an exclusive list but I suppose I have always been attracted by those projects which appear effortlessly to blend the environmental solution and usually many of the other aspects of the design into a total whole.

An early attempt to integrate services with structure and produce a coherent architecture can be found in Birmingham University's Mining and Metallurgy Building. Here a cluster of four precast columns define and enclose the vertical service routing and support in turn four square precast floor panels. The gaps between the panels provide a horizontal service routing space. Secondary service routing is provided by depressions within the floor panels themselves.

Another early influential experiment can be found at Loughborough University where planning rigour and discipline controls the interaction between structure, services, and architecture. It could be argued it is a prime example of co-ordination rather than integration. Structure, services, and architecture are separated and defined by the planning grid. The grid on plan defines the location of structure and partitions within the minor

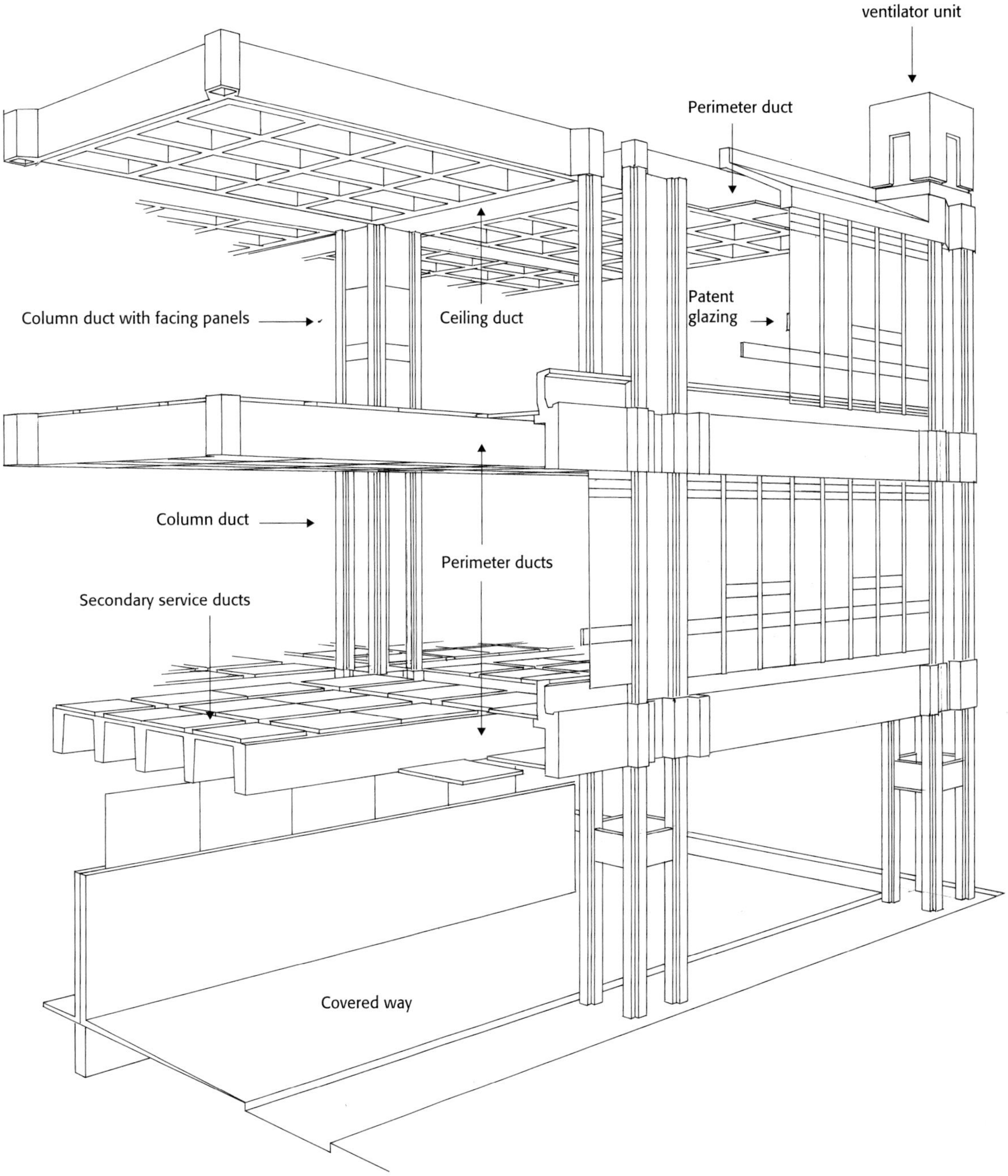

ventilator unit
Perimeter duct
Column duct with facing panels
Ceiling duct
Patent glazing
Column duct
Perimeter ducts
Secondary service ducts
Covered way

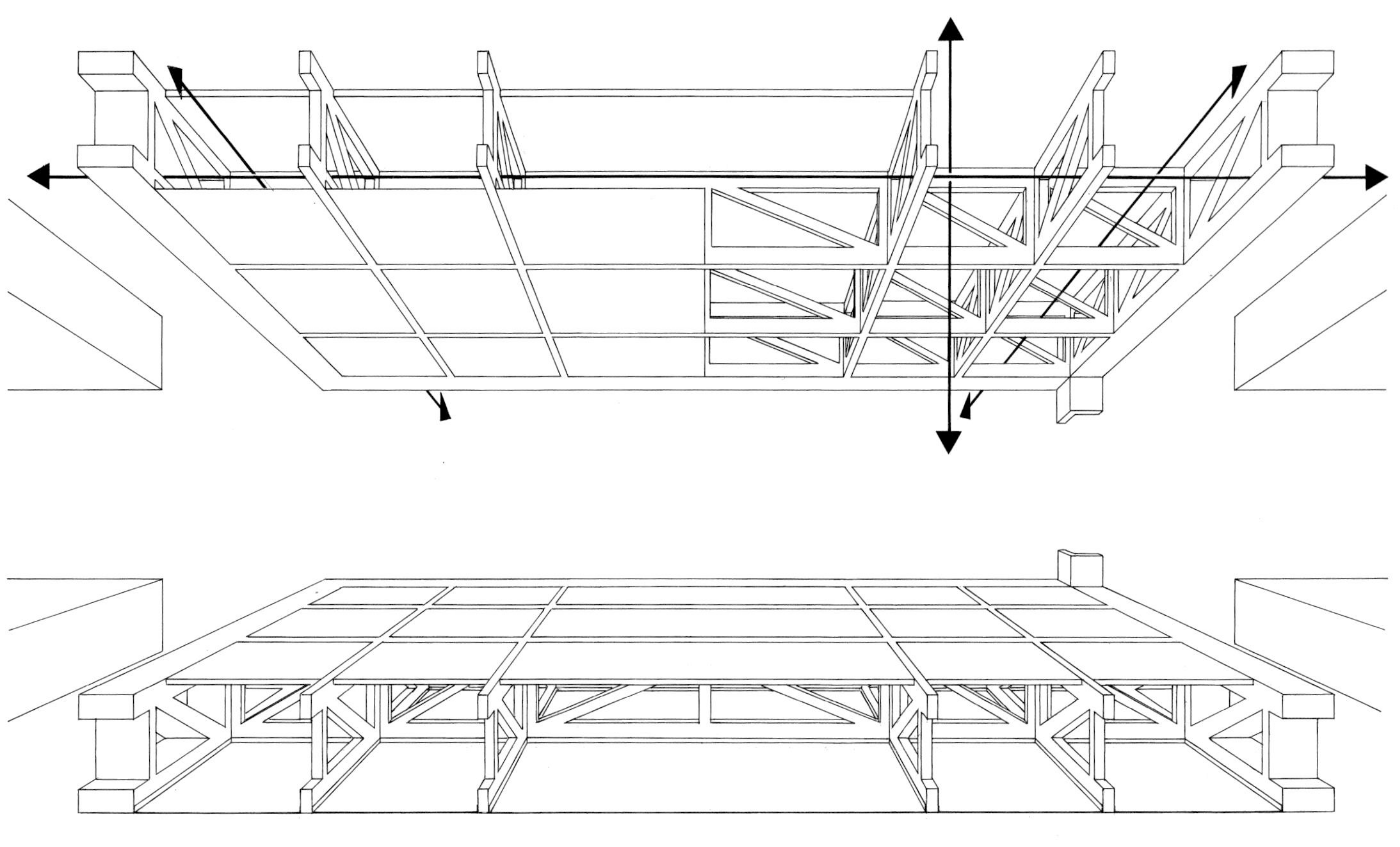

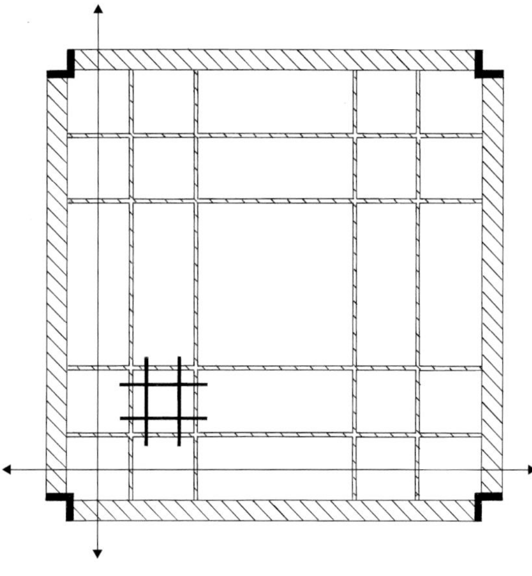

University of Loughborough. (above) Diagrammatic section: zones for services distribution integrated into the structural system. (left) Relationship between all grid networks.

grid and services outlets within the major grid. Structure and services are separated in the horizontal plane and interstitional service floors are provided between each laboratory floor.

This same planning rigour can be found several years later at the Prince Philip Dental Teaching Hospital in Hong Kong and the Glaxo Medicines Laboratories at Stevenage. A clear advantage of this approach in very heavily serviced buildings is that design can proceed relatively rapidly within an established set of guidelines, and client changes during or post design can be undertaken within the same guidelines. Therefore the design itself permits a certain amount of flexibility and adaptability.

A somewhat different approach emerged in the wake of the 1973 oil crisis with the design of a headquarters building on a greenfield site for Lucas Industries. The concern was to produce what would now be called a green building which minimised its energy consumption, particularly compared with then current alternatives. A four-fold strategy was developed which:

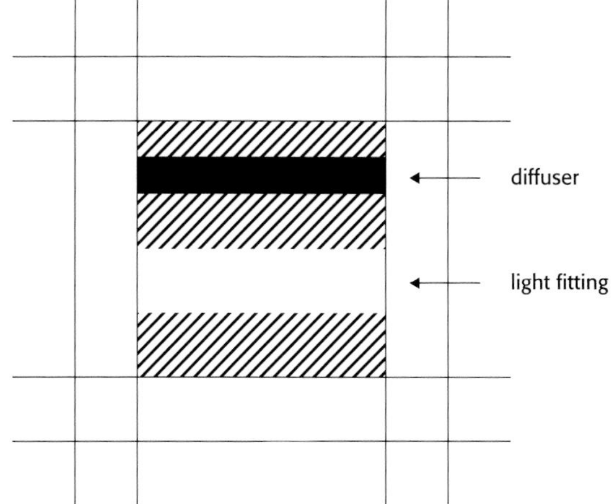

Prince Philip Dental Teaching Hospital, Hong Kong.
(Top left)
Planning grid showing relationship between minor planning grid for partitions and major planning grid for services outlets.
(Top right)
Completed laboratory

- minimised internal gains through the use of task lighting and the maximum use of natural lighting

- minimised external gains through high roof and wall insulation values and solar shading to any glass surfaces

- had ample exposed thermal mass to help dampen rapid changes in internal gains and thus temperatures

- only cooled occupants at their task level and used the stack effect of high spaces to reduce cooling loads further.

The architect used the opportunity to create north-lit domes whose size encompassed a working group and whose form help to delineate the territory of the same working group.

Some years later, Cable and Wireless Training HQ used several of the same techniques, but the wave form of the roofs encourages a natural ventilation effect through the teaching spaces.

Kansai Airport Terminal, on the other hand, uses the waveform roof in a completely different manner. Here the geometry of the roof is determined by the need to keep an airstream attached to it over nearly an 80m distance, and thus ensure adequate ventilation and air mixing over the entire occupied space from one air jet nozzle.

The Inland Revenue Building at Nottingham returns firmly to the totally integrated theme. Here the prefabricated high thermal mass of the external walls and the slabs contribute both to the speed of erection of the building and its energy equation. The vaulted structural panels are self-finished and at the same time reflect light evenly throughout the space. The glazed stair cores also double as thermal flues and encourage cross-ventilation through the office floors.

The attraction of these projects – and many others not discussed here – to the practising engineer is that their success is dependent on a full contribution from him at all stages of the project. He must comprehend and sympathise with the architectural intent in order to do this but he must also grasp the opportunity, understand the client's brief, appreciate the cost constraints, look carefully and comprehensively at buildability, and most importantly communicate clearly and unambiguously to all the parties involved.

Many projects conceal their true engineering from the untutored eye, or even some of the engineering may be appreciated at a superficial level whilst its more important contribution is concealed or ignored. The Lloyds Building in the City of London is an excellent example of this. Whilst the clearly articulated external structure and the co-ordinated vertical service risers can be easily appreciated, other equally important engineering contributions are not so evident. The high thermal mass of the exposed coffered horizontal floor structure serves to dampen rapid changes of internal conditions, whilst the horizontal space above the coffers acts as a distribution space for electrics, sprinklers and extract ductwork. The false floor, as the final layer in this sandwich, provides the means of distributing task ventilation and air-conditioning, small power and most importantly nowadays, communications.

Embankment Place.
(Above) Tapered steel beams during erections.

(Below) A night view across the river with the bow string arches illuminated.

(Opposite) Perspective section showing the integration of services and structure. Note the principal air-conditioning ducts perpendicular to the tapered beams and adjacent to their support where the beams have minimum depth.

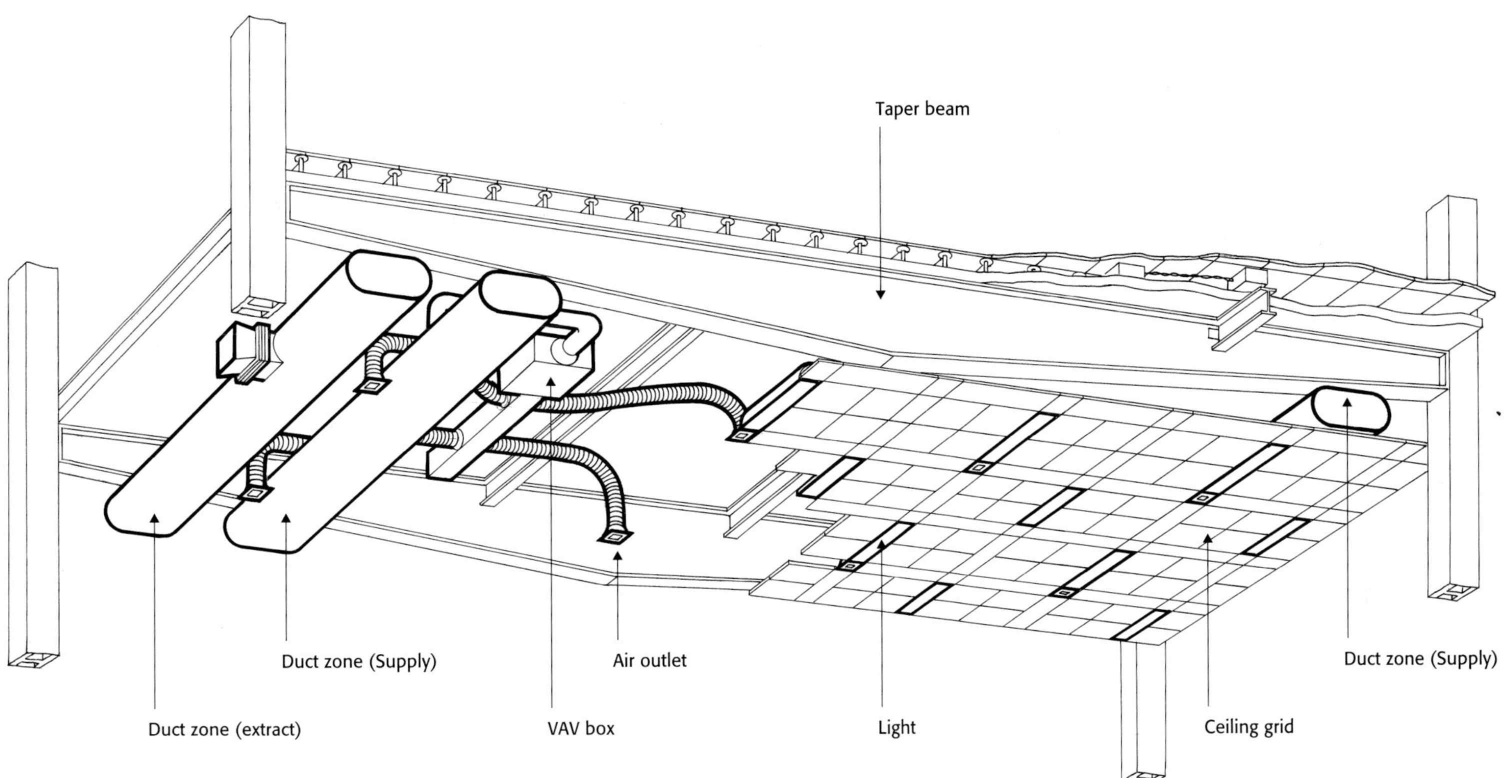

The whole environmental system is integrated with the triple glazed façade where air is passed between the glass panes so that external gains are minimised.

Embankment Place, built over Charing Cross Station, posed a different challenge. Here it was to provide a structure which spanned the railway floors and provided the maximum number of highly serviced dealer floors within a minimum height. A clear-span tied arch spans the tracks at the highest level of the building and lower floors are suspended by vertical hangers from this arch. The hangers are at relatively large spacing to give column-free space in the dealing rooms. The individual floor sandwich thickness of structure and services is reduced through the adoption of tapered steel beams. The shape of the beams roughly follows that of the imposed bending moment which requires maximum depth at centre span for strength and minimum depth adjacent to their support. The additional space provided below the minimum depth of the beams can then be used to advantage to allow transverse distribution of air-conditioning ductwork without greatly increasing the overall depth of the sandwich.

Bracken House, adjacent to St Paul's Cathedral, posed similar but different problems to Embankment Place. Bracken House was purchased by the Obayashi Corporation and was promptly listed. Eventually a compromise was reached which retained the two book-ends of the building whilst the central former printing works of the *Financial Times* were demolished and rebuilt. Again planning requirements, in this case sight lines from the river to St Paul's Cathedral, would not permit a higher development. The challenge was to provide one more highly serviced floor within the overall height than the number that previously existed.

We found that a minimum sandwich could be achieved by placing the fire separating floor at mid-height of the beams, not a solution which would have been selected by a structural engineer working in isolation. Secondary services distribution is between the beams. Above the slab air-conditioning and below the slab extract ducts, sprinklers

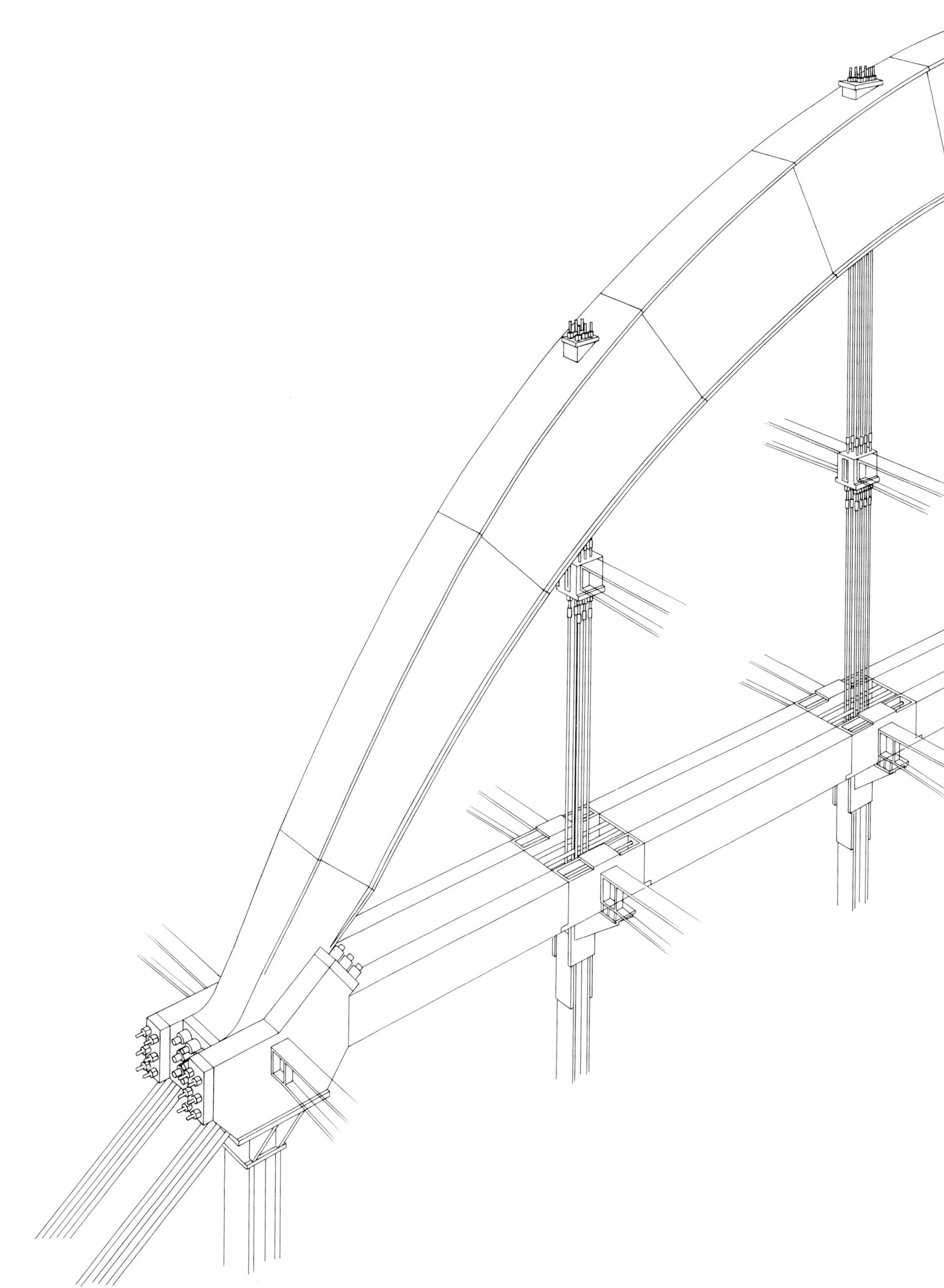

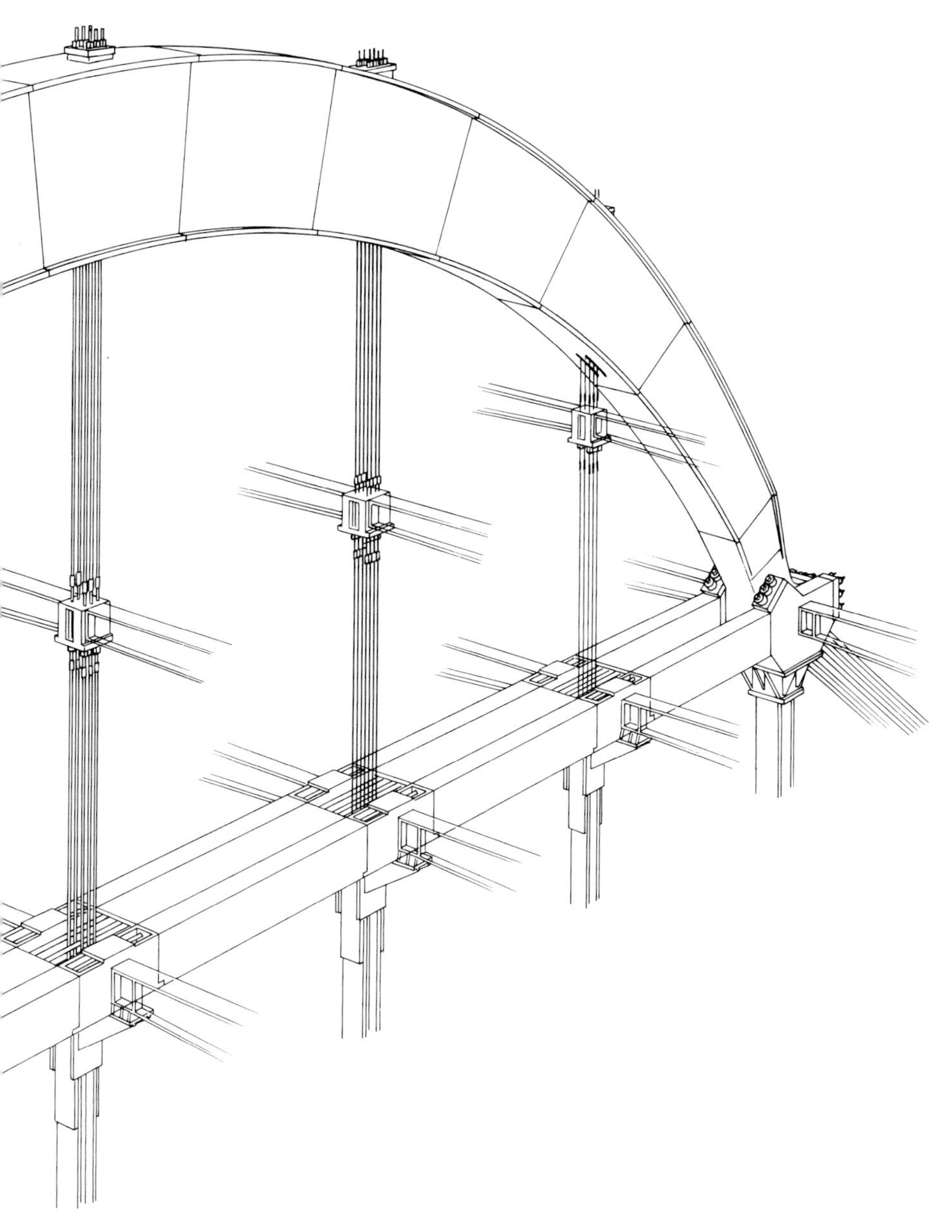

Embankment Place.
The two-storey high bowstring arch supports the entire weight of the nine floors below. It consists of 2m deep by 1m wide steel box sections tied by Macalloy bars and weighs 60 tonnes. The arches are horizontally prestressed as load is transferred to them.

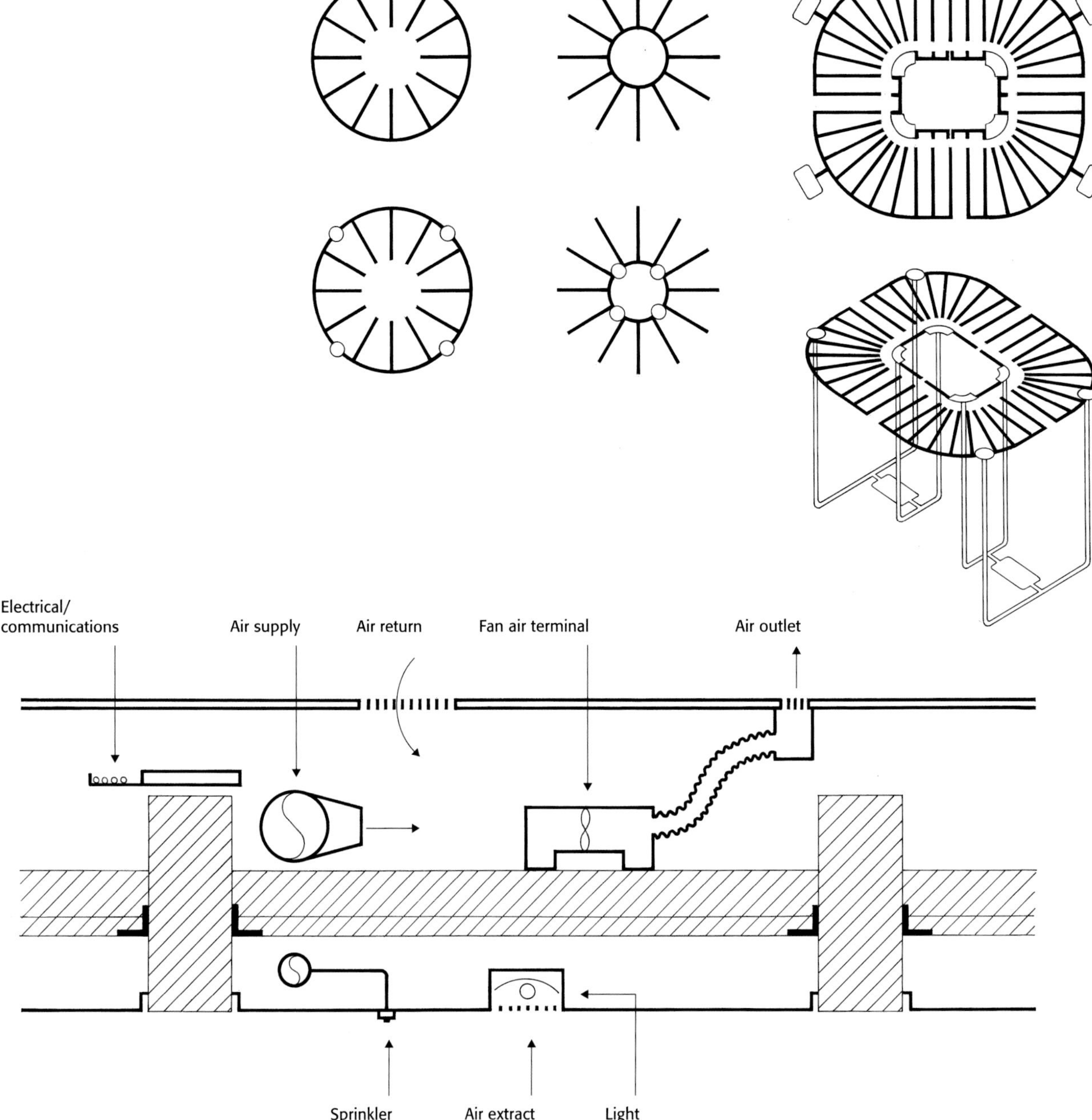

Bracken House.
Pinwheel diagrams of structure and services. (Clockwise from top left) Basic concept pinwheel; structure; air exhaust; air supply. (Far right) plan and perspective of air supply and exhaust.

(Above) Cross-section through Bracken House floor. Note 150mm thick structural fire separation floor at mid-height of precast beams.

and lighting distribute. A tertiary distribution above and transverse to the precast beam carries small power and communications. Primary service distribution is around the perimeter of the building and feeds laterally into the spaces between the beams.

Integration is facilitated where the engineer is sympathetic to the architectural intent, and the architect is sympathetic and responsive to the engineering contribution – but it needs more than this. It needs teamworking, where there is an interactive dialogue and a shared intent to deliver a coherent whole design. This does not (and nor should it) require the explicit expression of the engineering of the building, but it does require a recognition at an early stage of the design process of the engineering contribution to the whole. Integrated design cannot flourish where the design professionals are compartmentalised and establish rigid impenetrable boundaries between their areas of responsibility.

(Top) Metal deck being laid between precast beams ready to receive 150mm thick structural fire separation floor.

(Bottom) Occupied dealers' floor at Bracken House.

If the conventional model of R&D applied to the construction industries is misleading because it is based upon a mistaken analogy between construction and other industries, then, Groák argues, a process model should be preferred over the feedback model. R&D can occur as strongly in the development of the project as in conventionally understood research. It will then have a new role: as 'advanced scouts' for the exploratory work in projects; as 'troubleshooters' – the more conventional role perhaps – and finally a role in new forms of feedback about the solutions during their lifetimes.

Steven Groák

Project-related research and development

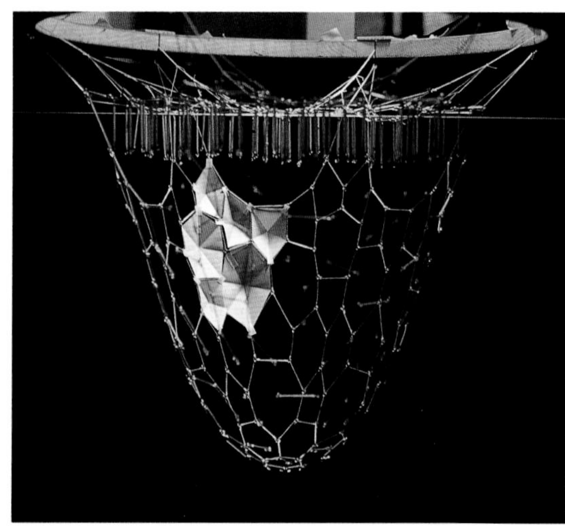

Form-finding structural modelling techniques have become a basis for exploration in projects and research in new technologies. Hanging chain models are used (in tension) to find compression structures (by inverting the derived form).

In the construction sectors of industrialised economies, we tend to identify various 'problems', whose solutions are seen as essential conditions for improving the overall process, the quality of the product, the service to clients, and profitability of the industry. Remove these barriers, remedy these problems (it is argued): only then can we move forward.

But there is an alternative perception: many difficulties and apparent anomalies are not 'problems'; they are characteristics of the industrial world in which we work. To complain about them is like a fish complaining that water is wet. They are endemic to the construction process. Many have been around a while.[1]

Take the familiar complaints about specialisation and fragmentation: Goldthwaite's work[2] on the building industry in Renaissance Florence reveals the 15th century's concern about subcontractors. In the UK, the Emmerson Report[3] of 1962 lists 13 'main criticisms' which 'are all too familiar' and are pretty much repeated in the Latham Review[4] of 1994 (simply the most recent UK version of this catalogue – albeit offering more sensible encouragements than Emmerson). We do not have to wait on their solution before we can progress significantly.

Construction lacks innovation?

One particular 'problem' which has loomed large in recent years is the view that construction stagnates because of its perennial lack of research and development (R&D).

Where has this analysis come from? It arises from more general industrial studies, which suggest that a manufacturing (or similar) industry needs to spend 4–8% of turnover each year to innovate, to maintain market position, to ensure a suitably forward-looking or even experimenting mentality. Construction is seen as committing only about 0.5% of turnover – and most of that in materials and component manufacturing, not in design or site-based processes. Hence, apparently, we cannot be innovating in the construction industries. Yet other industries look (sometimes with envy) at the ability of construction to adapt successfully to changing circumstances. Either the theory is wrong, or the R&D is actually going on – unnoticed. Construction is not a backward form of manufacturing. There is good evidence that R&D, innovation, etc, are going on – not in discrete programmes but very much on a project-led basis, often via a series of projects (not inevitably linked to one firm or a fixed set of firms). Historically, this has been the dominant mode of innovation in construction. Much of the work at Ove Arup & Partners provides good examples – often working with a variety of different clients, architects, contractors, manufacturers, etc. Not only is the innovation project-led, but it has emerged from work with a strong team-based design approach where design responsibility is genuinely shared.

Several of these jobs have resulted in theoretical output which any university research group – and the grant-awarding councils – would have been proud to claim. Those concerned were not on a research budget; they would probably be surprised to have themselves called 'researchers'.[5] To all intents and purposes, the research and innovatory work has been, at best, informal, at worst invisible. Because it does not fall into the normal survey categories of R&D, it is not captured and considered in industrial policy or education.

Today, this informal innovation infrastructure faces two new and significant limitations as the industry adapts again to change:

- The research knowledge (and practical know-how for its realisation) do not reach the public realm – including the general awareness of clients and users. To find out you have to hire the people concerned.
- It is not well co-ordinated with the 'formally funded' R&D of universities and other research centres, which are mostly organised on the model of research in the natural sciences.

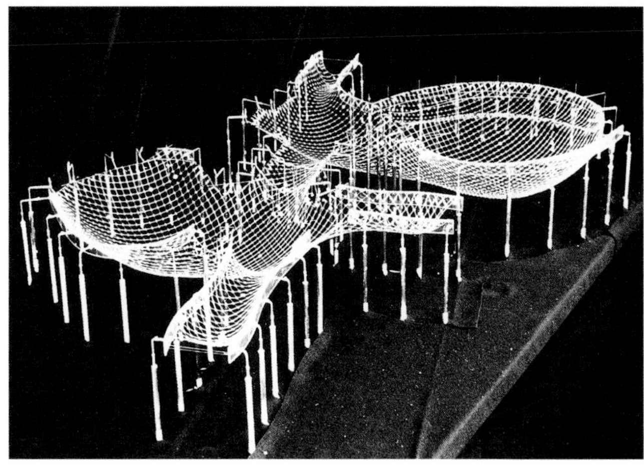

The Mannheim World Garden Exhibition (above and below) involved new exploratory methods for a unique project which led to a new understanding of large span grid shells made from small timber sections.

Construction has suffered from innovation?

Another response has been to suggest that, anyway, what little innovation and experiment we have done was disastrous (eg heavy precast concrete panel systems for public housing) and that we should instead return to tried-and-tested methods of construction which we know will perform satisfactorily. The implied principle is that useful change is brought about by steady development of what we have and do, by first commanding the conventional. Is it? After all, electronic calculators weren't invented by slide rule manufacturers.

In reality, we have to doubt whether we still retain a reliable basis of precedent. Many tried and tested traditional methods of construction are ill-adapted to new circumstances. In the UK, the most obvious example has been the inadequacy of traditional methods to perform in very new conditions of use, internal warmth, and humidity – giving us epidemic problems of condensation. What we might call our 'robust technologies' are becoming increasingly fragile – sensitive to errors of design, manufacture, assembly, or use.[6] These are the methods we thought we could rely upon, as the reference for education and training young professionals, the anvil on which we beat out the recognisable forms of good practice. Perversely, we are experiencing the curious failure of the familiar. Even history has been overtaken by time. In a sense, we could almost assume that every project today is 'innovatory' whether we choose it so or not. Moreover, the very innovatory may even be less risky than the slightly innovatory: the former may be fully recognised and be reflected in the project development – via prototypes, etc. We will have to organise ourselves accordingly.

Kansai Airport (top and above) provided an important project-led programme of fundamental studies in structure, air movement and fire safety.

Innovation in a project environment

Interesting changes are emerging:

- Major construction projects to which R&D studies are linked (eg the EC Joule projects, where public building contracts can include research studies, with reporting in the public realm).

- More deliberate technology transfer, or even 'technology fusion', where we bring in methods, materials, concepts from other industrial domains altogether. The evolving mixture of wider range of project and wider range of specialisms should encourage this considerably.

Throughout the construction sectors of the world, the constant sub-dividing of specialisms and its concomitant diaspora of technical know-how – including design know-how – mean that we have to develop new means of re-integration. Indeed, in this context, it is misleading to describe construction as 'an industry', capable of coherent management, development, or with distinctive sets of organisations. The 'industry' making box-girder bridges is completely distinct from that making 'intelligent buildings' on dense city centre sites or that repairing private houses. Increasingly, we borrow from outside 'construction'; and other industries draw upon construction services. Construction is distinctive because it has always been a sector based on site-based projects, realised through temporary coalitions of many organisations, skills, and people combining variously over time to deploy an extraordinary range of resources. It is this enduring characteristic – which is not a 'problem' – to which new forms of manufacturing may turn in understanding production systems which are moving to mass customisation, and to the projects and champions of current business literature.

Many developments in construction are now being driven by major clients, via their portfolios of projects. Much of the formally-funded R&D which proves useful to construction is not done in construction. 'Technology fusion' is an issue in many industries, and construction should be no exception. Some developments will come permanently from Information Technology and related industries. The locus of construction R&D/innovation is the project rather than the industry.

Changing concepts of practice

Against this background, what does it mean to command 'good practice'? How can any one designer develop this understanding of a technology, and its sound application? And where does project-based R&D fit into this very different 'industry', with its very complex pattern of what constitutes 'the client'? The necessary integration will never be solved by bureaucratic methods – forms of contract, admonitory notices, moral crusades, etc. Flexible team-based approaches offer the most fruitful model we have at present for redefining innovation and R&D which will most benefit the clients and users of our constructions. They will, however, have to become faster, more nimble, and more precise in their responses to clients, context, project circumstances – and certainly much better integrated.

This also redefines a role for those primarily (but not exclusively) charged with an explicit R&D role: to become advance scouts to assist designers and constructors, in addition to current troubleshooter roles. There will need to be different connections with formally-funded R&D groups, although they may wither as the concept of 'publicly-funded research' dissolves into generic information for public access.

The process to which R&D now refers has changed. With the evolution of the kinds of technical information and learning we require, and the new concepts of the process in which they find meaning, we also have to rethink the concept of 'feedback', whether these be lessons from technical matters, use, or contract management.

We now look not just at the building process, not just at 'building as a product' – because the modern conception of what a 'product' is has changed. We recognise that construction is a means to providing a service – part of which may be an investment, a physical environment, an aesthetic object. All the work in facilities management, etc, has shown that notions of construction ownership are changing significantly – as is the attitude to the static elements of ground works, structure, services, fabric, etc. Lifetime behaviours of people will provide a new research/innovation agenda.

So R&D and innovation in construction are now concerned with the necessary knowledge for that emerging process, a process of 'building as growth', in which we increasingly are charged with providing 'design as a non-invasive procedure', as an activity which recognises wider public responsibilities, which is sensitive to the social context as well as the physical environment. It amplifies earlier concepts of the practitioner-researcher, because designers and constructors will develop much further and more usefully these ideas of team-based project-led innovation. This will be heavily driven by the project, by the definition and, indeed, the invention of that project and its unfolding over time – by the client, the designers, the makers, the users.

Western Morning News building, Plymouth. (above, and right) The cladding brackets demonstrate a new succession of projects in which ferrous castings are revised to new purposes.

References

1. Some of these issues have been explored initially in Groák, S. *The Idea of Building*, 1992. Related ideas are given a fascinating review in Happold, E. 'The nature of structural engineering', *The Structural Engineer,* 70 (20), 20 October 1992. Sadly, Ted died just before this book went to press, mourned by many friends and admirers.

2. Goldthwaite, R. *The building of Renaissance Florence*, 1980.

3. Emmerson, H. *Survey of problems before the construction industries*, 1962.

4. Latham, M. *Constructing the team*, 1994.

5. Other chapters in this book (eg by Brian Simpson, John Miles, John Berry, John Thornton, etc.) illustrate variations on this theme.

6. See articles, for example, in *Architect's Journal* in 1986: 9 April, 16 April, 7 May. These demonstrate problems with many familiar forms of constuction – the most astonishing being the failure in a matter of months in lead sheeting roofing to churches, constructed in exactly the traditional manner!

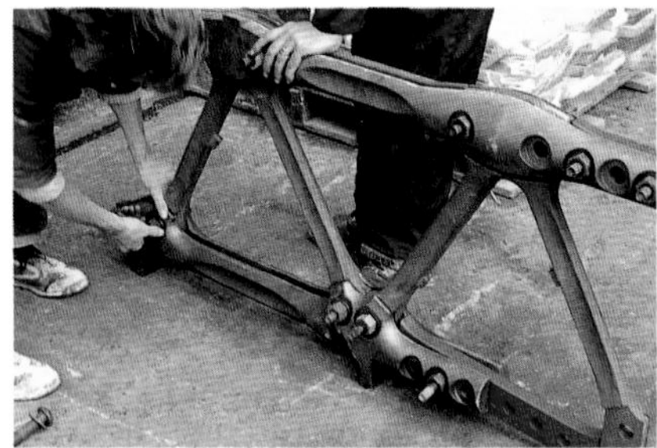

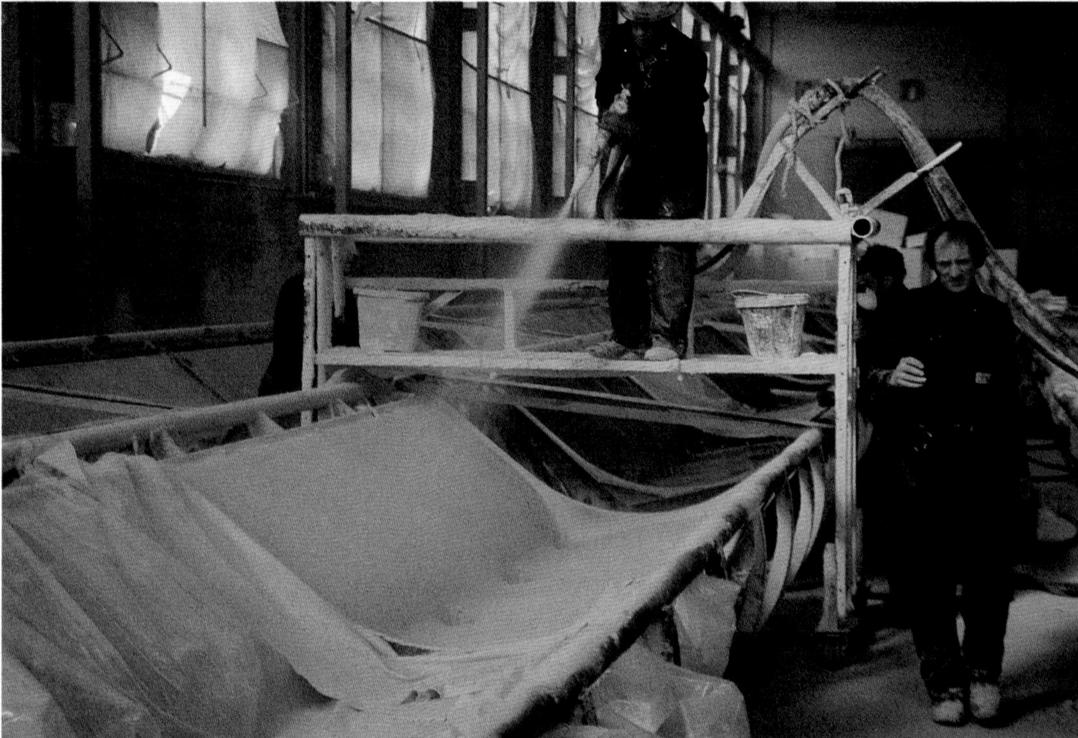

The Menil Gallery illustrates a process of investigation–which had all the systematic hallmarks of true research–into the structural use of brittle materials and the provision of gallery lighting which retained many qualities of natural light.

The crucial role of the materials scientist in engineering innovation, and in avoiding inadvertent innovation, is discussed and shown to depend upon the demands of clients for better buildings more quickly constructed. This has led to innovations in the production of materials; for example, brick panel construction necessitated the invention of mastics and sealants. However, prediction about new materials is difficult because the long-term cannot be simulated. With the integration of a wide range of skills, however a greater degree of holistic design becomes possible.

Turlogh O'Brien

Materials science in construction

The construction industry is a voracious user of materials. It needs them in abundant supplies, at low cost, and at acceptable distances from the site. Yet, despite this, there are few people specialising in the science of construction materials.

Before the 1950s, designers seemed generally to be content to work with the character of existing materials, following design rules which recognised their variability and other limitations. A change in approach began to emerge with the wider acceptance of the Modern Movement in architecture. Whilst the pioneers in the 1920s and 1930s had tried to develop designs based on ideas of machine-made components, design often seemed to become disconnected from the materials with which it would be realised.

The materials industry was called upon to develop products to meet new requirements created by the designers. A simple example of this trend can be seen with panel construction. In traditional brickwork, movement could be accommodated by very small strains at the large number of joints, formed with compliant mortars. Panel construction concentrated the movement at fewer joints. In addition, the relative movements were often greater, due to the nature of the panel materials, and so the performance requirements for the jointing material became more onerous. It became necessary to invent mastics and sealants.

The application of materials science to construction materials should enable better predictions to be made about long-term performance. However, as the history of the last 50 years of construction technology shows, these predictions are extremely difficult. Real time testing takes too long; accelerated testing is subject to great difficulties in relating the test regime to reality; interactions between materials and workmanship variables add further dimensions of complexity and uncertainty.

The change in approach to the requirements for materials was not only driven by designers. Contractors wanted to rationalise site processes and, in particular, to speed up some of them. For example, concrete could be made to harden more quickly, either by using chemicals added to the mixer or blended in with the cement. It was found that a variety of relatively cheap chemicals could be used to alter various properties of concrete.

Better long-term performance has also come to be expected. Exposed steelwork needs regular painting to be kept free from corrosion, with longer intervals between re-coating becoming a requirement. A large variety of products began to be offered by the paint industry to provide this improved durability. As traditional corrosion protection coatings were relatively slow to dry, the new faster-drying products were able to offer advantages during construction as well.

Whilst there is no specific date which can be said to mark the key point in the transition of the construction materials industry, it was clear that by the beginning of the 1960s, chemistry and materials science had come to the aid of construction. With them came increasing variety and complexity within what had formerly been relatively standardised categories of materials.

It is therefore essential that materials performance predictions are based not simply on observation and testing, but on fundamental understanding of the materials and the influences on them in service.

Since the early 1960s, the Arup approach to this type of

problem has been to have 'specialists' within the firm. Whilst materials science is just one among many disciplines, Arups were unique for many years in having materials scientists among the staff. Their prime role has been to act as technical advisers to the designers, but in practice the role has been much wider. They act as key interpreters within their discipline. They must be in close touch with the researchers, in universities and other laboratories, with the technical and research staff in the materials manufacturers, and with all information sources. They must be able to interact with these sources as fellow materials scientists. Then, when working with the designers, they must be able to synthesise and re-interpret the information in a form and language, free from materials science jargon, which is understandable to them.

However, there are some key differences in the approach of the materials scientist and the engineering designer. The latter is often content simply to know what a material will do. He does not seem to want to know why. To the materials scientist this is the crucial issue. Without understanding the 'why', the 'what' can be unreliable in circumstances which may differ only slightly from previous experience.

For understandable reasons, design engineers in construction place considerable reliance on codes of practice. These documents have done much to raise general standards. However, they also seem to have had a negative influence, in that they can be followed blindly, without real understanding of their relevance to a particular set of conditions. The materials scientist has few codes or standards to help him. He must work from his understanding of basic principles, as applied to the nature of the problem, using the information that is available.

There is one respect in which the materials scientist working in construction is closer to the engineer than to the scientist in his methods. At the end of the day, a decision must be made. In many cases the amount of reliable information will be less than ideal. However, a judgement must be made, with appropriate qualification and allowance for uncertainty, to enable action to proceed.

In its optimum form the interaction between designer and specialist adds to the creative process. The 'product' is improved; innovation can be undertaken with less risk; the designer learns more about an aspect of materials technology; the specialist broadens his understanding of the applications of his knowledge and of the context in which he works. With a large number of designers interacting with the same group of specialists, the potential for cross-fertilisation of ideas and feedback is considerable.

However, there are dangers. Over-reliance by the designer on the specialist can actually lead to a de-skilling process. The tendency to call for specialist help to retrieve a problem, rather than to help avoid it in the first place, can lead to specialists seeing so much that goes wrong, that they become over-cautious in their advice.

The increasing complexity of modern construction is leading to an increase in the number of different types of specialist who can make worthwhile contributions to design or construction. On the face of it, this would be expected to lead to less integration in design, and a less holistic approach. This would be contrary to the fundamental tenet of the Arup design philosophy. The paradox is that with a wide range of skills available in the one organisation, greater integration is actually possible.

Materials science, well applied, can enhance creative design. The following notes describe some prominent applications illustrating the relevance of materials science to construction.

Sydney Opera House, Sydney (1957–1973)

The roof structure is made up of precast concrete ribs, formed in manageable segments and bedded on the adjacent units with an epoxy resin adhesive. The units had been cast against each other, so the they would have matching faces, thus enabling a relatively thin adhesive bed to be used (approx. 1mm). The units were then stressed together and the epoxy adhesive provided a medium for even transfer of load from one unit to another. This was a novel application in a building structure in the early 1960s, but it had been used just prior to this for a bridge outside Paris.

The covering to the roof structure is of panel form. Ceramic tiles were placed face down in a mould and a 'ferrocement' mix cast on the back, with integral stiffening ribs. The ferrocement is a mixture of mortar and galvanised steel reinforcing mesh, with a thickness of 30mm. To enhance the resistance of the panels to wind-blown salt ingress, a specially formulated epoxy resin composition was poured into the joints between each of the tiles, so as to form a 1mm thick barrier layer. The roof panels, or 'tile lids' as they were called, vary in size, but all have the same basic geometry. The joints between them were subject to complex movement patterns under the influence of changing temperatures. A high performance sealant was required to fill these joints. The technology of such sealants was still in its infancy at this stage.

In each of these cases, a high degree of materials science input was required from the design team in order to get the best possible formulations for the relatively new products that were being used in an unusual application.

Centre Pompidou, Paris (1973–1976)

A key feature of this building is external exposed steel frame on the two long sides. Through the application of fire engineering principles and various protective measures in the façade, it was possible to leave the steel structure without applied fire protection. The corrosion protection system had to be of a high order, so that requirements for long maintenance intervals could be met. In addition, the system was chosen so that when maintenance is required, complete cleaning back to bare steel would not be necessary. This was achieved by using

Sydney Opera House
(Top) Adhesive joints join structural elements together; sealants contribute to watertightness.

(Left) Centre Pompidou
Pre-applied corrosion protection systems must withstand the construction process.

a base layer of flame-sprayed zinc, followed by epoxy and urethane coatings.

The particular choice of coatings and the build-up of the layers differed from standard practice in France. It had, therefore, to be presented to the relevant authorities to ensure that the decennial insurance arrangements would still apply.

Bush Lane House, London (1974–1976)

This building also features an external steel structure. In this case it takes the form of a diamond lattice, supported off major columns. The same issues of corrosion and fire protection were involved as with Centre Pompidou. However, here, the solution to the fire protection was different in that it involved a circulating water system within the structure itself.

The use of a similar corrosion protection system was considered. However, under the London Building Acts, the structure would only receive a temporary licence, which would have to be renewed every 10 years. This was not considered appropriate for a building of this type and in a location in the City of London. The solution adopted was to form the structural elements out of centrifugally cast stainless steel. This was the first major application of stainless steel for a building structure, although it borrowed heavily on the technology from the brewing and process plant industries.

The choice of the composition of the stainless steel to be used was the subject of detailed study. With the availability of specialist in-house advice, the design team were able to achieve a greater match between the material characteristics and the detailed form of the structure. By having access to the same team of materials scientists, the designers of this structure were able to gain the benefits of the experiences of a different design team, in a different country, on the Paris project.

Credit Lyonnais (1974–1978)

The design development of this building produced the need for complex moulded cladding panels, which would be light in weight and have a surface finish which was sympathetic to the natural stone of surrounding buildings. A solution was developed based on the use of thin skins of glassfibre reinforced cement (GRC).

The idea of being able to reinforce cement with non-metallic materials has long challenged materials scientists. Asbestos fibres, which have an extensive history of use for relatively utilitarian applications, are no longer acceptable.

Centre Pompidou, Paris.
(Left) Cast steel components offer flexibility of form, but protection systems must be maintainable.
(Right) Protective coating systems must withstand all the influences on them.

Al-Shaheed Monument
The sculptural form adds geometrical complexity to the performance requirements for the cladding.

The mechanical properties of glassfibres have made them seem attractive. However, they are susceptible to deterioration in the strongly alkaline environment of a Portland cement matrix. Experimentation with different glass formulations and with different surface treatments to the fibres has sought to overcome this limitation.

Suitable materials became available in the early 1970s. However, considerable detailed work was required to interpret the likely long-term behaviour from the test results that were available, and to develop design rules appropriate to this new material. Of necessity, considerable factors of caution had to be used. The Credit Lyonnais building is the most successful of the early projects using GRC in cladding.

Al-Shaheed Monument, Baghdad (1981–1983)

The winning concept for this large scale sculpture and museum features two 45m high half domes, clad in blue ceramic tiles. The design was developed by one of the large Japanese construction companies. Given that the half-domes were pure sculpture, having no useable spaces inside them, the key issue was how to construct and clad the complex geometry. The solution for the structure was to form it from structural steelwork.

Ceramic tiles are normally applied to a continuous supporting surface with an appropriate adhesive material. There is a long tradition of their use in Islamic countries. The solution for this project was anything but traditional. The idea that had been used on the Sydney Opera House, of tile-faced ferrocement panels, was developed further. The backing for the tiles was a cementitious mix, incorporating a combination of lightweight fillers, which was reinforced with carbon fibres and hardened in an autoclave process.

The particular combination had been researched over the previous few years by the Japanese contractor, but had never been used on such a scale. Carbon fibres have attracted considerable interest for their combination of strength, stiffness and weight. However, cost has restricted their use in construction. A novel form of carbon fibre, derived from pitch by a cheaper process, opened up the possibilities of wider use, even though the properties are not as spectacular. These fibres were used in this application.

Even though the detailed technical assessments showed that the combination of materials being used to form the tile-faced panels should give the required performance in the harsh environment of Baghdad, a fail-safe system was

Al-Shaheed Monument
Prefabricated panels of ceramic tiles, backed by a carbon fibre/cement mix, provide a high performance solution.

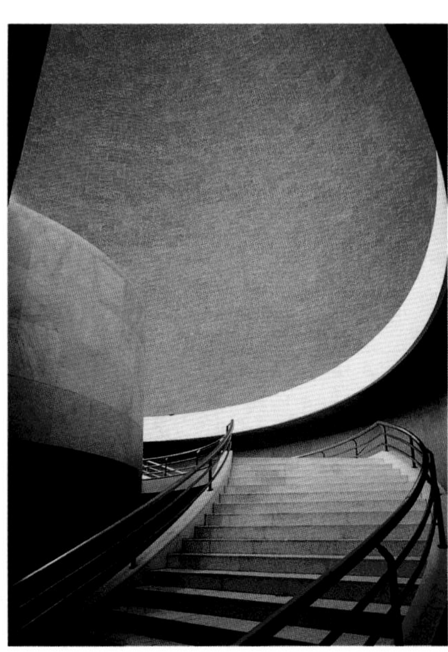

judged to be prudent. The backing to the panels therefore incorporated a light stainless steel reinforcing mesh to which the fixings were connected.

Hong Kong Bank, Hong Kong (1983–1988)

The external elements of the steel structure of this building are wrapped with a fibre protection blanket and clad with aluminium panels. The cladding, whilst designed to exclude most water, allowed any that penetrated to be drained to the outside. Inevitably this meant that the corrosion protection to the steelwork had to allow for damp conditions for an unknown proportion of the time. Any maintenance of the corrosion protection would require removal of the cladding, and would be extremely expensive.

Various options of conventional corrosion protection systems were evaluated. However, it was concluded that the type of protection afforded by concrete encasement would give better long-term performance. The problem was that the normal thicknesses used for concrete encasement to provide corrosion and fire protection could not be accommodated within the zone available. In addition, the added weight of concrete if cast onto the steel off-site would be significant for some elements.

The innovative solution was to adapt technologies which had been used for various other applications in construction. A gun-applied mortar was sprayed onto a carefully prepared steel surface. The adhesion of the mortar was enhanced by the use of a polymer, which also reduced the water loss during the curing period and reduced the permeability of the hardened product. To improve the toughness of the system, particularly for the handling phase, hardened steel fibres were incorporated into the mix. The resulting system, which uses a combination of the low permeability of the mortar and its alkalinity when wet (to passivate the steel) to provide the necessary protection, is a good example of the results that can be obtained from a deeper understanding of material behaviour.

Al-Shaheed Monument
A lattice steel structure provides the framework to which the novel cladding is applied.

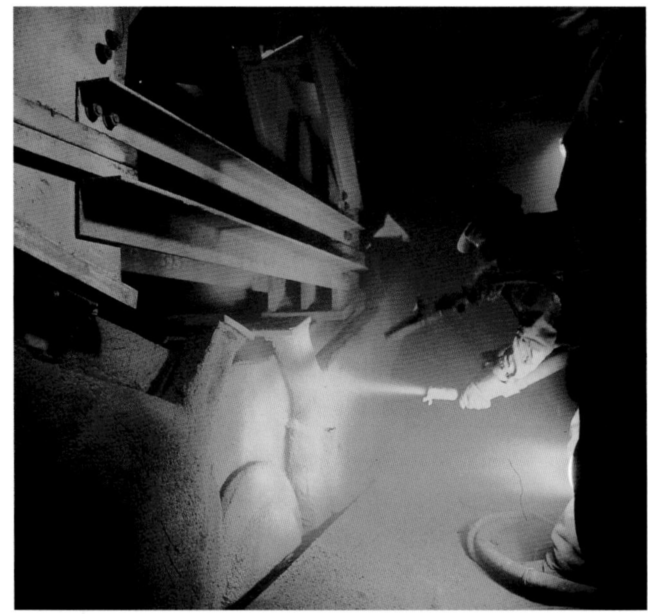

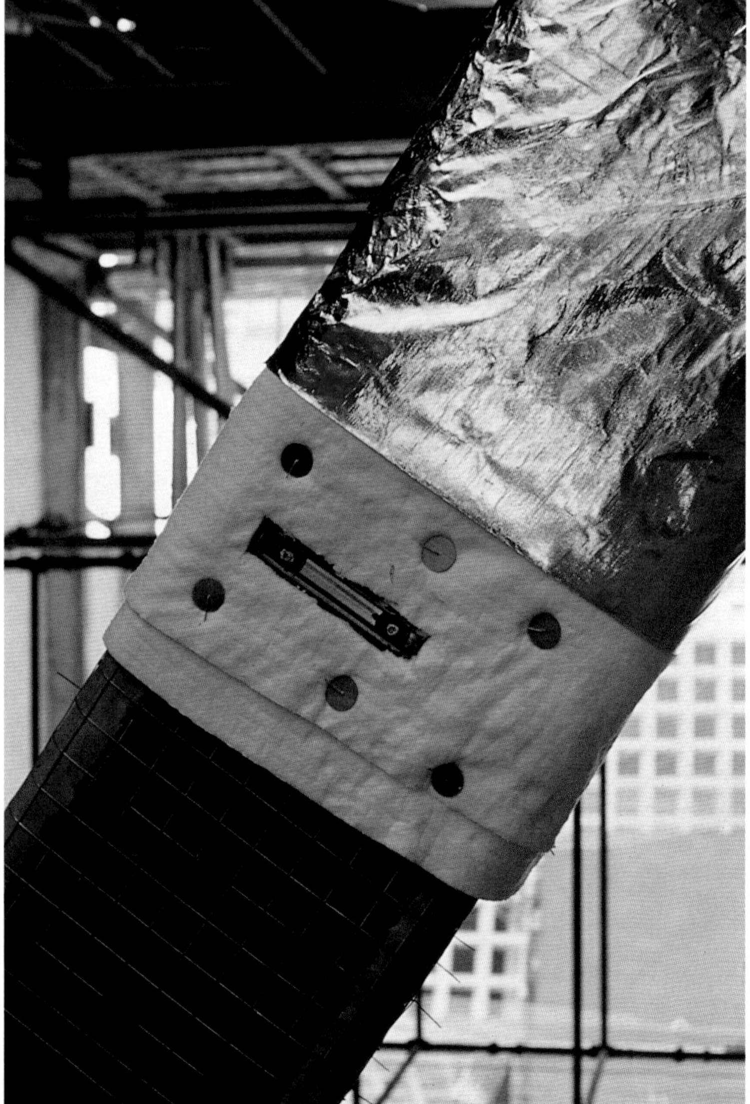

Hong Kong Bank.
The corrosion protection system has to be able to deal with a variety of complex details.

Analytical perspectives

Risk is the frequency of an unwanted event multiplied by the consequences of it. New methods of analysis are described which enable clients to make more rational decisions concerning capital expenditure to minimise risk. This is particularly true of structures housing industrial processes. Analysing risk requires an engineering understanding of constructional, industrial, and legal requirements. It is not just a form of extended insurance policy.

Charles Milloy

Designing with assessable risk

(Opposite)
BP Magnus Platform.

Daily professional life does not normally require us to put a monetary value on human life. It is sometimes argued that human life is beyond price but life assurance companies take an opposite view. Should any value be related to domestic responsibilities? Is £5 million a good offer?

Acute life and death decisions are usually only made by medical practitioners who, thankfully, do not yet consult an expert system running on their PCs. It is believed that there are no software packages which can take account of all the quantifiable and unquantifiable factors. However, planning the Health Service is another matter and the cost-effectiveness of every medical procedure is closely examined. In engineering, we too have to make similar decisions and there are a number of techniques which we can use to help us decide if a process is too risky or, conversely, if we are mis-spending money on unnecessary safety precautions.

There are various kinds of risk. The most emotive is loss of life and much of construction industry legislation is directed at saving life. However, in many of our projects, loss of life is not an issue and our client is concerned with financial loss in all its various forms. Other common risks are environmental risk, loss of function and loss of reputation.

The Piper Alpha disaster in 1988 cost hundreds of lives because the structure failed due to fire and blast loadings beyond the design basis. The subsequent cost to the North Sea oil industry has been enormous. The recent collapse of the tunnels under construction at Heathrow cost no lives, but will cost considerable sums in repair, lost revenue, and has dented the reputation of the UK tunnelling industry.

Codified design

Much of engineering design is guided by UK and European codes of practice. In the main, these codes have evolved over time and have been adjusted to take account of recorded failures and changes in technology. For the UK land-based construction industry, buildings designed and fitted out to the UK codes of practice generally protect human life and ensure the building performs its design function. In general, our codes serve us well and the UK has a good safety record for completed buildings. As we know, very few, if any, of the loads and capacities of a building's structural, mechanical, and electrical systems are known exactly. Good design strives to keep the capacity greater than the load in all reasonable circumstances and, if the capacity does remain greater than the load, all is generally well. Some of these loads, such as the building self-weight, are known relatively accurately at the time of construction but may vary during the structure's life. Others, such as wind loads, vary considerably and much effort has been expended to be able to predict wind loads.

Our codes take account of the more common variations, mostly by the application of partial safety factors for loads and materials. An overt statistical approach is adopted in the case of wind loads but most of the partial safety factors (factors of safety) have evolved over time and have been found to be suitable for well-documented materials and environment conditions.

When the load variations are less well understood, or there is not time, or it is too dangerous to let the codes of practice evolve, it is necessary to examine the risks directly. For example, fire and wave loads on North Sea

platforms were virtually unquantified 20 years ago. For these, and other unprecedented design conditions, it is necessary to develop a risk-based design approach; it is abundantly clear that the evolutionary approach to fire safety was unacceptable at Piper Alpha.

Risk definition

It is worth remembering that risk is defined as the product of two variables, being the product of the frequency of an unwanted event and the consequences of that event. The frequency is usually stated in events per year, or per million years, and the consequences in terms of human loss or financial loss. It is possible to have high frequency, low consequence events – like bumping the family car – to low frequency, high consequence events, such as the collapse of a fully occupied multi-storey office block.

The methodology of a risk assessment is to find out all the ways a system can fail, estimate how often each failure can be expected to occur, and assess the consequences of each failure. In this sense, 'system' is used to describe any structural, mechanical, and electrical system, or even a management system.

In many cases, the circumstances leading to the failure can be identified by inspection, especially if the system components are set out in a linear form and there is no redundancy. The obvious analogy is that failure of any link in a chain supporting a load will lead to 'system' failure. Such simple linear systems rarely exist and it is much more common to have systems where there is redundancy and single component failure does not lead to system failure. For complicated systems, it is rarely possible to spot all the failure combinations by inspection, so techniques such as Fault Tree Analysis have been developed to help ensure that a rigorous and comprehensive failure identification can be completed.

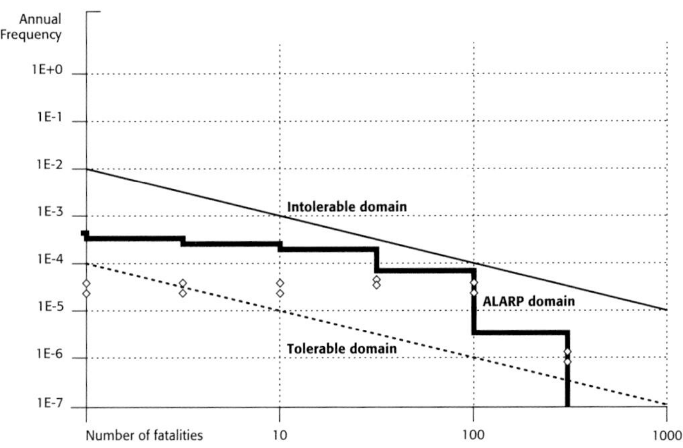

Figure 1. A typical risk acceptance criteria diagram.

Qualitative and quantitative risk assessment

Risk analysis can be both qualitative and quantitative. All good engineers carry out a qualitative assessment, often subconsciously. They assess the risk and design it out if at all possible.

A quantified risk assessment predicts the failure frequencies and consequences mathematically and clearly requires more resources. It can only be absolute when there are reliable historical failure data for the system components.

A qualified assessment must be carried out before proceeding to a quantified assessment, so this effort is never wasted. Semi-quantified assessments, which are largely based on engineering judgement rather than on formal failure data, are often an excellent use of resources which take full advantage of the limited component failure data which are normally all that is available.

The estimated risks need to be compared with the risk criteria. For risks to life, the Health and Safety Executive have recently quantified tolerable risks to individuals and to society at large. Legislation covering the chemical, nuclear, and off-shore hydrocarbon industries has been enacted which requires formal quantified risk assessments to be prepared and submitted to the Health and Safety Executive. The *Construction (Design and Management) Regulations* have recently brought the same ethos to the UK land-based construction industry.

The results of a typical quantified risk assessment are shown on Figure 1 which also shows the risk criteria. The upper zone is the intolerable region; risks in this zone are too dangerous and the activity should not take place. The middle zone is the ALARP zone (As Low As Reasonably Practical). The ALARP principle is an excellent concept as it encourages the risk reduction to be cost-effective. Further risk reduction measures should be investigated and costed and, if cost-effective, should be implemented. It is at this point that the cost of statistical life comes into play. Spending between £0.5 million and £5 million to save a statistical life is worthwhile according to current wisdom. If the cost was approaching £10 million, as it was at one stage in the nuclear industry, it is likely the money would be better spent elsewhere.

A typical risk profile for an offshore platform is the risk domains shown on Figure 1. The profile can be seen to be mostly contained in the ALARP zone but extends through into the intolerable zone at one point. No matter the cost, additional fire protection must be added to take the high point back into the ALARP domain and then the cost-effectiveness of further measures to reduce the risks more must be investigated.

The use of these techniques on Arup projects is best illustrated by two examples; the Magnus North Sea Platform and the Øresund Link between Denmark and Sweden.

BP Magnus North Sea platform

As mentioned above, the Piper Alpha disaster resulted in new legislation governing the safety of UK offshore platforms. Part of the legislation covers the preparation and submission of a Safety Case for each platform. The legislation is not prescriptive, in that it does not stipulate which safety systems should be installed, but suggests risk targets which should be met in a manner which best suits the platform operator. To satisfy the Health and Safety executive who administer the regulations, it is necessary to submit a quantified risk assessment with each safety case.

The failure behaviour of a platform structure in a fire is very complex and it would be unwise to predict the precise behaviour of every conceivable fire. Typically, an oil or gas leak will start at a flange on a high pressure pipeline. If the leak ignites, it is probable that an automatic deluge system will operate; the emergency shut down valves will close and the pumps will stop, allowing the system to depressurise. If the leak is large, the deluge system may not extinguish the fire, which can grow into a pool fire. The heat of the fire will depend on the amount of oxygen available and structure's temperature rise will vary accordingly. In addition, the times to failure of exposed steel columns will vary with their actual load and with associated reductions in strengths and stiffnesses. In other words, there are a lot of variables.

It is also important to predict the mode of failure and decide whether the collapse of one part of the structure will bring down adjacent fire walls, allow the fire to spread into the next compartment, and escalate towards the Safe Refuge where the platform staff have mustered to await rescue.

It is meaningless to solve the problem deterministically; there is no code of practice which has been tried and tested over many years. Only a probabilistic solution has any merit.

The studies on the Magnus platform included fire and structural engineering to model the fire sizes and temperature rise of the structure, and the prediction of failure modes and failure times. Using the Event Tree approach illustrated in Figure 2, a large number of leak sizes and positions were investigated which took account of the reduction in leak rates as the system was depressurised.

The airflow through each module was modelled using computational fluid dynamics. The results were checked later by measurements on the platform. The effects of the deluge and of the smoke were also included to construct behavioural matrices of temperature rise rates for critical structure for a number of leak sizes, leak positions, fire types, and external wind conditions.

The structural engineering initially had to understand the primary structural loadpaths and differentiate between the critical and non-critical members. For the critical members, it was then necessary to estimate the actual load in the member at the time of the leak. As the modules are designed for several load cases, including transport and erection, some members were under-utilised to a fair degree and could accept significant reduction in strength before failing.

Most structural engineering is driven by the compulsion to avoid failure so there is comparatively little knowledge about the actual failure modes of real structures which often have a lot of reserve strength. A mathematical model of an entire module on Magnus was created and run numerous times with a selection of members missing to simulate the loss of a critical member due to fire. This analysis provided a good background to making difficult decisions about the probability of module failure in a certain time, the way in which failure would occur, and the damage the failure itself would inflict on the fire compartment walls.

The results of our work were combined with chemical engineering studies to estimate the frequency with which the Safe Refuge might be impaired.

It is very important to state that the probabilities are recognised as estimates, and the analysis method is based on engineering judgements of uncertainty and not on pass/fail criteria from codes of practice.

There is the inevitable, and entirely justifiable, suspicion that it all turns into a great numbers game and the statistics can be massaged to show just about anything. A more positive viewpoint is that the methodology forces experienced engineers to use their skills to understand the behaviour and allows them to use their best judgements to reflect, and to record, their own uncertainty about failure times and modes.

Having assessed the risk and compared it with predetermined risk criteria, various options are open to the team responsible for the platform. For example, further non-linear analysis may show that the structure is stronger than indicated by a simple elastic analysis. Or, the addition of more gas detectors may detect the leak earlier allowing a more effective emergency shut-down. By estimating the cost of a number of remedial measures and the associated statistical lives saved, sensible decisions can be made about spending on increased safety.

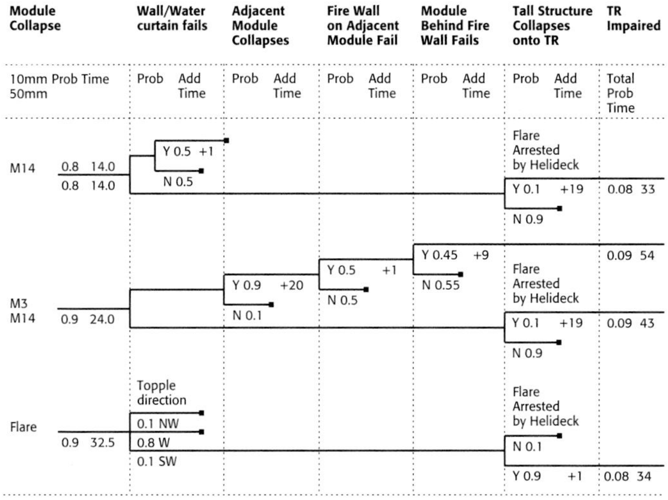

Figure 2.
A typical event tree.

Øresund

Another interesting use of designing with assessable risk has concerned the design of the Øresund Link between Copenhagen in Denmark and Malmø in Sweden. The planned link across the Øresund is a glorious mixture of submerged tunnel, artificial island, low approach bridges and a high bridge over an important shipping lane. The 16km link leaves the Danish shore in a 4km long submerged tube carrying a rail track and a two-lane motorway in each direction. The link is submerged at this stage to stay clear of the approach to Copenhagen's Kastrup Airport and to avoid obstructing the main shipping channel which passes adjacent to the Danish coast.

The next section is an artificial island taking most of the spoil from the tunnel and bridge foundations. The artificial island is adjacent to but does not interfere with the Island of Saltholm and its bird sanctuary. Approach bridges of approximately 100m span then rise from both the artificial island and the Swedish coast to meet at the 490m span cable-stayed bridge over the Flinterenden shipping channel. The risks are varied and numerous.

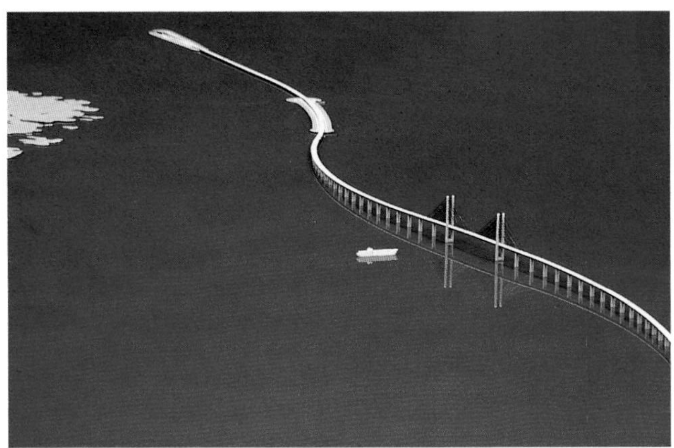

(Left and opposite) Øresund high bridge: proposed scheme.

They include train derailments and collisions which could demolish vital parts of the bridge. They include substantial fires as the result of accidents involving hydrocarbons being transported across the link. They also include explosions which could so badly fracture the submerged tube sections that the tunnel section would flood, and they include impact by large jets, straying from their approach to Kastrup Airport.

All these risks and more have been investigated and quantified. As a result, the ability of the bridge to survive without a truss diagonal or a couple of cable stays has been made a requirement. Fortunately, the frequency with which a fully-laden Boeing 747 can be expected to crash into the bridge has been found to be tolerably low as no economic bridge design could withstand such huge impact energies.

By far the most contentious issue is that of ship collision with the bridge piers. Historically, there have been only a limited number of large ships colliding with bridge piers and the circumstances of each are so unique that the data cannot be used directly by Øresund. A complex numerical model of the shipping collisions in the Øresund has been developed and the model has predicted the incidence of large ships hitting the piers and disrupting the link. It has prudently been decided to protect the six piers near the main shipping channel by forming artificial islands around their bases; both to protect them physically and also to reduce the incidence of hull damage leading to oil spillages. To provide artificial islands for the remaining 60 piers would be a great expense, but it would also significantly reduce the cross-sectional area of the Øresund which may restrict the flow of the North Atlantic into the Baltic.

The ship collision model has been used to determine the design impact force for the piers but this would result in very large pier bases and considerable additional expenditure.

At present, the expenditure on increased bases to give a lower frequency of disruption due to ship impact does not appear to be cost-effective, so the collision model is being re-examined to see if the initial assumptions are valid. An attempt will be made to validate the model against the history of collisions and groundings in the Øresund.

Conclusion

As with many statistical techniques, risk assessment in design can be a very valuable tool in the right hands. In the wrong hands, it can be used to obfuscate a problem and provide a false sense of security.

Used properly, it obliges the designers to understand fully the implications of their design and how the function will be used throughout its life. In extreme cases, it can show that a system is too dangerous for the benefits it provides. In more usual cases, it can highlight the risks which are not always evident and can help to select a level of safety which is neither too much nor too little.

There is sound evidence that formal risk assessments will be required more and more, both in design and construction.

Fire spread diagrams.

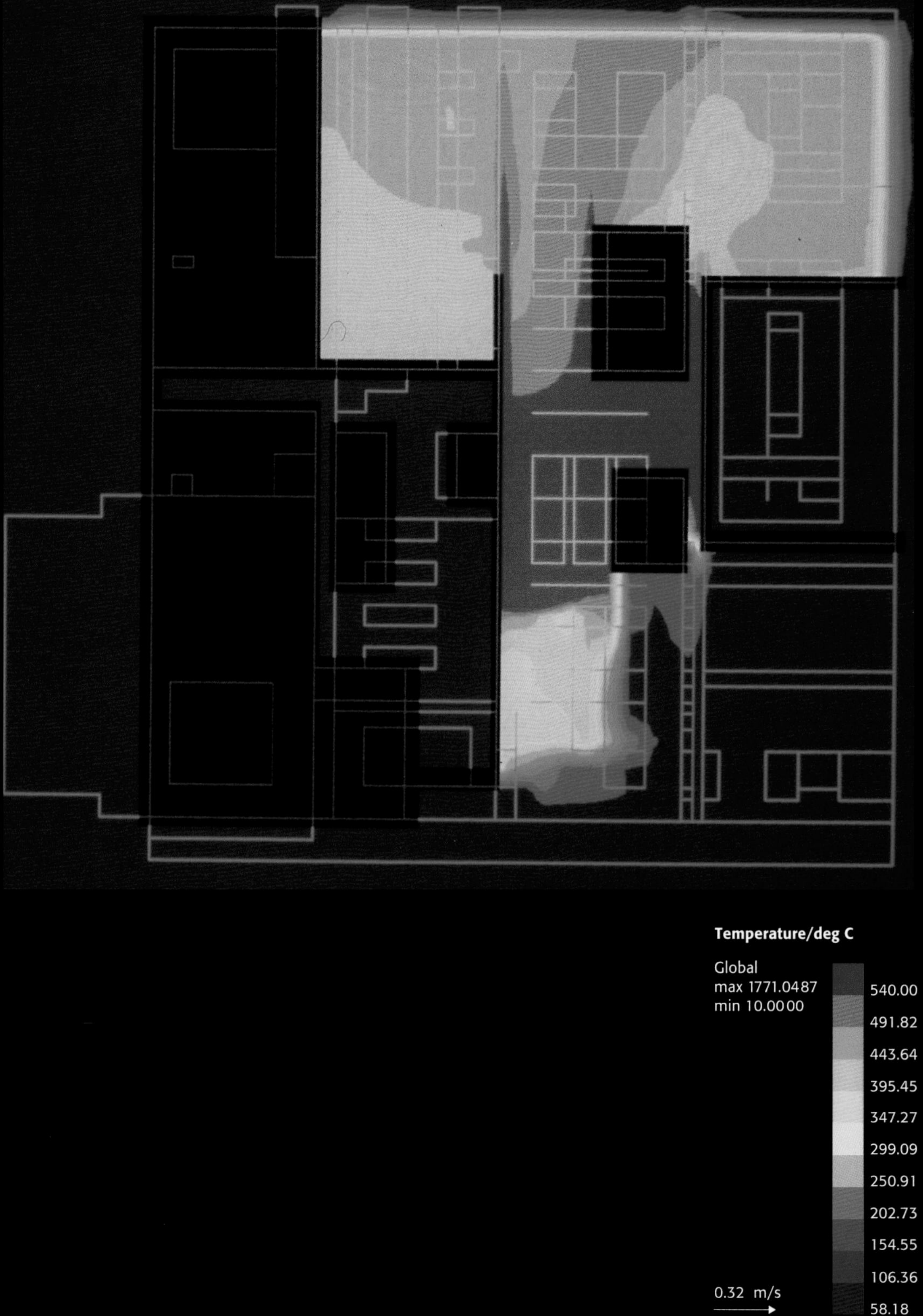

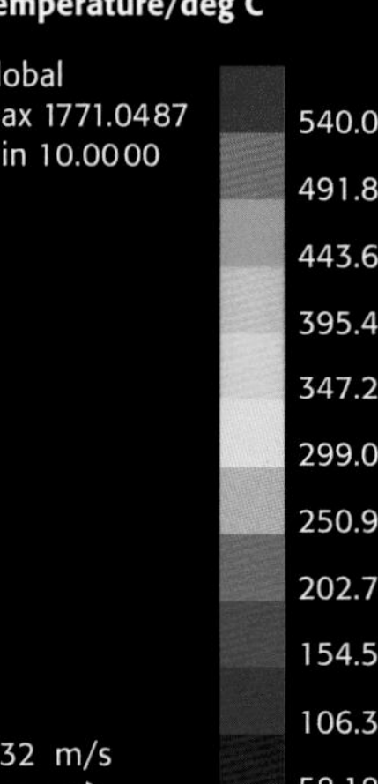

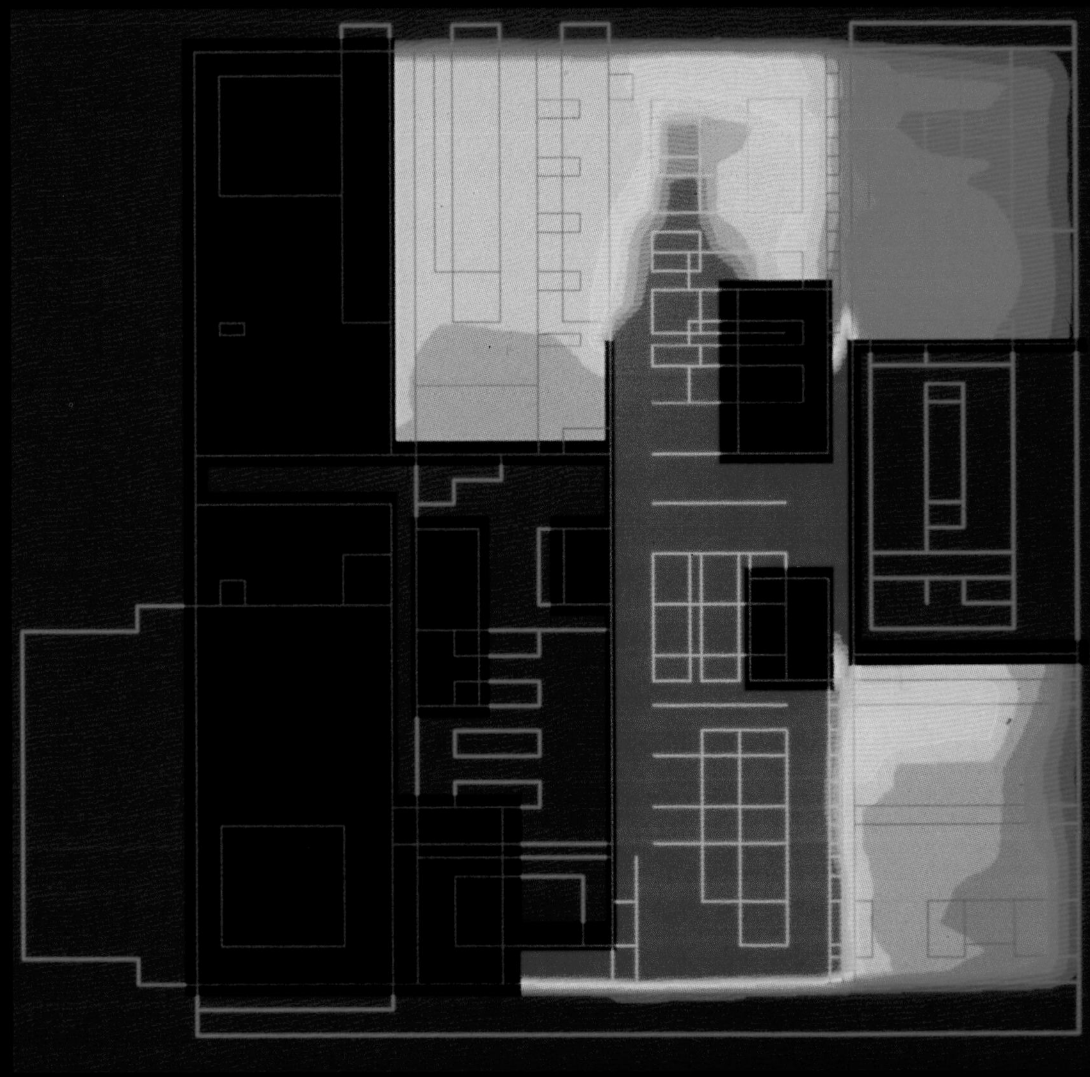

Temperature/deg C

Global
max 926.4881
min 9.9999

540.00
491.82
443.64
395.45
347.27
299.09
250.91
202.73
154.55
106.36
58.18

13.75 m/s

In the design of airport terminals and offshore structures, codes covering fire safety cannot be satisfactorily applied. New concepts have been used to make these structures safe from fire: the 'Cabin' concept, the use of CFD to calculate smoke movement, modelling movement of people, and scenario-based design. Investigation of these in particular cases has led to new ways of thinking about safety from fire.

Paula Beever

Design for fire hazard

In common with more familiar building engineering disciplines, fire safety engineering is about better design. 'Better' in this context may be interpreted very broadly: more architecturally satisfying; easier to use; more commercially viable; more environmentally beneficial; and so forth. Fire safety design does not therefore contribute only to safety: it can be used intelligently to influence all of the aspects which improve building design. In other words the value of fire safety engineering is to enable designers to explore how to deal with fire safety in a rational and flexible way.

In order to illustrate the current position, projects chosen for discussion here fall outside the normal categories of buildings and occupied structures as envisaged by building regulations.

Transport terminal buildings, with potentially long escape routes in large volume spaces, demand an understanding of fire development and smoke spread in order to keep people safe as they move away from any fire incident before it is fought.

Offshore structures, particularly in relation to the safe use of combustible materials, require an understanding of how to restrict fire growth and keep operatives safe until a means of rescue arrives, possibly after the fire is extinguished. Use of quantified risk assessment is also needed to prioritise modifications to existing structures.

Each of these situations needs a quantifiable understanding of the physics of fires applied to the structure, service systems, and functional planning of the buildings and installations involved. They also require an understanding of human response, pedestrian traffic engineering, and the pre-planned response of the emergency services, in order to complete equations of safety for personnel and (where significant) property.

The essence is to quantify and make sound engineering judgements about the particular objectives of designing against the effects of fire. It may be surprising to realise that many of the fire safety-related aspects of the building regulations are subjective and do not necessarily have an engineered basis. As we continue to develop our discipline, especially in areas outside the scope of current regulations as described below, so we will be better able to influence on a rational basis the design of all buildings.

Transport terminals

In recent years airport terminal buildings and railway stations have increased very significantly in size. Designs have changed to permit greater user convenience as well as accommodating large areas of retail, office, and passenger service facilities. Strict compliance with the fire safety codes of many countries would frustrate these goals and result in less efficient designs. These buildings have characteristic features in common, and the problems which have to be tackled illustrate how an integrated approach may be used to develop a soundly-based strategy. A number of calculation techniques come into their own, both simple and sophisticated, and covering a range of technical disciplines.

In order to ensure the convenience of large numbers of people using a terminal building, it is frequently necessary to have very large public spaces. This removes the usual protection of fire-resisting compartment walls to limit fire and smoke spread. In the event of fire, people may have to move long distances across possibly unfamiliar territory in order to reach a place of safety. In international terminal buildings the escape procedures may be complicated by the need to maintain the landside/airside boundaries, and the problems of communicating to people in foreign languages. However on the positive side, transport terminals are designed specifically to promote a smooth flow of people. There tend to be open spaces for circulation which are largely free of combustible material. In addition there may be a high ceiling over all or much of the terminal which can act as a smoke reservoir. It is the task of the fire safety engineer to exploit the positive features of this kind of building and to propose additional fire protection measures which will ensure that people are at no more risk in the proposed building than in one that can comply with the letter of the regulations.

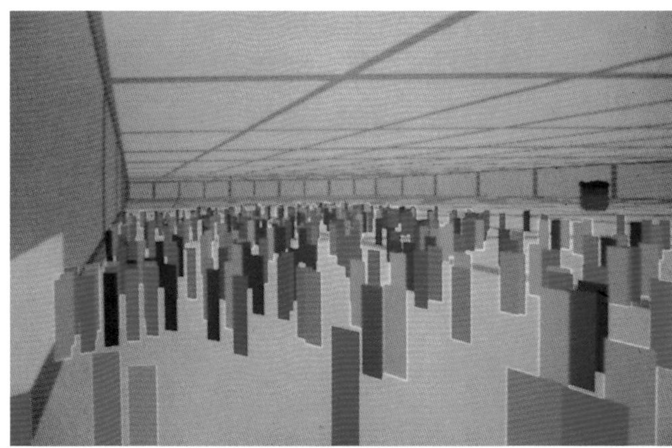

Combined modelling of flows of people and smoke movement are used to demonstrate safe evacuation.

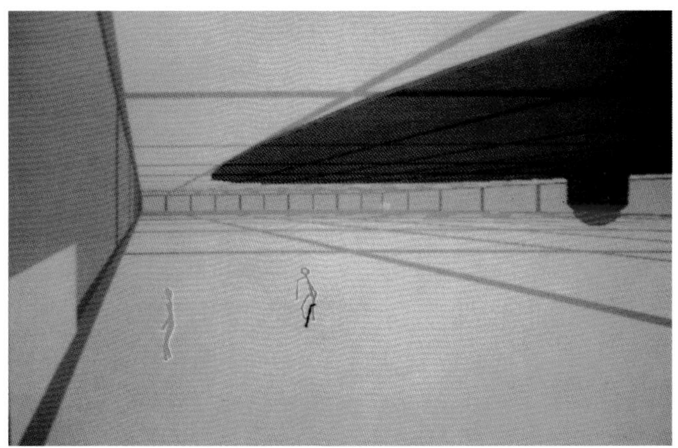

Cabin Concept

The new terminal building at Stansted Airport opened in 1991 provides a good example of how a range of techniques may be brought together to form a coherent fire safety strategy. It was decided that people should be protected from the effects of a major fire as they made their escape by fitting all the areas of high fire load (for example shops) with a local roof to carry sprinklers and a smoke extraction system. The required extract volume for smoke was calculated using standard techniques based on the volume of air entrained into a hot plume, and assuming that all the heat from the sprinklered fire goes into the hot smoke layer. The use of this conservative approach ensured that smoke would be contained within a limited area, even without fire-resisting walls. This was termed the Cabin Concept. This essentially simple but very useful idea has found wide application in, for example, the stations on the New International Lantau and Airport Railway Hong Kong and at Kansai Airport in Japan.

Smoke movement

Outside the Cabins, an unsprinklered fire might occur in seating or in passengers' baggage. The ceiling of the building is some 12m above the floor and it was argued that this provides a reservoir to keep smoke above head height in the event of a fire in the circulation or seating areas. To study this in detail, a computational fluid dynamics (CFD) analysis was carried out to assess smoke flow across the ceiling. This technique divides the three-dimensional space under consideration into a large number of cells. The mass, momentum, heat, and turbulence conservation equations are solved for each cell for each short time step. By considering a fire suddenly starting at a given point in the building, it is possible to study the flow of smoke away from that point as a function of time.

This was almost certainly the first time that a CFD analysis had been used in demonstrating fire safety in a building, and with current advances in computing speed and graphic output this is a technique which has enormous potential in the future.

CFD models are used to study the interaction of wind and smoke in a semi-enclosed building, 40m high.

Evacuation

Calculations were additionally carried out to assess how people might evacuate the space. Measurements were undertaken specifically to determine actual walking speeds as individuals moved through an airport baggage reclaim area. The average speeds, around 1m/s, were in accordance with those quoted from a number of studies elsewhere, but with a range from 0.2m/s to 2m/s. The flow of people towards and through the building exit doorways, as a function of time after escape started, was estimated by assigning the measured distribution of speeds to the maximum population within the building, and undertaking a time-dependent analysis of people movement.

By combining the smoke modelling with the people flow modelling, it was possible to demonstrate that people would be able to move freely over the escape period with large margins of safety. Use was made of graphic display techniques to visualise the movement of people compared to the flow of smoke.

Regulatory frameworks

A similar set of problems was tackled in developing the fire safety strategy for Kansai Airport, recently constructed in Osaka Bay in Japan. The issues are related to the fact that this is a huge uncompartmented space within which people might move for many minutes whilst a fire is detected and confirmed, and evacuation completed. The fire safety strategy depends on the Cabin Concept and exploits the high roof space once again as an effective smoke reservoir. Detailed calculations were prepared in support of the proposed scheme.

The attitudes of the regulatory authorities to the presentation of a fire safety strategy which is based on engineering calculations rather than on regulations is variable. The UK *Building Regulations* have a flexibility which allows this kind of approach, though paradoxically this does not always make the path of the negotiations smoother. In Japan, where the regulations are less flexible, the building control authorities have the option to set up a committee of experts if they feel that they are not in a position to judge the scheme presented. Whilst this introduces an element of delay into the proceedings, the prospect of negotiating on an engineering level with this committee has definite advantages.

The Ministry of Construction in Japan has recently published a unique integrated guide to fire protection. This addresses a wide range of issues and provides calculation techniques for the sorts of studies that are relevant. The guide proposes design fires, sets alternative safety standards for escape routes, deals with smoke flow, suggests how fire resistance of steel may be calculated, and so on. The document is comprehensive, drawing on research work from all over the world and quoting all relevant equations and sources. It points the way to future developments in regulatory matters which will free the designer to adopt a range of approaches to fire safety. It should be said that the document has no legal status in Japan, but it was applied in developing the fire safety arguments for Kansai Airport, and its value in this was noted with interest by the authorities, and formed the basis for discussions with the expert committee. Documents with a similar scope are now starting to appear elsewhere in the world.

Fire spread

In the Japanese negotiations, the committee expressed concern that fire spread might occur across the circulation spaces and suggested that the absence of compartment walls required some other kind of additional protection. Estimates were made of the radiation from possible flames in a fully-developed fire in every combustible area, assuming that the sprinkler system had failed. It was shown that each area was sufficiently far from the next for the incident radiation on fresh fuel to be too low (<20kW/m^2) to cause ignition. This became known as the Island Concept. It was additionally shown that even if each passenger departed without their baggage, fire spread across the space via the abandoned baggage would not occur.

Fire resistance

Large compartment sizes and long escape distances are not the only issues which need to be addressed in terminal buildings. Architectural concerns are often of essential significance in major public buildings. It is frequently desirable, for example, to maintain as a feature exposed structural steelwork. It is then necessary to consider the thermal response of the steel under fire conditions and to assess how that affects the structural response of the building as a whole. At Kansai Airport, calculations on the response of parts of the exposed steel structure to fire conditions were combined with structural calculations to show that collapse of the structure was not a safety problem. Similar calculations were carried out for parts of the arched roof structure for the new Lille TGV Station. In

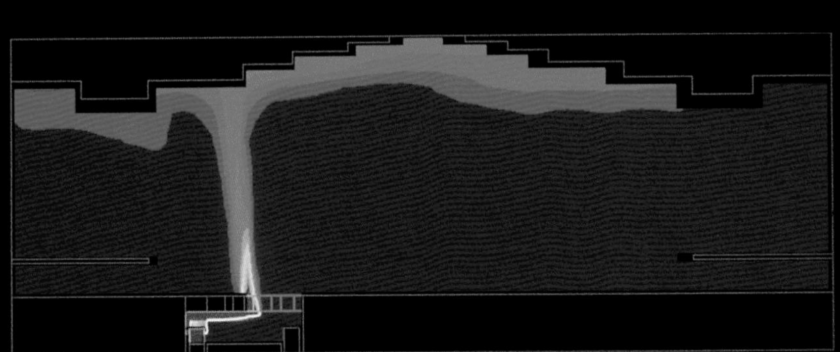

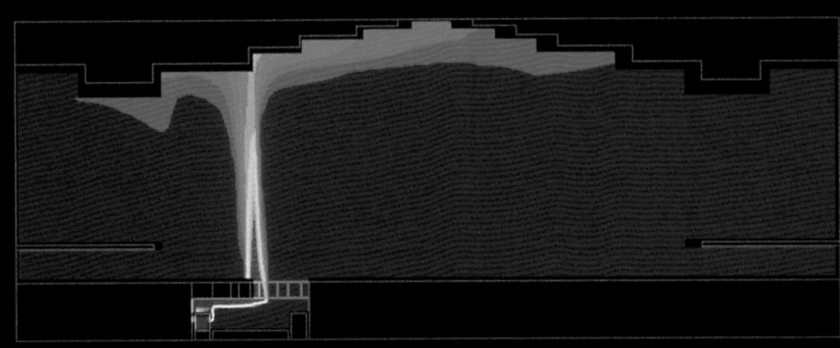

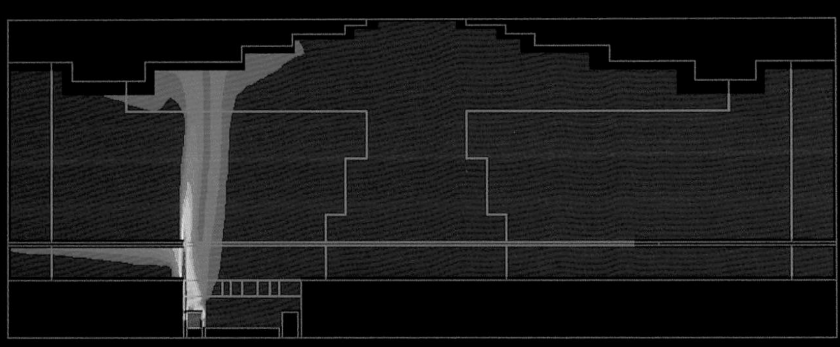

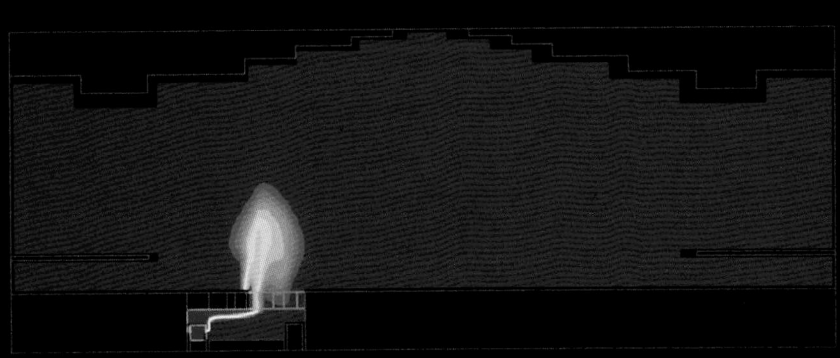

this case the steel thickness of some of the columns was increased internally to maintain architectural uniformity, whilst achieving greater fire resistance in certain elements. This made the elements stronger, with a greater thermal mass, which allowed survival of the design fire chosen. In both cases, the limited amount of combustible material in the building, together with the reserve strength in the structure, allowed arguments to be made for leaving the steel bare.

Offshore fire engineering

The involvement in offshore fire engineering work arose indirectly as a result of the Piper Alpha disaster. This occurred on the night of 6 July 1988 when an explosion on board the Piper Alpha platform resulted in a fire that escalated to involve the main oil export line. 20 minutes after the initial incident a second explosion in a gas pipeline led to a massive intensification in the fire. Of the 165 fatalities, 79 occurred inside the accommodation unit where personnel had mustered.

The report from the public inquiry, headed by Lord Cullen, took over two years to complete and was eventually published in November 1990. In the interim period all offshore operators questioned their safety procedures and problems were highlighted, particularly in relation to the use of living accommodation as a place of safety.

Safe muster stations

It was during this time that Unocal approached us with the brief to investigate their existing Personnel Living Quarters (PLQ) on the Heather Alpha platform. The aim was to protect personnel who had mustered inside the PLQ for the duration of any likely fire. Escape could then be effected after the fire if necessary. Extra process safety equipment was to be provided which would reduce the risk of secondary fires occurring.

The PLQ was a three-level 'Armadillo' unit built up of repetitive modular sections which were effectively cells with plywood and timber framed walls, floors, and roof. The plywood panels were skinned internally with plasterboard and externally with glass-reinforced-plastic (GRP). The side of the PLQ facing the oil and gas process equipment, shielded by a 6mm steel plate cover, had a fire resistance of 60 minutes when measured against the standard furnace test used to simulate a growing cellulosic fire. It was not designed to withstand a hydrocarbon fire, which have higher temperatures achieved over a shorter time period. Thermal shock and greater heat fluxes had to be accounted for when assessing passive fire protection.

The timber structure was clearly the main problem for fires outside and inside the PLQ. It was concluded that this should be fully protected from involvement in any likely fire. From experiment it was found that the timber, with a fire-retardant treatment, started to char and give off smoke at a temperature of approximately 250°C. This was chosen to be the critical temperature that the wood could endure before it was assumed to become involved in the fire.

Consequently, fire scenarios for the inside and outside of the accommodation unit were developed. Estimates of the likely severity and duration were made for each of the design fires. For internal fires it was found that the plasterboard would prevent the timber from reaching its critical temperature of 250°C. No major internal structural changes were necessary and only upgrading works to make good services, doors, and surface finishes were recommended.

For external fires the story was different. These included potential fires from process equipment, crashed helicopters, refuelling tanks, and from diesel generator storage tanks. Flame sizes and shapes were estimated, taking into account the effects of wind, and used to calculate radiated heat fluxes onto the accommodation unit. The worst fire considered had flames impinging directly onto the PLQ with a temperature of 1000°C and a radiated heat flux of 150kW/m^2. Even at a relatively low calculated heat flux the plywood/GRP interface would exceed the critical temperature of 250°C. Therefore it was necessary to insulate the wood.

It was evaluated that a 20mm sprayed-on application of intumescent paint onto the outer skin with suitable fixing details would be the most efficient solution to protect the accommodation unit. The weight and thickness of this material made it the most desirable (even though this involved application by abseiling down the side of the PLQ). This product had also been tested with hydrocarbon fires. Thus, the existing structure was maintained and a solution to upgrade the PLQ to protect against all credible fires was established.

Through negotiations with Lloyds Register of Shipping, both the internal and external protection systems were proved with full-scale panel tests. All PLQ upgrade work has now been successfully completed.

After the work had been carried out, together with other fire engineering studies on platforms in the Southern North Sea, the report by Lord Cullen into the Piper Alpha disaster was published. In a wide-ranging review of the offshore industries safety measures, Cullen recommended that existing fire regulations should be replaced by 'goal setting' standards. His aim was to introduce engineered solutions to fire hazards by means of 'scenario-based design' so that an integrated approach to safety could be developed. It is this philosophy combined with quantified risk assessment that points the way to the future for all types of complex structures that fire engineering teams adopt in all our projects both onshore and offshore.

Summary

This article has only touched on a few of the large number of major projects to which fire engineering techniques have been successfully applied in recent years. More and better tools are becoming available to us all the time. These, when coupled with more flexible attitudes on the part of regulatory authorities, may permit us to have an even greater influence on building designs of the future.

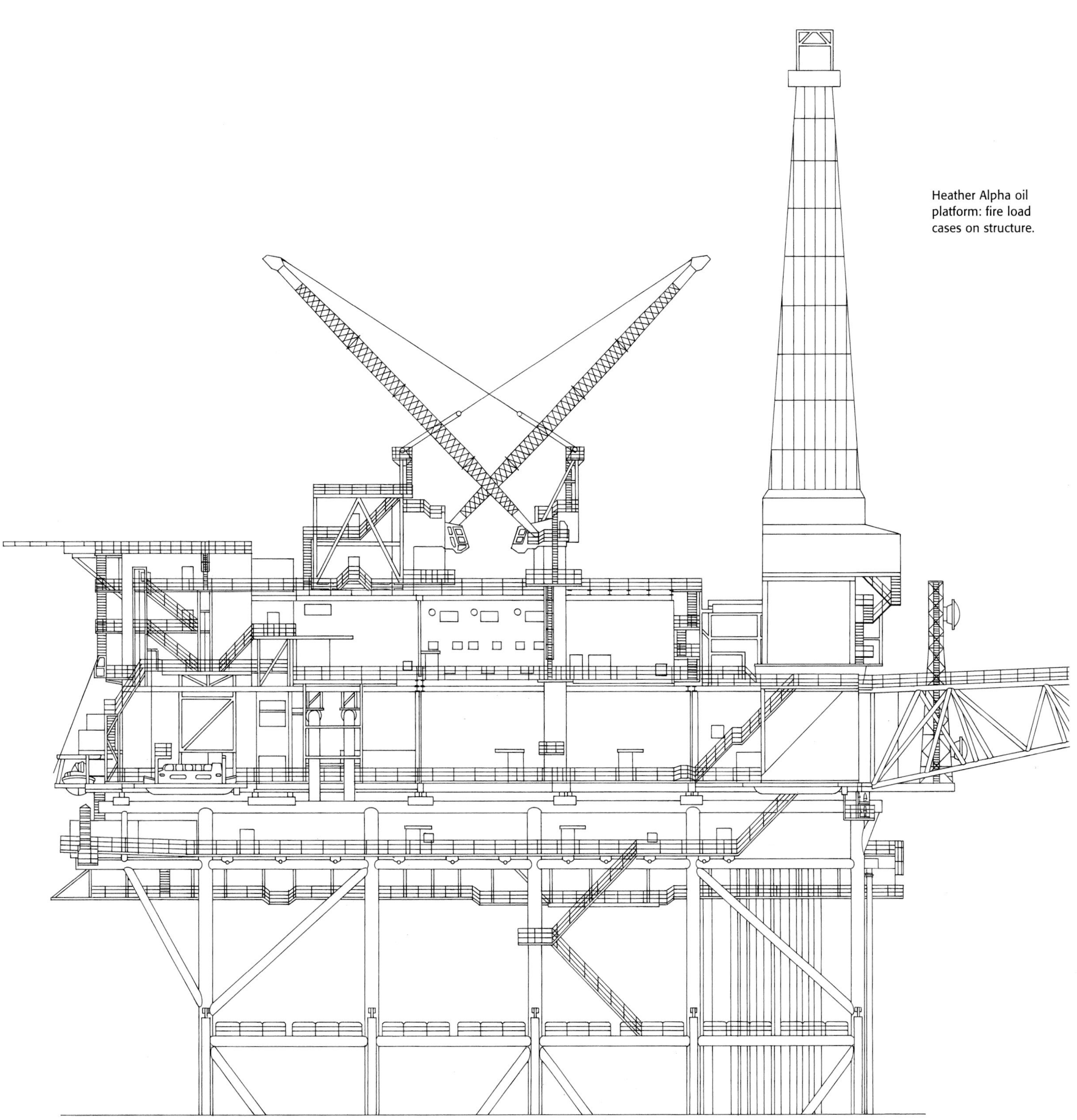

Heather Alpha oil platform: fire load cases on structure.

The change in attitude to road building, from the simple satisfaction of perceived need to arguments about the desirability of new infrastructures, together with concern about the pollution arising from mobility, means that new strategies need to be developed. These involve a larger political consensus as well as more reliance upon managing needs so that restrictions on the freedom to travel are considered within an economic and an environmental agenda.

Malcolm Simpson

People moving

Mans' desire to travel seems insatiable. We have always wanted to travel further, faster, and cheaper, and this trend is growing with increasing rapidity. The consequences are significant.

Traditionally, the desire to travel has been limited by the technology available, regardless of any environmental consequences. Improvements in technology have allowed us to be increasingly selfish in our use of it. We have grown to delight in the individual freedom it gives us without regard to the wider effects our actions have on others. Increases in car use have dominated movement patterns at the expense of other modes of travel. The movement of goods has also continued to increase rapidly, and suppliers are now truly global. We buy our beans from Kenya; meat and wine from Australasia; cars from Japan; and even manufactured goods are shipped halfway around the world for assembly. The volume of goods movement is increasing as dramatically as person movements.

The technological restraint on travel has determined the amount of movement, but this is now being superseded as a prime consideration by the unarguable need to preserve our local and global environment. The increase in travel has affected the towns in which we live, and they have been turned inside out with peripheral developments for retail and business parks replacing traditional town centre activities. This has largely been driven by our overwhelming desire to travel by car.

The trends that have developed will produce a wide range of unsatisfactory consequences unless they are not only slowed but reversed. New political and social attitudes must be developed to produce acceptable and sustainable transport systems in order to recognise the full environmental, social, and economic consequences of a lack of action.

Transport development

When Ove Arup was born 100 years ago, transport desires and the ability to cater for them were totally undeveloped by today's standards. Travel was by foot or horse, with public transport being provided by tram and rail systems. 1896 was effectively the dawn of the car age, when drivers were allowed freedom to travel further and faster following the abolition of the law that required a red flag to be carried in front of the vehicle and speeds to be limited to 2mph. Average travel by individuals was probably one to two miles per day. Early development of transport infrastructure was entirely private sector-led; road development was carried out by turnpike companies, and canal, rail, and tram systems were all developed by private companies. This led to a lack of integration in transport systems.

50 years ago, when the firm was founded, had seen the advent of cars and aircraft, but the average travel distance was probably not more than five miles per day. No environmental factors were considered and there was a continuing desire to expand travel horizons. New towns were planned with massive road infrastructure and their location often followed the prewar metropolitan railway thinking where the major locational criterion was good transport access.

By the 1960s we believed that travel would continue to expand and the visions of the time included supersonic travel, urban helicopter travel, and people movers of all sizes and shapes. This decade saw the beginning of motorway construction in the UK; the first motorways were built with little consideration of environmental effects but were nevertheless hailed as engineering triumphs and as a portend

Figure 1. U.K. Passenger transport by mode

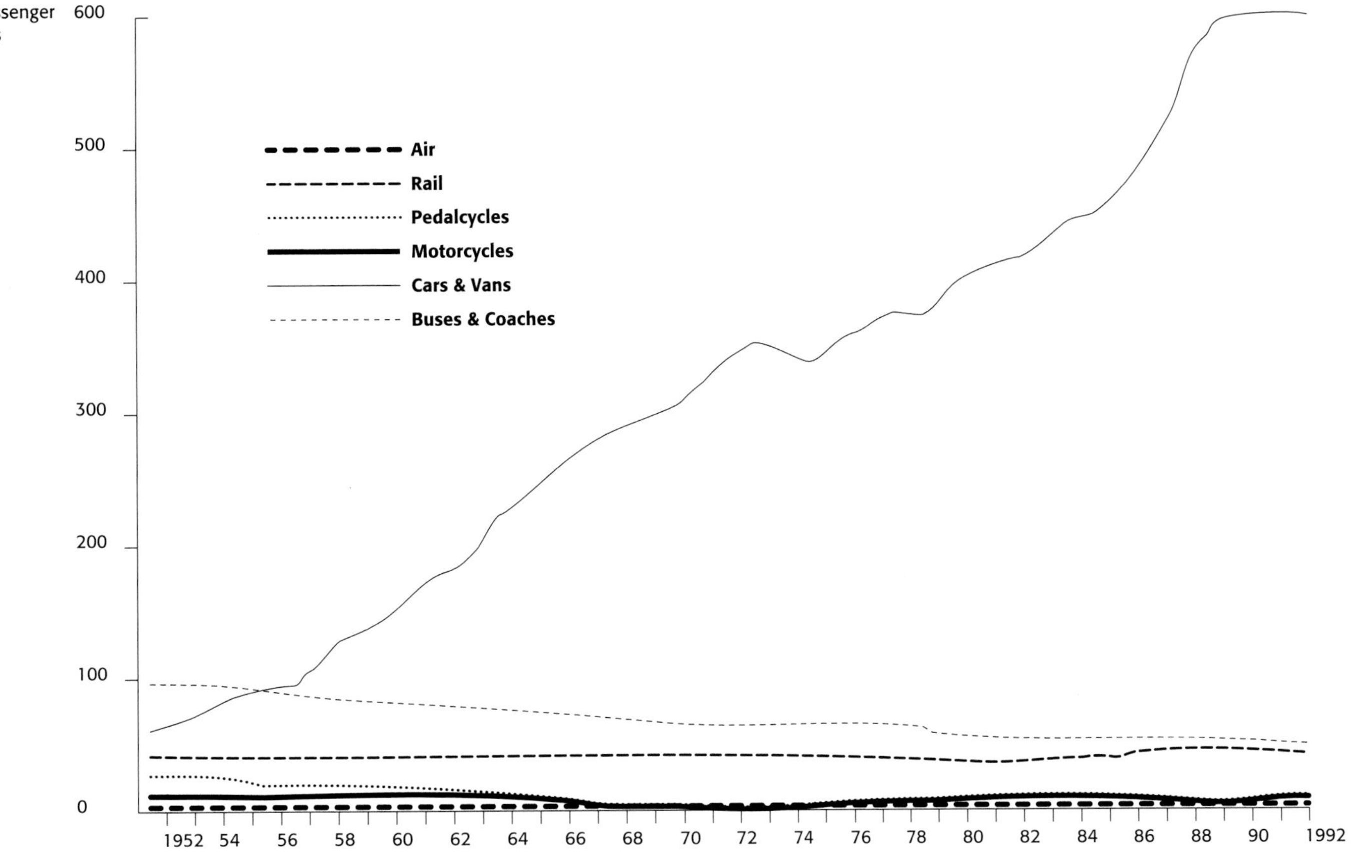

Figure 2. U.K. Freight transport by mode

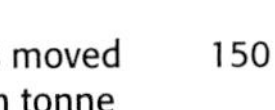

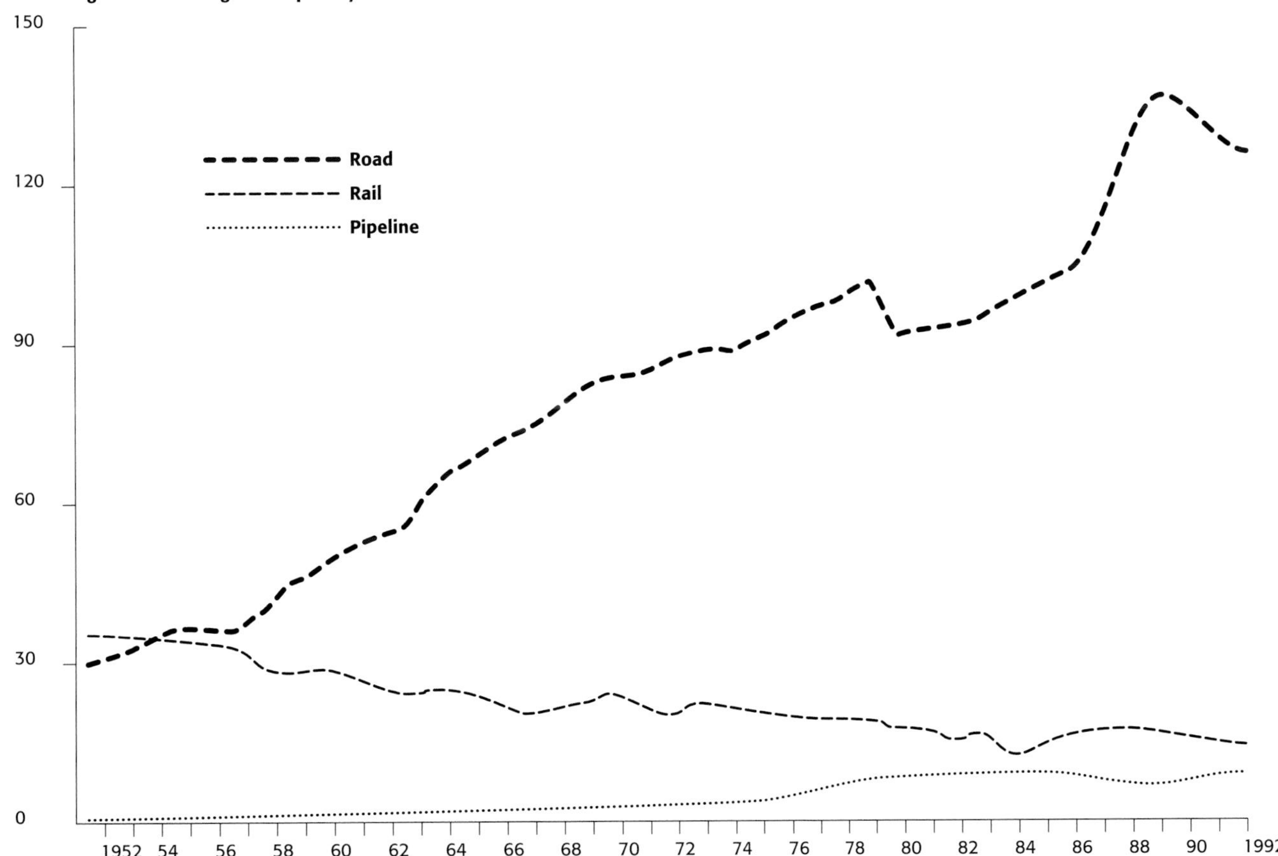

(Above) Edinburgh buses. A guided busway will provide the flexibility of bus operation with a high speed segregated route linking the airport and major developments in the city centre. (right) The US solution to moving people.

of the future. Colin Buchanan's landmark report, *Traffic In Towns,* did indeed suggest the creation of environmental areas, but at a cost of constructing major urban motorways with widespread destruction of existing urban fabric. One of his three study areas included Fitzrovia with Tottenham Court Road being replaced by an eight-lane urban motorway.

The increase in travel distance grew rapidly toward the current level of some 21 miles per person per day. There was still no consideration of integrated travel patterns or of environmental effects and widespread rail closures followed the Beeching Report. The trend for increased car use and less public transport use was almost universally hailed as a desirable ambition and was encouraged by governments until very recent times.

Traditionally, Ove Arup & Partners and other consultants and public sector authorities undertook inter-urban road building in the UK on the basis of satisfying projected demand for road travel, the only constraint being the funding available for the road programme.

Current situation

Recently, there has been a significant change in attitude in many sections of the community. It has been realised that the continuing and rapidly increasing trend of more and faster travel cannot be satisfied because of limited financial resources and environmental considerations. This is no longer in dispute, but personal desires continue to lead to higher travel demands for work trips, business trips, and all types of leisure trips.

Other factors are coming into play. The global mobility now available through the vastly improved technology means that commerce and industry compete on a far wider stage than has existed to date. An essential requisite of establishing world-class commerce and industry is that there should be good accessibility for employees, customers, and goods. Without it, our industry and commerce will suffer at the hands of other countries which provide a higher-grade transport system.

It is clear that the transport infrastructure cannot keep pace with the demand. Over several decades we have been investing far less in our infrastructure than the rest of Europe, and the ability to provide transport infrastructure has been limited by financial constraints and the development density of the UK. Introducing new infrastructure has also been an extremely slow process, with any new road or rail link taking anything up to 30 years to develop. The rapidly increasing demand and little new capacity inevitably means that an equilibrium level will be reached nationally, as it has in our larger conurbations with high levels of congestion and pollution. These are both economically and environmentally unacceptable, as has been demonstrated in the United States.

The way that we approach transport infrastructure is currently the subject of widespread debate; with obvious failings in the mechanisms that we have, the procurement methods that we have traditionally adopted, and the environmental consequences of our actions. It is apparent that radical new thinking is required on the way forward.

We need to integrate transport planning and investigate wide-ranging transport options, as we have done in West Glamorgan and Edinburgh. The study in West Glamorgan was an investigation of a range of options for travel in the area, and proposals for an integrated package of road and public transport

(Above) Tuen Mun Light Railway System, Hong Kong, which combines the benefit of rail operation with the convenience of on-street running.

measures. The Edinburgh study was for a guided busway, which combines the advantages of a segregated, higher-speed, light rail type facility with the flexibility offered by bus travel.

Private sector funding is essential for any major new developments and although this method of procurement was appropriate in the early days of transport infrastructure, it may no longer satisfy the current need to optimise land-use travel patterns and make the best use of our current infrastructure. We can no longer consider simply linear travel demand, but need to plan a network on a regional or even national scale.

The environmental and health consequences of our love of the car are now being realised. Effects range from local noise and air pollution, through regional and national impacts, to global impacts of emissions which lead to climate change.

We have no strategic transport planning. Piecemeal planning exists, with privatisation of public transport leading to increasingly fragmented services. Our assessment methods for road and public transport investment are not compatible, one being on an economic base, the other financial, which does not allow a fully-integrated transport assessment to be carried out.

Local authorities have little power or finance to promote transport improvements, in stark contrast to their counterpart institutional structure on the European mainland.

As we cannot satisfy travel demand growth, traffic restraint methods are being considered, and our work for Avon County Council is among the first studies investigating options including road pricing, parking restraint, park and ride schemes, and improved public transport systems such as the Tuen Mun LRT system we designed in Hong Kong. As an example of the challenge facing us, Figure 3 shows how bus use must change in order to meet the Royal Commission on Environmental Pollution report transport targets, a formidable task in the light of the continuing trend since 1955.

It is evident that there are wide-ranging differences in attitudes, policies, and investment criteria which must be overcome if we are to maintain an efficient transport system as part of a buoyant national economy consistent with the sustainability aims set down by Government.

The future

We must change our attitudes to the demand for travel and its achievement, in full recognition of the environmental, socio-economic and planning consequences.

We need a clear framework strategy for the future. The continuing debate must be focused and actions must be taken, even if they are experimental and prove to be ineffective. We are moving into a new era and it is impossible to predict the consequences accurately, so risks must be taken. We need to establish the consensus view on how to develop, but if this becomes self-defeating then a more streamlined decision-making process must be established. Any strategy must be vertically and horizontally consistent; local policies must integrate with regional and national strategies and inequalities between adjacent localities must be avoided to maintain a balanced land use and economic strategy.

The future of transport development in the country cannot be allowed to continue as a transport issue in isolation. The economic effects of industrial and commercial development must be fully recognised. The environmental consequences must be considered, not only in terms of local impact but

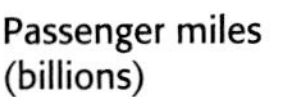

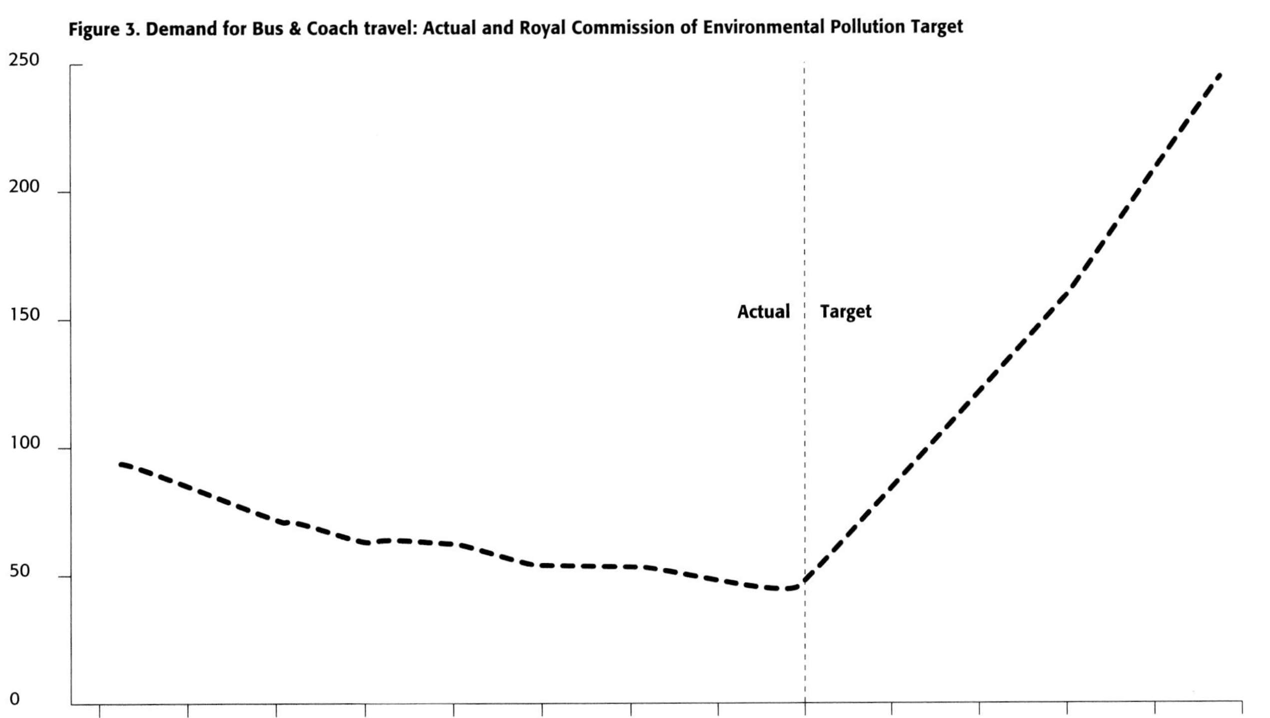

also at a strategic level, and technology must be developed in a controlled fashion so as to be compatible with other aims rather than the dominant force it has been in the past. The debate must be led, not by technology satisfying uncontrolled demand, but more by sustainable environmental and economic considerations. The need to manage the transport resources we have will also become increasingly important. The motorway control systems and pricing mechanisms for motorways and urban areas will inevitably occur and full integration of all modes is needed to produce a seamless transport system. The organisational and institutional aspects, especially those relating to local authority powers, must be reviewed. A coherent policy framework must be developed to cover public and private transport modes, including funding options. These may include charging by means of taxes, tolls or development costs; strategic and local environmental assessment, land use planning, including such aspects as development locations and parking policy, and identifying the priorities and constraints that should be applied to different types of journeys. A consequence of all of this may be that there will be more restrictions on personal freedom of movement, either by physical or financial means.

We need to have the confidence to develop this new way of thinking which will have a style and sustainability to enable our transport systems to develop in the future without the current threat of self-destruction that now hangs over us.

The emphasis on transport planning must broaden from the pride in technological achievement which has led our thinking for the last 100 years to include other skills on an equal footing.

We are well placed to be at the forefront of this new transport thinking, by fully integrating and developing the economic, planning and environmental skills as an integral part of the transport process. We can further enhance our reputation in the complex and challenging area of transport by applying our innovative thinking in a manner which will enhance our physical, social, and economic environment in the next decades.

The 1980s saw the emergence of dynamic façade design – the interactive façade. Curtain walls had grown out of the window industry relying upon cheap power to condition the interior. Now the façade is a mediator between energy flows of many kinds with much innovation and research going into this area.

Neil Noble

The interactive building façade of the future

Traditional buildings have evolved as passive response to their environment using building orientation, material and forms to moderate the external environment to produce comfortable conditions within.

In considering future trends in building design and construction, it is important to reflect on examples from the past, both good and bad. Whilst the development of new technologies can provide a boost to the technology or industry concerned, they can sometimes have a detrimental effect on other components of a building, which get left behind in the pursuit of the new goal. Contemporary building construction contains many examples of components designed and manufactured in accordance with narrowly defined objectives, whether these be technical, visual, or commercial. 'Total architecture' or fully co-ordinated, multi-disciplinary design, is still the exception rather than the rule in the building industry. Compare this with the automotive or aerospace industry where each component is designed to react with another, to carry out more than one function, and to be integrated in the most cost-effective and space-effective way.

The building façade, in particular, seems to have suffered from this narrow approach over recent decades. Lessons learned over many centuries, using building form and building mass, with ventilation using the natural forces of wind and heat, were forgotten with the advent of cheap energy and the invention of mechanical air-conditioning. The façade became the curtain wall, its main functions being to provide the image of the building, and to keep the wind and rain out. The curtain wall grew out of the window industry and adopted many of the principles used in domestic window manufacture. It had nominal thermal performance and relied on boilers, chillers, and air-handling units buried deep in the building with cheap electricity to drive them, to create the required internal environment. Over recent years, new materials, particularly glass coatings, have been developed which have improved the performance of the basic system. New technologies have also been developed in the design and manufacture of curtain wall frameworks to improve their weather resistance, including overlapped drainage routes and pressure-equalised cavities. Performances of façades incorporating these materials and principles have improved gradually over the years, but there have been very few examples of façades being much more that static curtain walls, keeping out the wind and rain.

In the future, however, the building façade is destined to play an increasingly important role in buildings. It must be used, not simply to present an image and barrier between external and internal environments, but to moderate the flow of various forces through it, playing an integral part in the overall form and function of the building. It must be considered in its full three-dimensional form rather than as a simple two-dimensional component, and its ability to store and channel energy must be developed.

Some of the flows which the façade needs to moderate are:

Water	rain, humidity, condensation
Air	wind, draughts, ventilation
Light	sunlight, glare, artificial light
Temperature	solar radiation, internal heat gain, heat loss
Sound	desired, undesired
View	in and out, public or private
Fire	flames, heat, smoke
Pollution	gases, particles
Security	breaking in
Safety	falling out
Explosions	from within, from outside.

In extreme environments (e.g. hot and humid, hot and dry, cold and dry) the requirements of the building envelope are reasonably clear. Although the conditions are not so extreme, temperate environments present other problems, where conditions are constantly under change. These changes include seasonal variations, daily variations, and variations between façades facing in different directions. The façade of the future must not only keep these factors in balance and provide the required internal environment, but it must also utilise these natural forces, and their inherent variations, to create healthy and secure conditions within an energy-conscious building. Natural daylighting should be maximised, but glare avoided. Solar heat should be harnessed in the winter but rejected in the summer. Natural ventilation should be increased, but draughts, noise, and air pollution should be excluded. Good views to the outside should be provided, whilst security and safety are maintained.

The interactive façade of the future must be capable of all of this: it must be capable of responding to and utilising the changing external and internal environments, thereby creating the living and working conditions we all desire. It is this 'adaptability', coupled with the potential of modern materials and computer technology, which is likely to differentiate the façade of the future from the curtain wall of the past. But by 'adaptability', we do not necessarily need a machine, an assembly of moving parts, a 'dynamic' façade. Whilst this may be appropriate in certain cases where external conditions permit, or maintenance procedures are planned for, the interactive façade is designed to interact with its environment in the most appropriate, simple, energy-conscious, and cost-effective way.

Other sectors of industry, for example the automotive industry, have given the building industry a lead. Motor cars of the 1950s and 1960s were large, fairly basic and consumed vast quantities of energy, both in manufacture as well as in running. Factory assembly with production-line methodology led to increased output, improved quality, and reduced relative costs. The energy crisis of the 1970s saw a radical change in direction. Cars became smaller, dramatically more energy-efficient and aerodynamically refined. Through the 1980s and 1990s these trends continued, with improved designs and manufacturing methods, coupled with extensive research and development into new materials and systems. The average motor car of the 1990s is a vastly improved, energy-efficient and cost-effective machine when compared to its 1950s predecessor. Unfortunately, the same cannot be said of the average façade of the 1990s. Manufacturing methods have not changed significantly over the last few decades, and, with the exception of some types of glass, very little progress has been made in the development of new materials. Air-conditioning, the equivalent of the 1950s five-litre, V8 'gas-guzzler', is still seen as the only way of providing the required internal environment.

When almost 50% of the world's energy is consumed by the construction and running of buildings, we should take the subject more seriously. We should take a lead from other industries such as the motor car industry, in providing built quality, and cost-effectiveness. The interactive façade is one response to the industry's needs for buildings in the future.

What should the specific aims of the interactive façade of the future be? In Europe there is a growing desire to maximise the following attributes for the façades of commercial buildings:

natural ventilation
natural lighting
good views through clear glass
energy efficiency

Natural ventilation is of course not new, and has been traditional in hot climates. Natural lighting and good views lead to larger glazed areas, and energy efficiency requires optimum use of the sun in winter whilst controlling solar gain in summer.

What would these façades be made of? How would they perform? How much energy would they save? Let us first look at materials.

Glass

As mentioned earlier, glass technology has been one area where progress has been made in recent years. Glass for passive control of heat and light flow is widely used, in particular reflective glass and body-tinted glass. Reflective glass contains a coating which reflects the long wavelength component of direct sunlight and reduces heat gain. Whilst transmitting visible light, it has a reflective external surface,

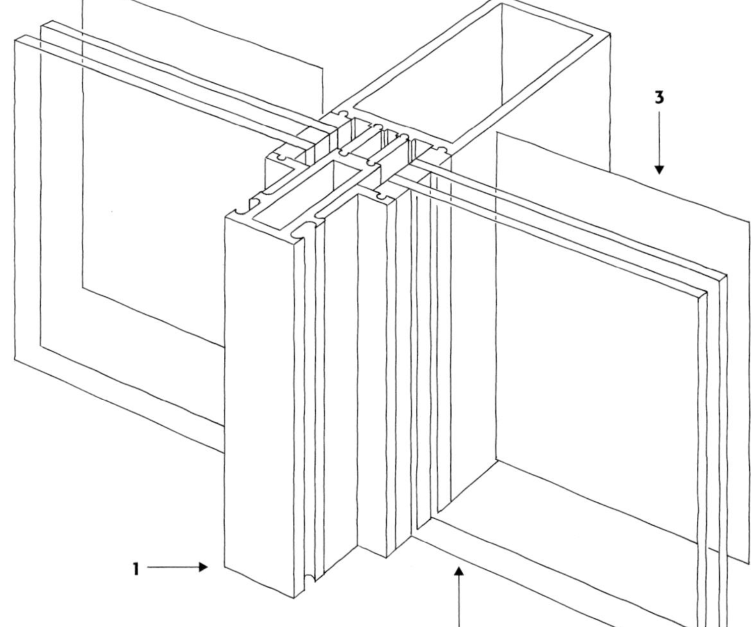

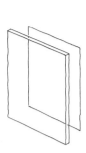

1. Thermally broken aluminium mullion.

2. Clear double glazed unit.

3. Internal blinds.

Combinations of different glass types can be used to improve thermal performance without compromising transparency.

which may not be desirable. Body tinted glass is coloured through its thickness and absorbs heat. The darker the colour the more heat it absorbs, and this is subsequently released as the glass cools down. Fritted glass has a ceramic frit or etching baked onto the surface, and is used to cut off direct sunlight and reduce glare. In addition to these existing products, new, adaptive glass products are becoming available for use. Switchable glass contains an LCD film laminated between glass sheets, which can be switched from clear to white by applying a current. It has been developed for partitioning systems. Photochromic glass contains a coating of silver halide which changes from clear to dark due to the sunlight and is used in some sunglasses. It is being developed in larger sheets for architectural use.

Thermochromic glass has a coating of vanadium dioxide which changes from clear to dark when the temperature rises between 25°–40°C.

Translucent insulating materials (TIM)

These are materials which have high insulating properties, but allow visible radiation to be transmitted. They vary in appearance from translucent to almost transparent. Sheets formed of polycarbonate honeycomb or acrylic tubes have good insulating properties and are translucent. Aerogels are silica-based materials with many small holes, having good insulating properties. They can be located in the cavity of a double-glazed unit, giving the units good thermal performance (U-value of 1.0W/m°C) and good light transmission (greater than 50%).

Photovoltaic solar cells (PV)

In contrast to the more commonly-seen solar thermal systems or hot water systems, photovoltaic (PV) systems convert sunlight into electrical energy. They are formed as an array of interconnected PV cells, which are semi-conductor devices, encapsulated in a laminated glass module. From their early use in the photographic industry, the technology has developed for use in the aerospace industry for power generation on satellites. The revolution in the computer industry, and the commercial development of improved semi-conductor manufacturing techniques over the last 30 years, has provided the spur to dramatic improvements in the industry and prices have fallen significantly over that period, whilst conversion efficiencies have improved.

By incorporating PV modules into building façades and roofs, electricity can be generated for use within the building itself. This power is governed by local climates, orientation, tilt angle, surface area, panel efficiency, and intensity of sunlight. Ideally, the system should be connected to the electricity grid and the power produced used to supplement that from the grid or, if produced in excess, fed back into the grid. Alternatively, 'closed' systems, with the power used for discrete local loads eg adjustable shading, are possible.

Thin film PV coatings are being developed which will allow greater architectural freedom, as well as achieving significant reductions in price. At present, efficiencies are low, but it is anticipated that these will improve and that coatings will be thin enough to be virtually transparent.

New materials such as these will form part of the interactive façade of the future. They need to be incorporated

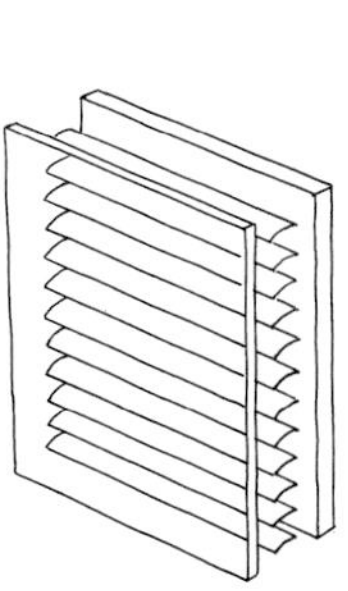

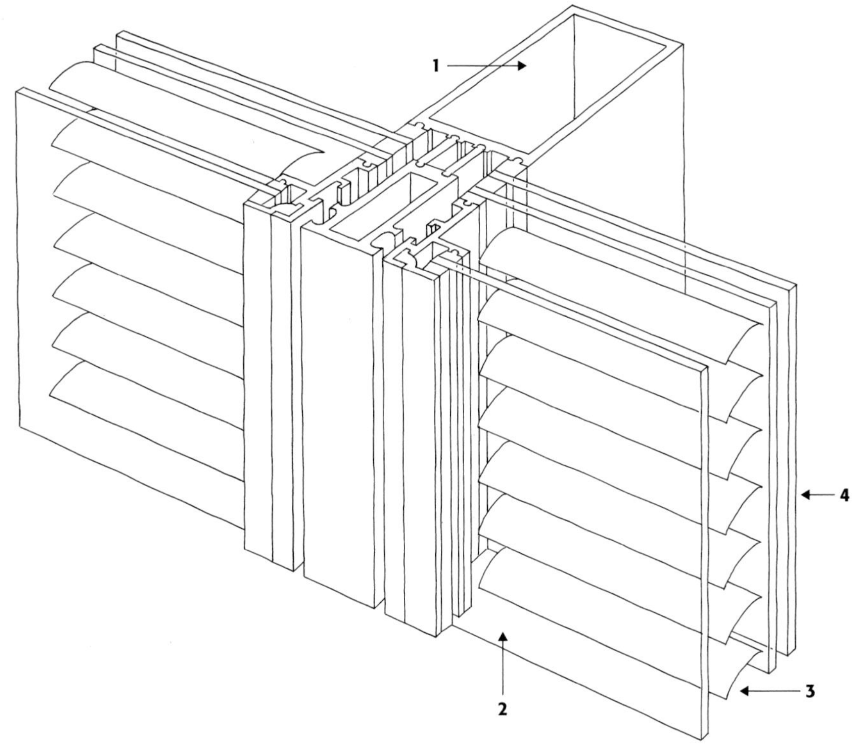

1. Thermally broken aluminium mullion.

2. Single glazed outer panel.

3. Retractable solar control blinds with ventilated cavity.

4. Internal clear double glazed unit.

into new systems which form the building enclosure and contribute to the overall performance.

PV cladding systems

The photovoltaic overcladding system is a version of the traditional rainscreen system using an economic, open-jointed construction to form a drained and back-ventilated rainscreen formed by the PV cells. Ventilation cools the PV cells, which increases the power efficiency.

Roof-mounted photovoltaic panels can be designed as part of a building skylight. Typically, the panels would be mounted into a system of sloping extrusions holding single or double glazing and orientated to receive maximum sunlight.

Roof-mounted systems are particularly useful for sites situated toward the equator where overhead sunlight is predominant. Mixed skylights, with PV cells directed south and glazing on the north façade, provide a combination of natural lighting and electrical power.

Curtain-walling systems can be designed to incorporate photovoltaic cells into factory-assembled double glazed units. These units comprise an outer laminated glass/PV-resin/glass pane and an inner glass pane with a sealed air gap between.

Shading systems

Research into the comfort of building occupants with respect to daylighting and views to the outside has led to the desire for large vision areas, incorporating shading devices. These shades can be fixed or adjustable and fabricated in metal or glass.

The incorporation of PV cells into shades has many advantages: moveable shades can maximise the incident solar energy; the surface area available for cells can actually be greater than that of the vertical walls; the use of shades in strips provides natural cooling ventilation for the cells, increasing the efficiency of power generation.

Fixed external shading can provide effective solar control, although systems where the orientation of the louvres can alter are most efficient. The Arab Institute in Paris (Architect: Jean Nouvel) has windows with variable apertures which respond to sunlight. However, moving parts are always vulnerable and the future may lie not with mechanisms but with computer-controlled systems with few moving parts.

The lightshelf is not a new idea, but is being rediscovered in some notable current buildings, for example the Inland Revenue Building at Nottingham and the New Parliamentary Building in London (Architect for both: Michael Hopkins & Partners). The lightshelf serves the double function of providing fixed shading while also reflecting daylight into the interior of the space. If the louvres within the shelf are variable, a range of external conditions can be moderated.

Blinds and ventilated cavities can be used to absorb and remove heat gains before they penetrate the façade.

Evaporative cooling

Fountains have traditionally been used to cool courtyards by evaporative cooling. They also help to ionise the air. The same principle was used for the UK Pavilion at Expo '92 in Seville (Architect: Nicholas Grimshaw & Partners), where one wall was cooled by flowing water over its surface, pumped by electricity generated by photovoltaic cells.

A similar principle is being employed for the Seawater Greenhouse project in the Cape Verde Islands (Promoter:

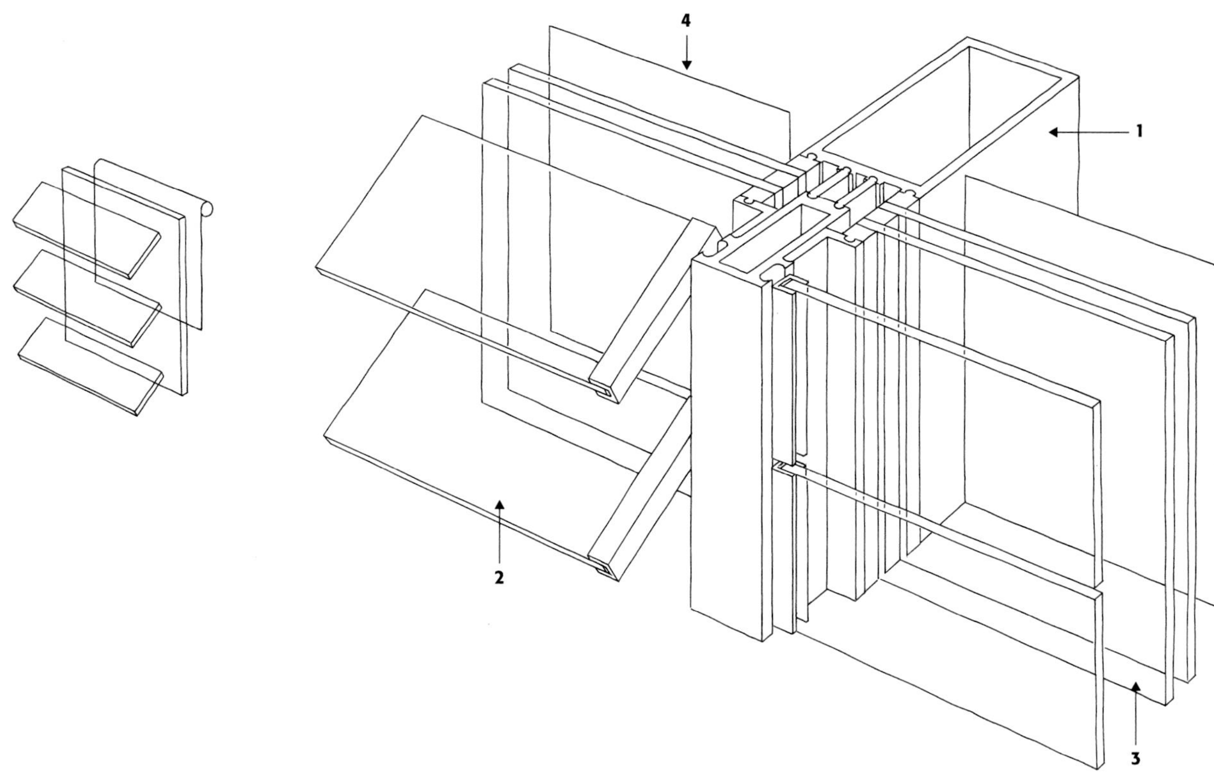

1. Thermally broken aluminium mullion.

2. Openable solar control glass louvres.

3. Clear double glazed unit.

4. Internal glare control blinds.

External shades or integral blinds can be accommodated within the façade to reduce glare and heat gains. Computer control maximises efficiency.

Light Works Ltd), where seawater is being used to humidify greenhouses for growing vegetables, while at the same time being itself desalinated.

Ventilated twin-glass wall façades

There is an increasing demand for buildings exhibiting high levels of transparency with large expanses of glass dominating the façades.

The benefits of an all-glass façade are obvious – high levels of natural lighting, maximum views to outside, maximum transparency of the enclosure. The disadvantages are equally obvious – high solar gains and heat losses, high levels of glare, and high heating or cooling costs.

The externally-ventilated twin-glass wall is one response to this challenge. Consisting normally of a double-glazed inner skin and a single-glazed outer skin, the wall includes a large cavity, containing shades or blinds, which creates a buffer between the external environment and the internal design conditions. When used on a fully air-conditioned building, the ventilated space will reduce heating loads on the inner skin in the summer by removing heat loads from the face of the internal glazing. In the winter, with ventilation to the cavity closed off, a warm buffer zone is formed. Whilst not achieving the lower levels of energy consumption of a traditional insulated building, significant savings of energy can be achieved over a simple, fully-glazed enclosure. Alternatively, by combining opening windows in the inner skin with a fresh air supply to the offices, the stack effect in the cavity becomes the driving force for a natural ventilation system to the building.

The internally-ventilated twin-glass wall consists of an external double-glazed unit, a smaller cavity containing blinds, and an internal single-glazed sheet. Internal air, passing through the cavity, removes the heat from the blinds and glazing, which can then either be recycled during winter months or rejected in the summer.

The design of such a façade is inextricably linked with the design of the ventilation system of the building, and the capital costs and running costs must be considered together. The optimum solution can only be obtained by a full multi-disciplinary approach.

Projects

In the proposed headquarters for GSW in Berlin (Architect: Lousia Hutton & Matthias Sauerbruch) the gap is approximately 1 m, forming a thermal flue through which air, warmed by the sun, rises up the building. In doing so, it draws air out of the offices and creates cross-ventilation. This is possible because of the narrow plan of the building which also increases daylighting.

The Green Building (Architect: Future Systems) has not been built, but is also based on a double-skin principle. The building form encourages airflow up the sides to the top. This flow, and the temperature regime resulting from it, has been analysed using computational fluid dynamics.

In the headquarters for Lloyds of London (Architect: Richard Rogers Partnership), the façade is triple-glazed system. The outer skin is a double-glazed unit and there is a gap between this and a single inner skin. Warm air from the room is extracted into the ceiling and pumped down the gap to warm the cavity in winter, after which it goes to the air-handling unit.

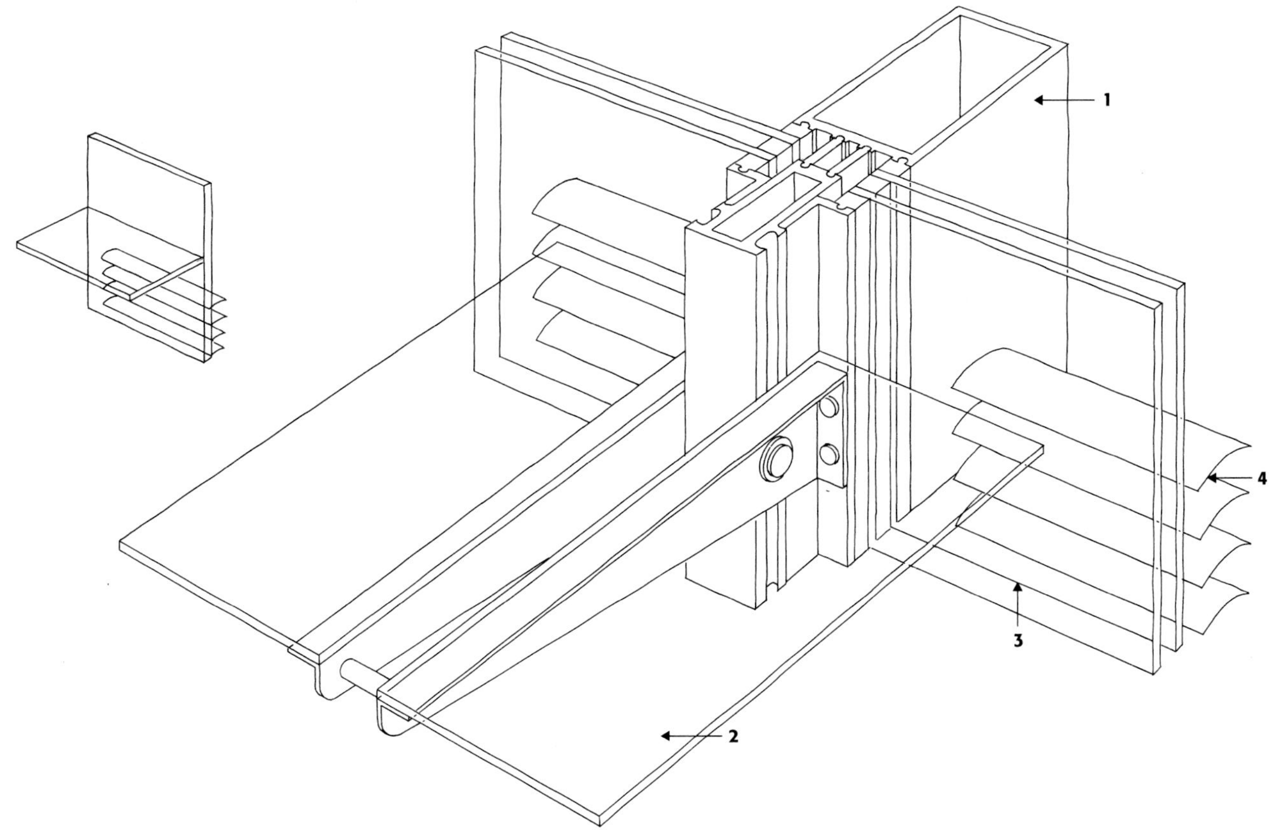

1. Thermally broken aluminium mullion.

2. Single glazed outer panel.

3. Retractable solar control blinds with ventilated cavity.

4. Internal clear double glazed unit.

Light shelves provide control of glare and heat gain whilst reflecting natural light deeper into the building interior.

In Kaiser Bautechnik's new Haus der Wirtschaftsförderung in the Duisburg Mikroelektronikpark (Architect: Sir Norman Foster and Partners), triple glazing is again used, but this time the inner skin is double-glazed and the outer skin is single-glazed. The air moves upwards through the gap and computer-controlled blinds are located in the gap.

For the New Parliamentary Building in London (Architect: Michael Hopkins & Partners), triple glazing is used, with a double-glazed outer skin and single-glazed inner. There are movable binds within the inner cavity and air can move from the room upwards through this cavity and then into an air plenum in the clerestory. The warm air then enters the vertical façade ducts and heat is either reclaimed via a thermal wheel in the roof or rejected through the roof chimneys.

Smart materials

Various materials are currently being developed which may have applications in façade systems.

An example of this is in the technology used in the US 'Stealth' fighter. Ferro-electric substances or polymer-based chiral compounds are built into the aircraft skin. The substances absorb incoming radar, making the aircraft radar-transparent.

A similar principle is used to make submarines sonar-transparent, this time using piezoelectric materials in the skin to detect incoming pressure waves. The skin then generates signals which are 180 degrees out of phase with the incoming signals, thereby cancelling them out. Similar systems have been used by Lotus in cars. This has possible applications in providing acoustic control in buildings and could radically alter the way we think about this subject.

The future of the interactive façade

It is essential that we change our view on the form and function of the façade on our buildings in the future. For too long, this vital component has been ignored and its contribution to the building, the well-being of its occupants, and the energy resources of our planet have been underestimated. New materials need to be developed and prototype systems tested. To progress this work, manufacturers need to be given proper incentives to carry out the necessary research. We can no longer ignore the fact that almost 50% of the world's energy is consumed in the construction and running of buildings.

I have only been able to describe some of the likely attributes of the interactive façade of the future, and some of the technical principles which may play a part in their development. There is a lot more research to be done.

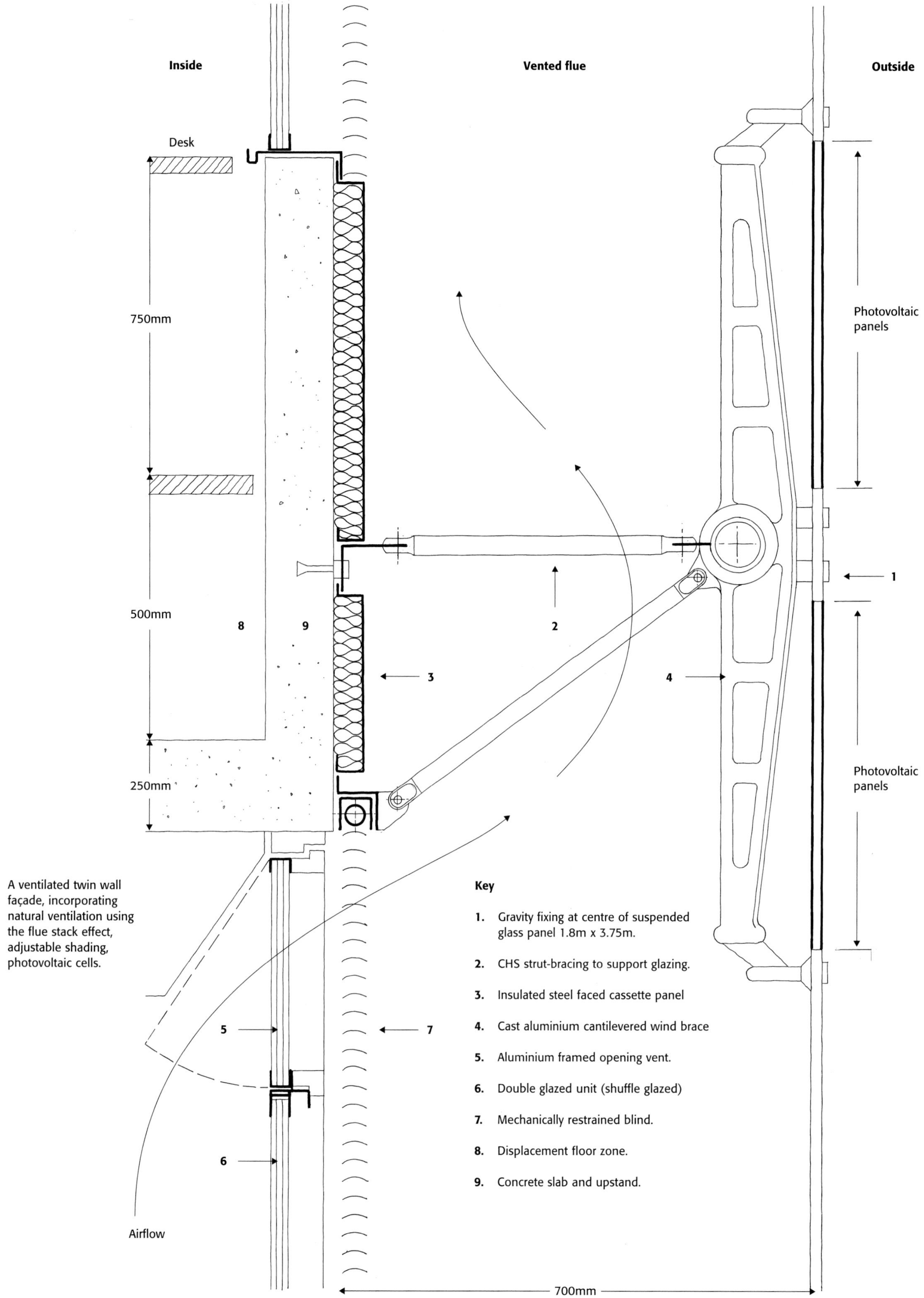

A ventilated twin wall façade, incorporating natural ventilation using the flue stack effect, adjustable shading, photovoltaic cells.

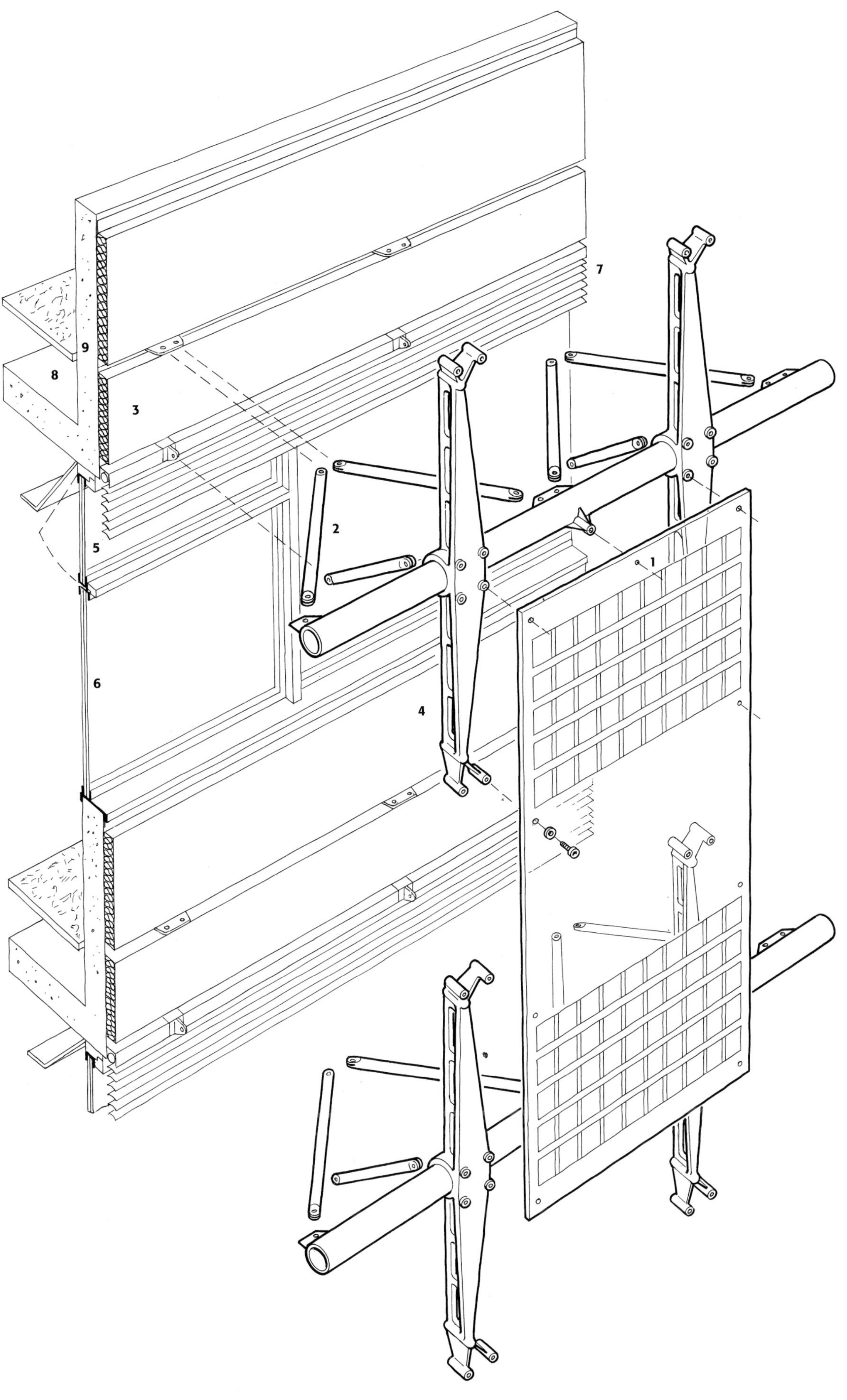
7
9
8
3
2
5
1
6
4

In geotechnics, stress has been the dominant concept but displacement is the physical reality. Soil behaviour is better explained using the concepts of non-linear plasticity – 'bricks on strings'.

Brian Simpson

Modelling ground behaviour

Almost all civil engineering structures depend for their stability and performance on the behaviour of the ground beneath or around them. It is the task of the geotechnical designer to understand the ground sufficiently to ensure stability, and to be able to predict the deformation and other features of ground behaviour.

Generally, the methods of geotechnical design have theoretical backgrounds, but they also depend heavily on empiricism and observation. This has sometimes led to debate about the nature of geotechnology: is it an art or a science – perhaps a 'black art'? I contend that geotechnics is founded squarely within the scientific method, requiring a perpetual combination of theory and observation. The term 'observational science' is perhaps most apt.

The behaviour of the materials in the ground – soil and rock – is complex and our level of real understanding is very limited. In order to proceed with design we must establish what can be known with confidence, and formulate models for important aspects of behaviour which we can only describe approximately. The 'models' referred to here are mathematical descriptions of the ground, usually represented in a computer.

Stress and deformation

Geotechnics has only been recognised as a discipline in its own right since the time of the Second World War. Before that, however, great pioneering work was done in understanding stability (Coulomb (1776) and Rankine (1857)) and the fundamentals of soil strength (Hvorslev (1936)). Terzaghi (eg Terzaghi (1943)) systematised much of the theory and, in particular, formalised the concept of 'effective stress'. In this, water pressure is subtracted from the ground's total stress to leave a measure of the stresses which act between solid particles, and which control ground behaviour.

In most of this early work, attention was concentrated on stress and strength. Later, when attempts were made to calculate deformation, linear elasticity was assumed because it was the only assumption leading to tractable calculations. More recently, carefully instrumented laboratory and field testing has led to major advances in understanding deformation properties, though the application of these is not yet routine. Burland (1967) neatly summed up a technical need which was completely in line with the layman's observation of engineering performance: Stress is a philosophical concept – displacement is the physical reality.

Displacement is cracking, an alarming tilt or unsightly bend, or sufficient movement to stop the lift working or prevent the function of delicate apparatus.

British Library, Euston

In the 25m deep basement for the British Library at Euston, designed in the mid 1970s, field measurement of ground displacements had been undertaken, as on many projects in London in the previous 20 years (Cole and Burland (1972), Burland and Hancock (1977), Butler (1975)). Stiffness parameters, assuming linear elasticity, had been derived from these observations. Attempts to relate the field observations to soil deformation parameters measured in laboratory tests on relatively 'undisturbed' specimens had shown poor success. It was fashionable to conclude that specimens obtained from the ground were so disturbed that nothing useful could be learned about their stiffness from laboratory testing.

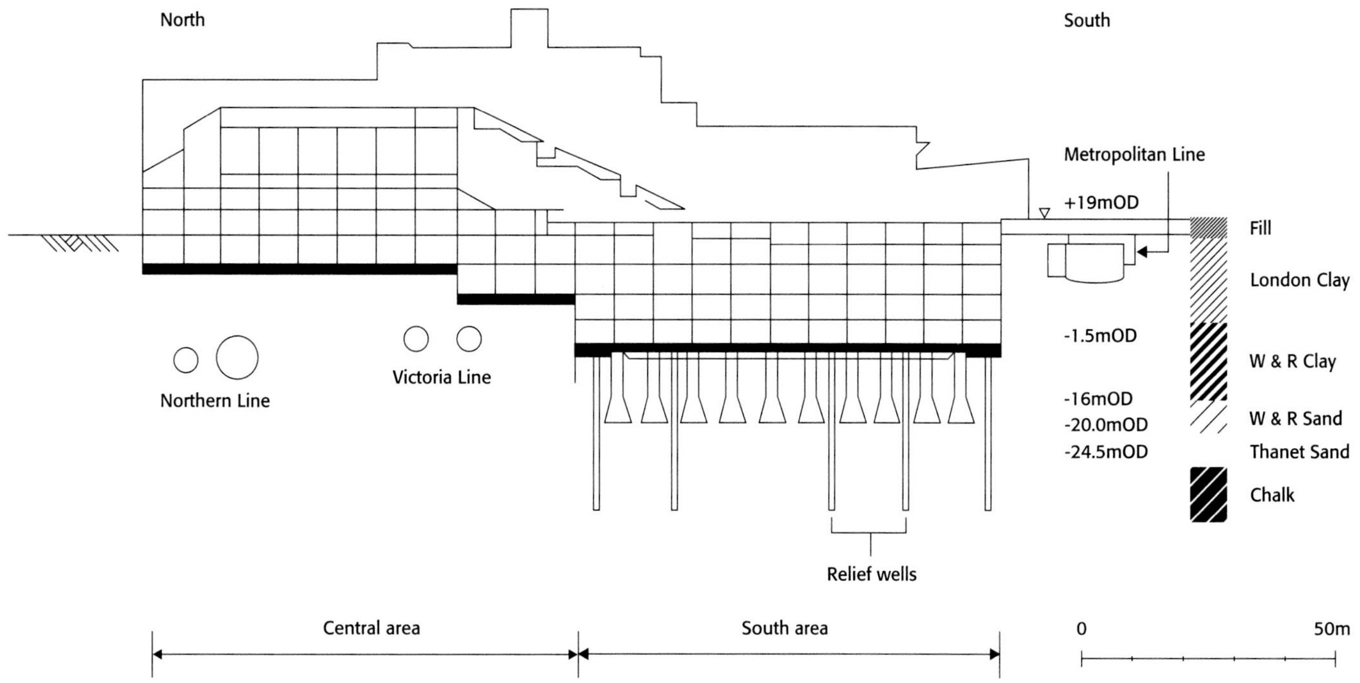

Fundamental laboratory tests made it clear, however, that the stress-strain behaviour of soil is rarely, if ever, linear or elastic. Thus there was a danger in extrapolation of field observations on the assumption of linear behaviour. There was available just enough evidence to suggest that the poor agreement between laboratory measurement and field observation might be due to the highly non-linear, and misunderstood, nature of soil behaviour. Much of the displacement occurring in the ground was derived from strains less than 0.1%, a range too small to measure reliably with laboratory techniques available at that time.

Simpson, O'Riordan and Croft (1979) published a model of the behaviour of London Clay which was somewhat speculative, but which incorporated these features of non-linearity in a primitive manner. They showed that the model correctly computed the displacements observed at the deep underground car park at the House of Commons. Earlier work of Ward and Burland (1973), showed that the magnitude of surface settlement could also be computed fairly well using a linear elastic model. However, the new model gave a more realistic pattern for the ground displacements outside the excavation, including the tilting of Big Ben. For the significantly larger proportions of the British Library excavation, however, the two models yielded markedly different results, the non-linear model indicating considerably more displacement. It was clear that these features of soil behaviour had practical significance; fortunately, the design could proceed without excessive conservatism whilst accepting the more severe predictions of the non-linear model.

Laboratory testing at Imperial College, London, designed to

(Above) The basements of the British Library at Euston.
(Right) Measured and computed settlements at the House of Commons underground car park.

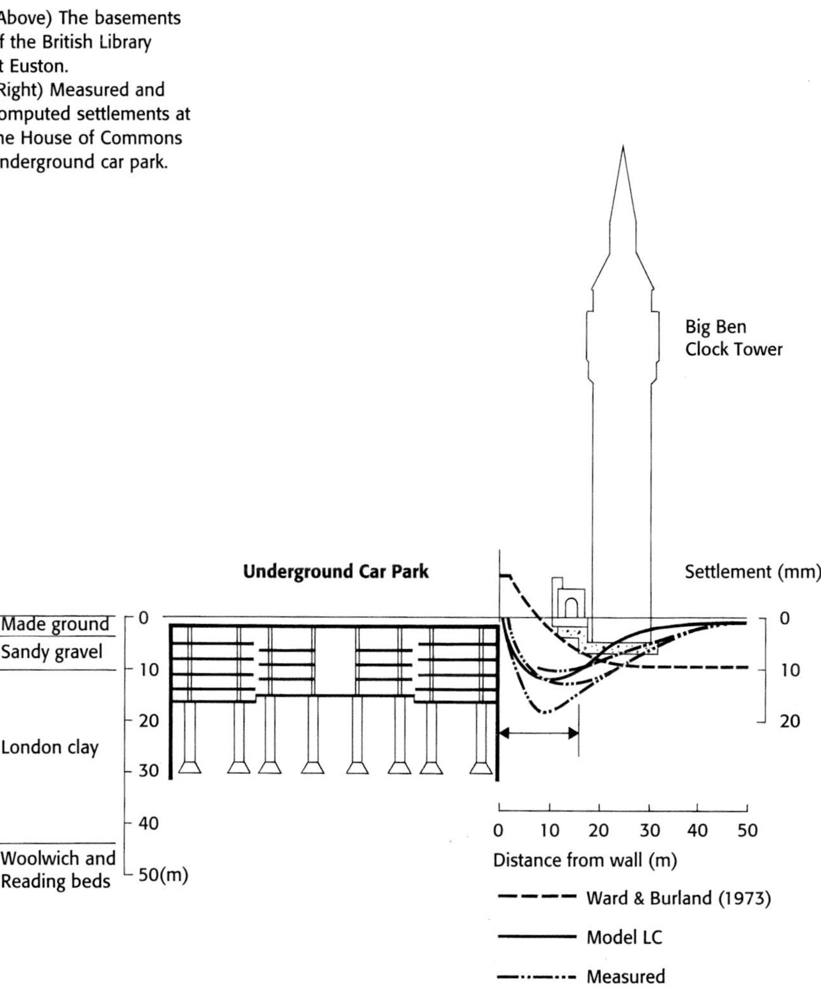

(Above) Soil can be simulated in a computer as a large set of discs.

(Opposite) Real soil has complex interactions between many tiny particles.

measure behaviour at very small strains, supported the development of ideas during the 1970s, eg Costa-filho (1979). Subsequently, many research establishments have established a clear framework of understanding, with notable publications by Imperial College, City University in London, and many others. As is often the case in geotechnical engineering, field observation had pointed the way to fresh understanding, which was confirmed and clarified by laboratory testing.

The work of Richardson (1988) and Stallebrass (1990) at City University was supported by Ove Arup & Partners through contributions to their CASE studentships. The ideas described below derived much from this research, and the writer has been pleased to work with City University in the capacity of Visiting Professor. The recent award of a Royal Society/SERC scholarship to Professor John Atkinson, of City's Geotechnical Research Centre, has strengthened this co-operation further. Based at Arups, he is to study the ways in which designers derive and use geotechnical parameters.

Soil behaves like 'bricks on strings'

Models of soil behaviour advanced beyond the concepts of linear elasticity by adapting ideas from metal plasticity, and this has probably had a restrictive effect. Attempts to describe soil behaviour more accurately have generally centred on deriving complicated sets of equations to fit phenomena observed in the laboratory.

Very simple soil, like rounded sand, deforms mainly by relative movements between particles, including both rolling and sliding. Although the soil may develop large strains before failing, it is clearly quite unlike a ductile metal. Research at Aston University, eg Thornton and Sun (1994), has shown that many of the properties of soil can be reproduced by a computer model which represents a moderate number of discs interacting. Such a representation of the discrete particulate nature of soil is vital to a proper understanding of the way it behaves.

Modern chaos theory may also hold a clue. This generally starts from a very simple mathematical statement about the nature of a phenomenon on a very small scale, then uses computers to predict what would happen if the same phenomenon were repeated very many times. The results may not be predictable from the start by analytical mathematics. Perhaps the rules of interaction between the particles could describe the small-scale phenomenon, the implications of which would be derived by computer.

These thoughts, in conjunction with the framework of understanding known as 'critical state soil mechanics', led the writer to propose that 'soil behaves like bricks on strings'. The idea was that a physical analogue, which had some similarities to the interaction between soil particles, could be used to make predictions of how soil would behave as it deforms. The analogue was that of a man, walking around on a flat floor, and pulling behind him a set of bricks on strings of varying length. (Most of the theory came together during one day at the end of a conference in Florence: the forecourt of the Cathedral was an ideal practice ground for the imaginary brick-man!)

As the concept developed, it became clear that the man was walking in a space defined by strain axes. His position represented the strain in a sample of soil, and movements of bricks represented the plastic components of strain. Thus, if the man moved in a direction which caused no brick movement, all deformation would be elastic and the soil was very stiff. At the other extreme, if he moved so that all bricks moved with him in the same direction, the soil was at failure. The model readily achieved the realistic feature that very little deformation of soil is truly elastic. The details and further developments are described by Simpson (1992a,b). The model had some features in common with others using nested series of yield surfaces; an important and unusual feature was its expression in strain space. Stress is a philosophical concept, but strain is a measure of deformation and of the displacement taking place between the physical particles of soil.

To the writer's surprise and delight, the computer representation of this physical model reproduced several of the most important features of soil behaviour, including shear failure at a constant angle of shearing resistance. It also predicted the constant ratios of principal stresses (K0) exhibited by soil during virgin consolidation, a well-recorded phenomenon for which the only theory, that of Jaky (1944), is much in question. Though the intention was to reproduce the behaviour of the stiff clays of London, the model was equally useful for computations in soft clays.

The BRICK model was tested by using it to compute the deformations of the ground around the British Library excavation, which had been measured during the 1980s. The building was built in very stiff London Clay – almost a soft rock. It was interesting to find that the same model, with virtually identical parameters, could represent the behaviour of the basement of the United Overseas Bank in the Singapore Marine Clay – a material so soft that it squeezes through the fingers even when taken from many metres depth.

The BRICK computer model is in constant use, and has recently been found to be particularly helpful in understanding the behaviour of the ground around tunnels. It appears that another feature of soil behaviour must be taken into account, however. The anisotropic nature of London Clay – its different stiffness in different directions – is unimportant in many situations but critical to behaviour around tunnels. Again, there is empirical evidence that this phenomenon exists: recent field testing by the Building Research Establishment (Butcher and Powell (1995)) and laboratory testing at City University seem to confirm the form of behaviour demanded by computation to explain observed ground movements. Perhaps this is the next advance waiting to be made.

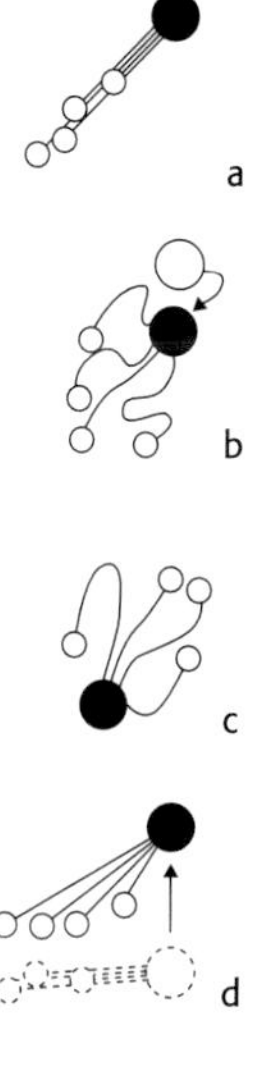

(Above) Imagine a man, walking through strain space, and pulling along a series of bricks on strings.

Ground-structure interaction

The usual reason for wanting to model ground behaviour is to predict the performance of a structure bearing on it. To achieve this goal, it is not always necessary to adopt complicated or innovative descriptions of soil behaviour. In predicting the effects of tunnel construction on overlying buildings, it is conventional to perform simple calculations based only on empirical observation with no theoretical background. However, predicting the effects of such deformations on buildings is a further challenge.

Cross-section through the United Overseas Bank project, Singapore

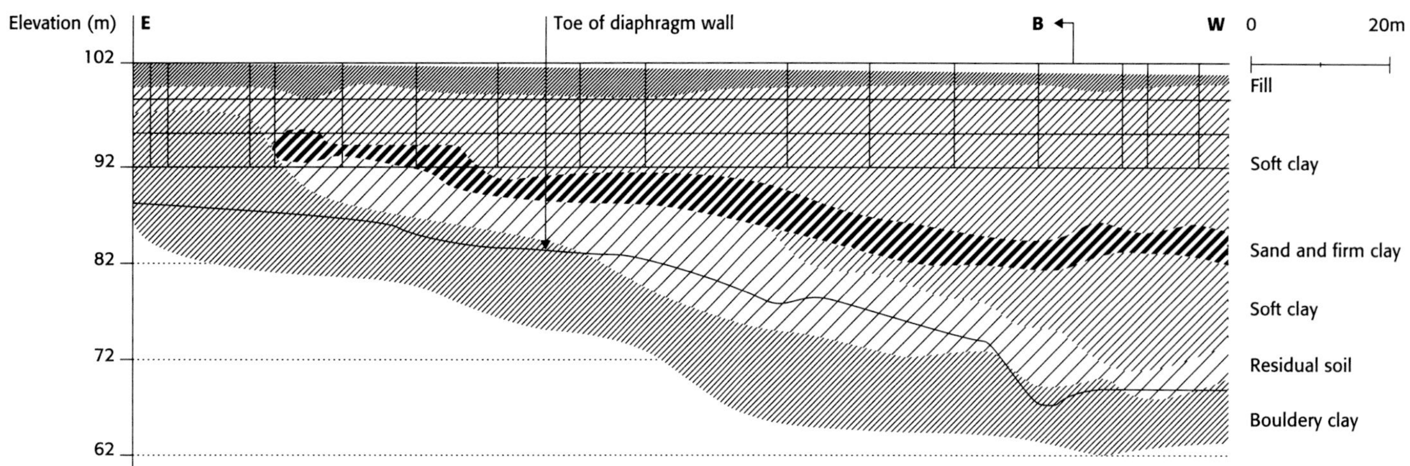

BRICK computed ground displacement compared with measurement: British Library

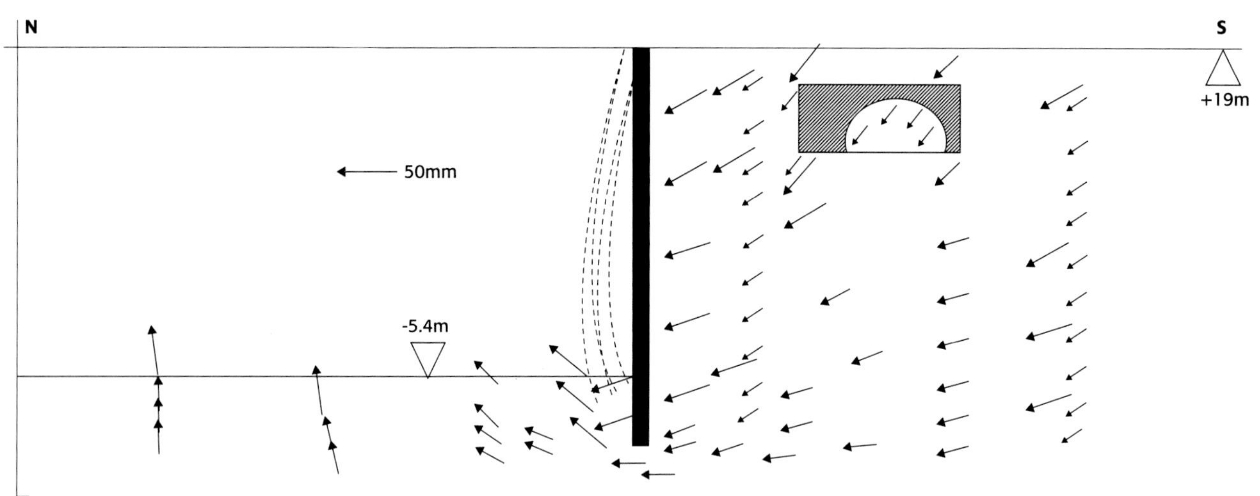

The proposed nearby construction of the CrossRail Underground station raised the spectre of damage to the protected façades of Britannic House, on Moorgate. For this purpose, ground deformations computed using an empirical formula were applied at the base of a computer model of the façades, through a 'cushion' which represented the soil around their foundations.

The act of making and using a computer model often educates the modeller and his colleagues. In this example, the reinforcing effects of the floors of the building were demonstrated, and the sequence of cracking computed, starting at the base of the façades and working upwards, was the opposite of previous intuition. Whilst a computer model does not prove what will happen, it sometimes has the important effect of suggesting possibilities that had not been suspected. Those unexpected suggestions are perhaps the most valuable contribution of soil and structure modelling, and the greatest source of pleasure for the modeller.

Measured and computed wall displacement for United Overseas Bank, Singapore

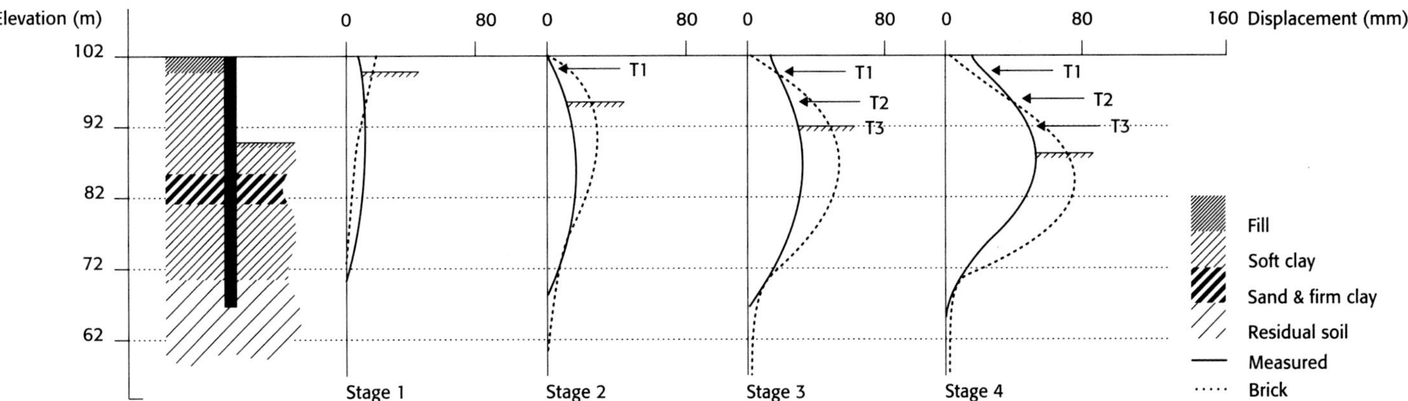

Computed settlement and damage to a masonry façade

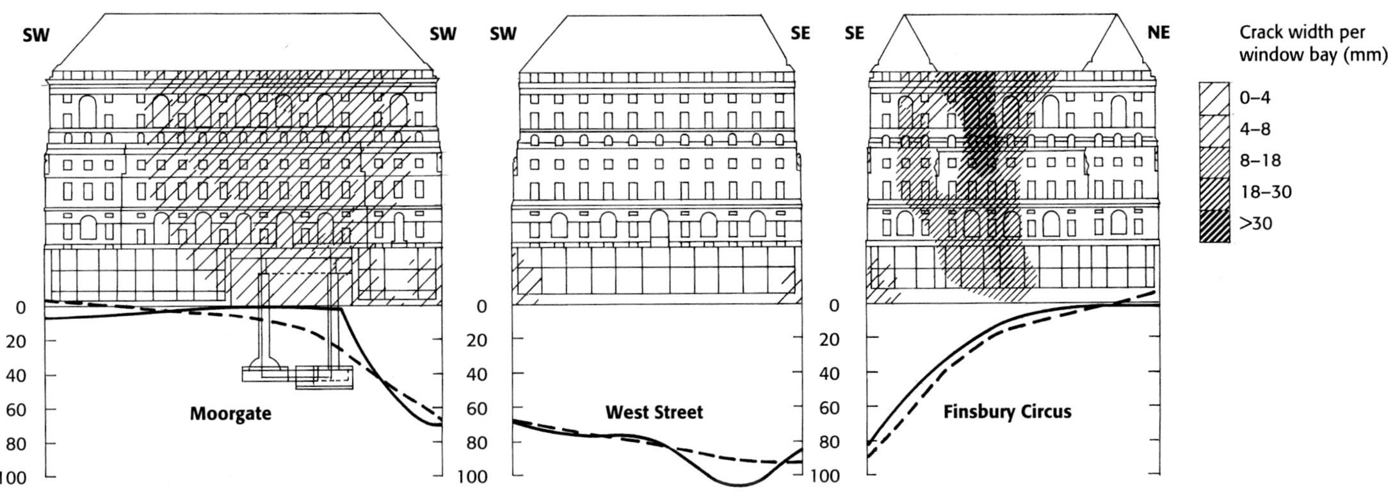

References

J. B. Burland, *Deformation of soft clay* (PhD thesis, University of Cambridge 1967).

J. B. Burland and R. J. R. Hancock, 'Underground car park at the House of Commons: Geotechnical aspects'. *Struct. Engr.*, 55, (1977), p. 87-100.

A. P. Butcher and J. J. M. Powell, 'Practical Considerations for field Geophysical Techniques used to Assess Ground Stiffness'. *Advances in SI practice*, ICE, (March 1995).

F. G. Butler, 'Heavily overconsolidated clays', *Conference on Settlement of Structures*, Cambridge, Review Paper, Session III. (London: Pentach Press.1975).

K. W. Cole & J. B. Burland, 'Observations of retaining wall movements associated with a large excavation', *Proc 5th Euro Conf Soil Mech and Found Eng*, 1, (Madrid 1972) p. 445-453.

L. M. Costa-filho, *A laboratory investigation of the small strain behaviour of London Clay*. (PhD thesis, University of London,1979).

C. A. Coulomb, 'Essai sur use application des règles de maximis et minimis, quelques problèmes de statique, rélatifs, à l'architecture.' *Mémoires de Mathématique et de Physique présentés,* 7, (l'Académie des Sciences, Paris, 1773), p. 343-382.

M. J. Hvorslev, 'Conditions of failure for remoulded cohesive soils', *Proc. Int. Conf. Soil Mech. and Found. Eng.*,1,(Madrid, 1936), p. 39-46.

J. Jaky, 'The coefficient of earth pressure at rest'. *J. Soc. Hungarian Architects and Engrs.*, 78 (22), (1944), p. 355-358.

W. J. M. Rankine, 'On the stability of loose earth', *Phil. Trans. Roy. Soc.*, 147 (2), (London, 1857) p. 9-27

D. Richardson, *Investigations of threshold effects in soil deformations*, (PhD thesis, City University, London, 1988)

B. Simpson, 'Retaining structures: displacement and design', 32nd Rankine Lecture, *Géotechnique*, 42, No. 4., (1992a).

B. Simpson, 'Soil behaves like bricks on strings'. *Arup Journal*, Winter 1992/93, (1992b), p. 15-17.

B. Simpson, N.J. O'Riordan and D.D. Croft, 'A computer model for the analysis of ground movements in London Clay', *Géotechnique*, 29, No. 2, (1979), p. 149-175.

S.E. Stallebrass, *Modelling the effect of recent stress history on the deformation of overconsolidated soils*, (PhD thesis, City University, London, 1990)

K. Terzaghi, *Theoretical soil mechanics*, (Wiley, 1943: Twelfth printing,1965)

C. Thornton & G. Sun, 'Numerical simulation of general 3-D quasi-static shear deformation of granular media', *Proc 3rd Euro Conf Numerical Methods in Geotechnical Engineering*, Manchester, ed. Smith, (Balkema, Rotterdam, 1994), p. 143-148.

W.H. Ward and J.B. Burland, 'The use of ground strain measurements in civil engineering', *Phil. Trans. Royal Soc.*, A27 4, (London, 1973), p. 421-428.

Forster explains the history of thinking in the field of lightweight structures and shows how the physical characteristics of models used by Frei Otto became numerically reproducible using the technique of Dynamic Relaxation. Computer software now allows testing of topological models, relative geometry and shape of support. He discusses generic types of lightweight structures, illustrating this with work in Europe and the Far East.

Brian Forster

Lightweight structures

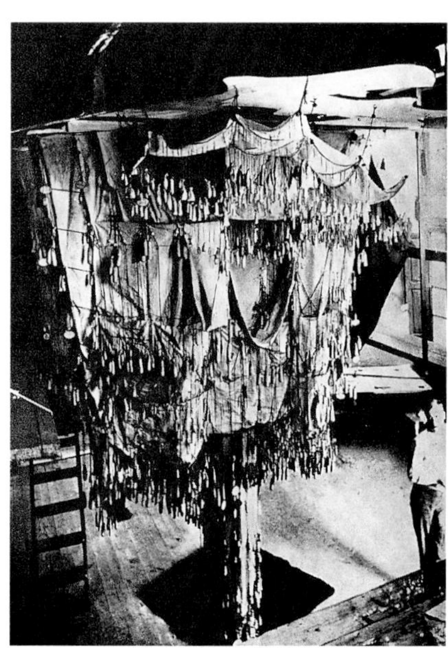

(Right) Antonio Gaudí used string models loaded with weights to develop the shape of the vaults for the Guell chapel.

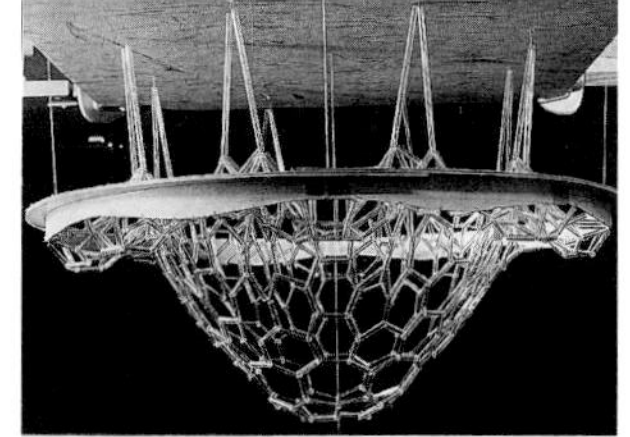

(Below) Frei Otto used hanging chain models to find the topology of reticulated steel domes.

'Lightweight structures' is a rather loose term covering a group of three-dimensional structures whose common link is that their effectiveness is primarily determined by their surface shape. The term includes tents, air-supported and pneumatic structures (generally made of coated textile materials), and cable nets. The origins of Western scientific interest in these structures commences in the late 17th century with the deduction of the catenary's equation by Bernoulli, Huygens, and Leibnitz. At much the same time, Parent, de la Hire, and Gregory investigated the mathematical conditions for equilibrium in vaults. In the mid 18th century Poleni illustrated this mechanically, following his study of St Peter's in Rome, but most significantly he applied Newton's principle of the force diagram to the field of statics and arrived at the explanation of the thrust-line as an inverted chain. His experiments also investigated non-uniformly loaded chains in which the loads were proportional to the weight of individual segments of the vaults.

This work formed the basis for Gaudí's physical modelling of vaults during the late19th century when he searched for a compressive configuration free of moment and shear, without resorting to the architectural devices used in the Gothic cathedrals. In effect, an equilibrium method was being used, albeit in an iterative way, in which judgements have to be made about the weights used (representing the mass of construction) and equilibrium shapes that they produced.

Arups' early opportunities to start designing lightweight surface structures coincided with Frei Otto's mature phase of work – the hanging roofs at Mecca, the Bundesgartenschau at Mannheim, The City in the Arctic, The Berlin Olympic Stadium, and Government Buildings in Riyadh. This period coincided with the presence of gifted engineers such as Peter Rice and Alistair Day; the latter being the originator of Dynamic Relaxation (DR), a mathematical technique which could reproduce the behaviour of non-linear/large displacement structures. This technique became our primary intellectual tool for calculating the shapes of structures, and analysing and predicting their behaviour.

A major constraint on the production of large-scale surface structures (either

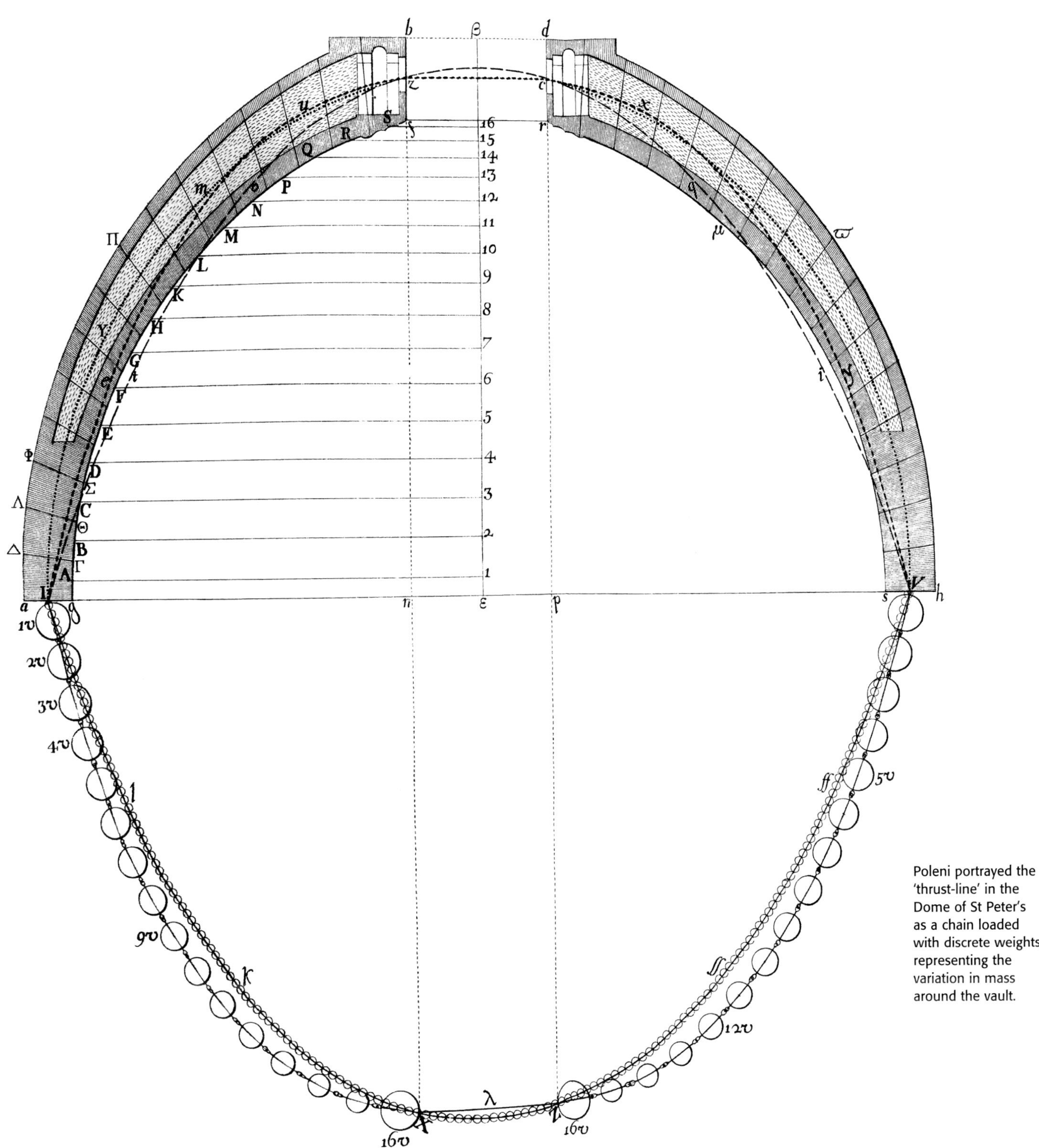

Poleni portrayed the 'thrust-line' in the Dome of St Peter's as a chain loaded with discrete weights, representing the variation in mass around the vault.

compressive or tensile) was the accurate definition of their spatial geometry. This was necessary for analysis and justification purposes, as well as for detailed drawings and cutting patterns for construction. In these respects physical models were slow, laborious, and prone to measurement errors.

The breakthrough for us came in using our own software, based upon DR, to reproduce the shapes of a variety of domes and vaults that Otto had been developing with hanging chain models, following a method similar to Gaudí. One of these structures was to span 90m and errors in measuring the physical model were shown to have very significant effects.

When a chain net is casually laid out on a flat surface, such as a table, its intended form cannot be perceived; it is only once it has been allowed to hang freely that the form is revealed.

Rather than start from an approximation to the final form derived by measuring the model, the imaginative step was to see that a precise statement of the net's topology (ie which element is joined to which and the lengths of each) would be sufficient input data, and that DR could easily compute the large displacements involved in starting from an arbitrarily chosen position.

The programme then simulated the net's free-fall and its damped vibrations as it came to rest – in fact just as a net does in reality.

If the analogue of the vault is the hanging chain, then the soap-film is the analogue for the continuous membrane surface. A soap-film is essentially a figure of equilibrium produced by surface tensions, rather than by distributed masses as for the vault.

Soap films show very clearly that a membrane's surface shape is conditioned entirely by the relative geometry of its support points and edge boundaries.

Membranes have very low mass and surface stiffness. Load carrying is largely determined by the alignment of the weave with the directions of principal curvatures, as well as the degree of these curvatures. By using doubly-curved surfaces, both inward and outward loading can be supported. Such surfaces are prestressable without change to their geometry; prestress contributes significantly to stiffness, due to the opposing curvature components interacting to constrain what would otherwise be severe deformations typical of flat or singly-curved surfaces.

Advantages of speed and accuracy were gained by computerising the 'form-finding' process for membrane surfaces, in which tensions are specified in zero stiffness elements connected to specified boundaries.

By changing the relative values of tensions and the relative geometry of the supports, different surface geometries result. This can be advantageous in accommodating the dominance of a particular service load, for instance, and in improving areas prone to ponding or inversion.

In the structural analysis that follows the form-finding stage, cable and membrane elements with real stiffness values are used; particular elements whose tensions go to zero are automatically switched out during the analysis to simulate a slack cable or buckled membrane. The element meshes used are aligned with the directions of weave and anticipate the seams between the individual panels of fabric that will make up the whole surface. For material economy in the cutting pattern for each panel, the seam lines ideally are controlled to follow geodesic paths.

However, such computerisation does not diminish the real virtues of physical models – they communicate directly in three-dimensions to their constructor, giving direct feedback on ill-conditioning and load paths. They also give immediate opportunities for adjustment, and indicate new directions of enquiry.

Working with physical models also educates the mind's eye in the process of conceiving three-dimensional shape. The explorative, divergent side of the

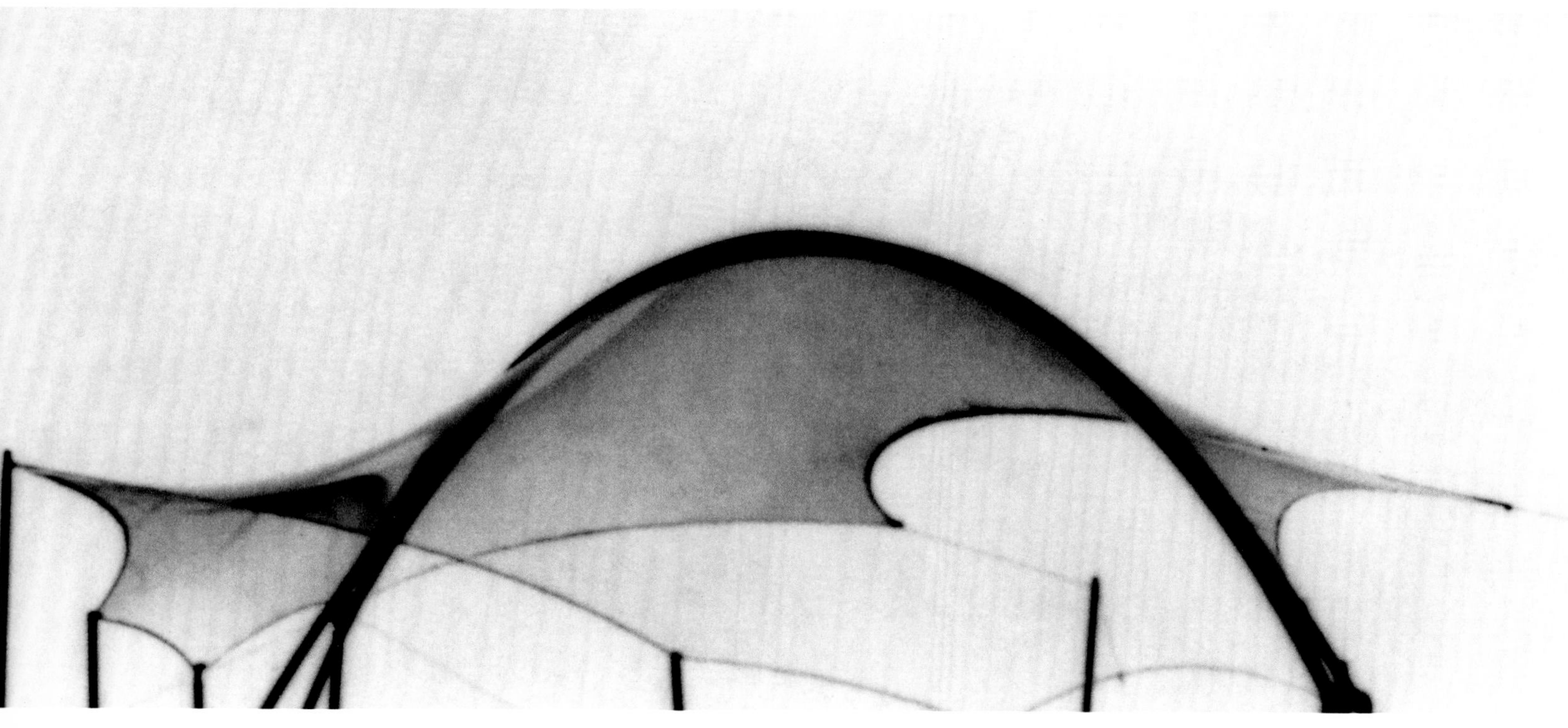

(Above) Soap film study model by Frei Otto.

brain comes into play in the process of presupposing supports and boundaries, and by extension, conceiving a supporting framework which complements it. Thus surface and skeleton can be seen to be totally interdependent geometrically; to achieve economy of means requires them to be conceived simultaneously.

The notion of integration between tensile surface and supporting framework is obvious in domestic objects such as the lampshade and tennis racquet, and also the bicycle wheel. Looking at structures built in the last 15 years, we can see several generic types.

The Lord's Mound Stand and the Sussex Stand at Goodwood have internal skeletons which connect to the membrane skin at discrete points, whereas the stadium roof at Bari, the Deliberatif at Marseilles, Thomson LGT and the Hong Kong Park Aviary all have rib cages against which their skins are stressed. A further aspect of the integration of skeleton and skin is that since cable nets and membranes rely on being prestressed, designs from the very outset have to evolve with a means of prestressing embodied in them. Ultimately this leads to connection details which provide essential load paths, as well as rotational freedoms permitting assembly in the unprestressed state (sometimes in a jumble due to site restrictions), thence moving towards the final geometry required at full prestress.

On large structures we have made calculation studies, together with the contractor, to show the efficacy of the proposed construction sequence as well as giving the distribution of forces at the various stages and the displacements involved.

For the Hong Kong Park Aviary the contractor proposed a sequence in which the cable nets would be stressed at first to values higher than those ultimately required; the underslung wire mesh would then be assembled and attached; the cable net would then destress slightly, and in the process induce prestress into the mesh.

Computer analyses were therefore carried out in advance of construction to test this proposition. It proved to be a satisfactory one and was adopted in practice. The calculation study also provided valuable information against which to monitor the actual movements of the arches, including when and where they would have to be freed from their temporary supports. Acceptance of the structure was made on the basis of measuring cable forces and checking the position in space of certain key elements. This is one of several projects where connecting details were standardised, yet were able to suit a wide range of geometric configurations in which each could undergo subtle changes during the different phases of assembly, prestressing, and load-in-service.

The capacity to produce structures with slender ribs followed from the perception of how the buckling of slender elements can be inhibited by tensile stays and bracing. This led to a second generation of our non-linear analysis program, still based upon DR but capable of including beam and strut elements. It might be said that this was an example of the perception of a structural phenomenon leading to the creation of a tool which could represent its behaviour satisfactorily. The arch ribs within the roof framework at Bari have fans of bracing rods which increase the in-plane buckling capacity of the apparently slender rib.

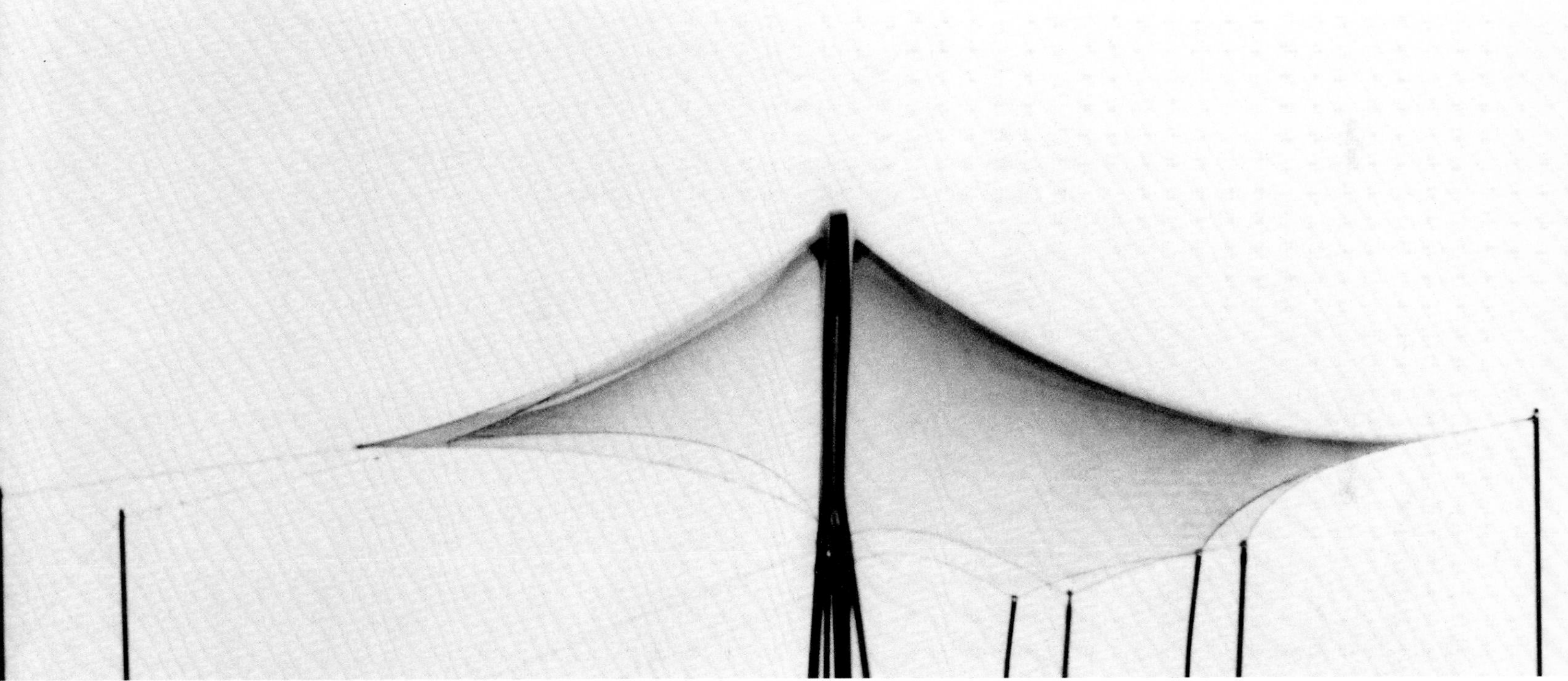

The cable nets forming the surface of the Hong Kong Park Aviary brace the principal arches both in-plane and out-of-plane. A key element in designing these two structures was to obtain the right balance of stiffness between the tensile and strut/beam elements. Another is to have a programme which automatically accounts for the reduction in axial stiffnesses of struts brought about by their deformation, and the change in structural stiffness brought about by the complete destressing of ties – both are phenomena which defeat orthodox linear analysis methods,

and as a consequence are beyond the scope of most building codes.

Other structures which, whilst falling outside the earlier definition of lightweight structures also rely on these analytical techniques, are the Tours de La Liberté in Paris, the glass roof of the transport Interchange at Chur, Switzerland the Louvre courtyard glass roofs, the TGV stations at Lille and Charles de Gaulle Airport, and the wings of the Kansai Airport passenger terminal.

For the stone arched screen at Seville our DR-driven software was used, not to calculate the geometry, but to study the formation of hinges between the stone elements so as to establish the load magnification required to produce instability – an extension of the work of many minds since the original enquiries made by Parent, de la Hire, and Gregory.

In conclusion, these structures, because of their geometry, behaviour and action, and the construction strategy required, provide fertile ground for ingenuity and innovation. The software revolution has had significant effects on what is now achievable. All engineering, regardless of time and place, concerns itself with the making of tools, the understanding of materials, and the exploitation of both. It achieves this through a unique combination of imagination and reason, and as such is considered an art as much as a science.

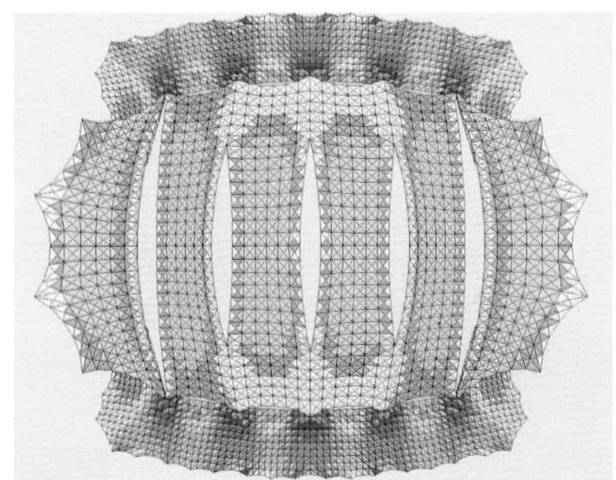

(Left) The erection and tensioning of the membrane roof of the Schlumberger building at Cambridge was planned using both physical and numerical modelling.

(Above right) A stress distribution plot of the Amenity Building, Inland Revenue Centre, Nottingham.

(Above, this, and opposite page) The roof of the San Nicola Stadium, Bari, was constructed as a series of discrete 'rib-cages' onto which membrane skins were tensioned. The slender intermediate ribs are stabilised against buckling by tensile fan-bracing.

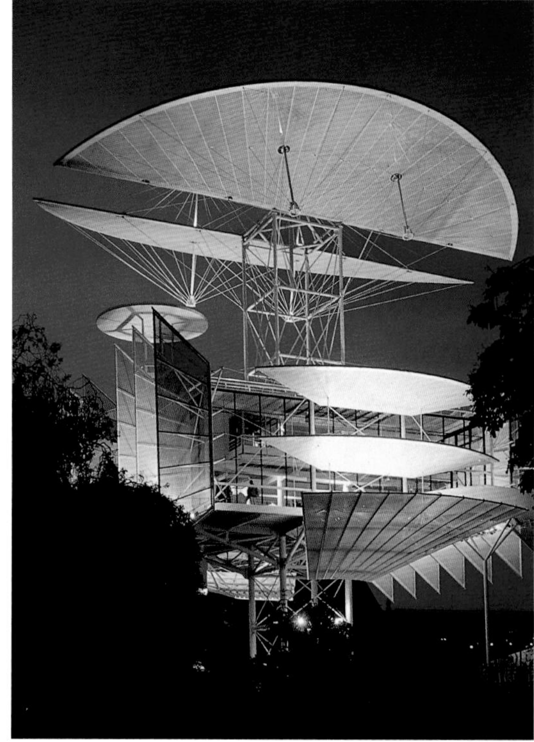

(Above) The 32m span 'wings' of the Tour de la Liberté are stabilised entirely by tensile bracing.

(Left and above) Membrane services are 'tailor-made'. The roof of the Lord's Mound Stand involved the calculation of 130 different cutting patterns.

(Above, left and right) Tensioned cable-nets interact with slender arches to support the 5000m² mesh enclosure of the Hong Kong Park Aviary.

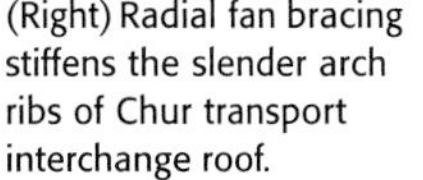

(Right) Radial fan bracing stiffens the slender arch ribs of Chur transport interchange roof.

(Below left) Thomson LGT, Conflans Ste-Honorine.

(Below right) Hôtel du Département, Marseilles.

Over half the consumption of energy in the world is accounted for by buildings, and with such usage comes a degree of environmental damage. Green issues therefore are coming to the fore in questioning the desirability of air-conditioning, for example. But a greater reliance upon passive factors implies a greater involvement of engineers and architects because the construction itself becomes the most important passive factor. At the same time the definition of comfort conditions is being seen as a balance of factors, not a formula.

John Berry

Green buildings

Buildings account for more than half of the energy consumed in the industrialised world today. Little has stemmed this demand during the last 20 years and every indication is that it is more likely to increase rather than decrease. The increase is likely to come from the growth in consumption of the third world, as it progresses toward industrialisation and per capita incomes on a par with the industrialised world.

At present fuel is relatively cheap and plentiful, so why should we be concerned? Surely the scientist has the answer in solar power, wind power, wave power, or cold fusion to ensure that we can carry on as before? Obviously there are attractions in pollution-free sources such as solar, wind, and wave, but there is a price to pay for all sources in one way or another. The notion of great wind farms, unless offshore, is unappealing to those living within sight or sound of those giant rotating blades. Also the impact on our global environment from tampering with nature in any significant manner should concern us. Small-scale operations or experimental set-ups are fine, but how confident can we be that our climate will be immune from the heavy hand of man? Past attempts at major river diversions to irrigate land for crop growing, for instance, give little hope for the future. One has only to remember the disastrous impact of the Ukrainian experiment, which effectively reduced fertile land to a desert, to urge caution. So why should we be concerned? Yes, energy is plentiful and relatively cheap, but two factors dispel the comfortable feeling that all is well: Will it be freely available in the future? What is the impact on our environment of proliferation? Fears of an energy shortage following the 1970s oil crisis undoubtedly prompted the major thrust for conservation; cost was somehow secondary. Perhaps for the first time the ability to pay was not in itself a guarantee of supply? At the same time the realisation that the burning of fossil fuels could damage the environment was gaining credibility. Excessive carbon dioxide emissions have been shown to contribute to global warming – the greenhouse effect. Alarm was also being expressed at the depletion of the ozone layer which protects the planet from harmful solar radiation. It is generally accepted that CFCs, which are used in many building insulation products as well as the large refrigeration cooling systems found in many buildings, contribute to this phenomena.

The link between energy use and environmental damage is proven beyond reasonable doubt and it is this relationship which is causing unease. This new thrust for conservation policies is widespread and for once is not just linked to the standard cost-effectiveness arguments. People now realise that it is impossible to put a price on environmental damage. Even for those who believe that science has the answer to all our problems, which indeed it may, the conservation ethic, as a temporary respite at least, is strong. But it goes further than this, in that conservation of our natural resources and limiting potential damage to our environment must be beneficial. How can the subtleties and intangible nature of the situation be expressed in any meaningful way? Of course it could be defined in purely scientific terms but that would, by definition, only apply to a limited part of the argument. The general argument that we are slowly but surely undermining our planet's ability to regenerate

(Above) Nature copes well with extreme climates. The compass termites of Western Australia maintain an inside temperature of 31°C to within 1°C while the outside temperature varies between 3°C and 42°C.

and support itself requires a far broader statement. Necessity being the mother of invention, the Green Movement was born. The word 'green' has no absolute meaning; it is an expression of man's unease with the present and a feeling that we must do better in the future. Green is simply the concept of nature and something which the populace can clearly identify with. It is so widely known and understood that its success is self-evident; Greenpeace and the German Green Party are just two examples which feature highly in the public awareness. The same can be said for energy conservation or the harmful effects of CO_2; so in a word it is successful. In most people's eyes it is seen as representing a positive attitude to our environment, and definitely one for the better. It knows no boundaries. It can be global or local. Now a common base exists on which to move forward and explore the new language of green thinking. Sun, light, wind, mass, volume, and materials can positively influence our designs. Aspects of green thinking which, perhaps above all others, are potentially the most influential are harmony and balance. It is all to easy to lose sight of the implicit simplicity which should intuitively flow from a harmonious design. If it doesn't materialise then question the process!

Each building will have its own unique set of goals and objectives and these need to be determined at the beginning of a project. It should not be forgotten that the primary function of any building is that of providing shelter and comfort for its occupant, irrespective of task. A fine balance exists between energy conservation, ecology, and health, and these need to be addressed in a pragmatic way and not just on fashionable trends. Nevertheless accepted wisdom does need to be questioned; an open mind is essential. One of the more deeply-rooted current wisdoms is the perceived need for air-conditioning. Not only is this questionable in the temperate European climate, but in a good portion of the USA, come to that. The equally dogmatic stance that the cocooned air-conditioned environment, free from external stimulus, is somehow the role model against which others should be judged is also questionable. Public opinion and user studies do not support such a universal theory. When

(Above) Inland Revenue, Nottingham Typical elevation.

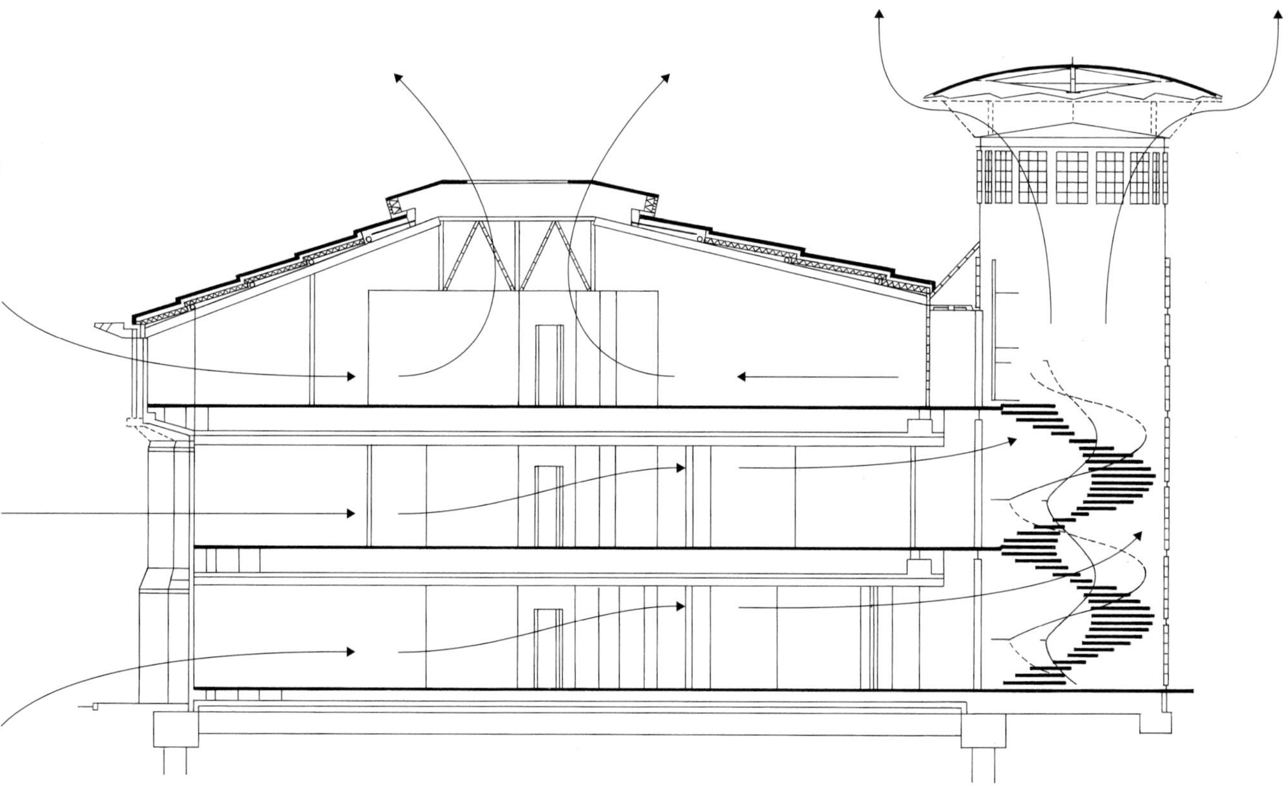

(Below) Principal air flows through the building. Note that the voluminous top floor is ventilated separately and not connected to the stair tower.

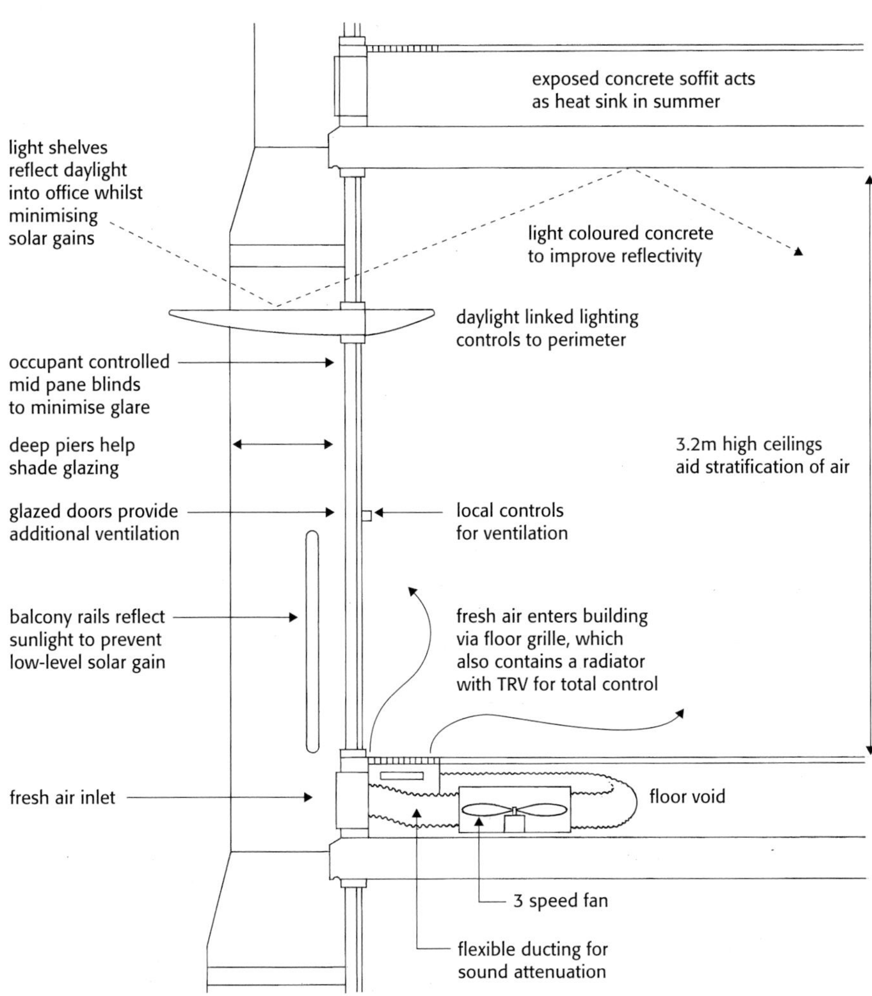
exposed concrete soffit acts as heat sink in summer
light shelves reflect daylight into office whilst minimising solar gains
light coloured concrete to improve reflectivity
daylight linked lighting controls to perimeter
occupant controlled mid pane blinds to minimise glare
deep piers help shade glazing
3.2m high ceilings aid stratification of air
glazed doors provide additional ventilation
local controls for ventilation
balcony rails reflect sunlight to prevent low-level solar gain
fresh air enters building via floor grille, which also contains a radiator with TRV for total control
fresh air inlet
floor void
3 speed fan
flexible ducting for sound attenuation

Inland Revenue, Nottingham. (Above left) Typical section showing the components of the high performance façade.

(Above right) Interior of typical office showing vaulted exposed concrete soffit (high thermal inertia).

asked which feature they most valued in a building 80% of the respondants of one survey voted for opening windows. That is not to say that 80% were arguing that environmental control should be by opening the window alone; they were simply expressing the desire to be able to do so. The notion of air-conditioning and opening windows goes together about as well as electricity and water. This is not an argument against air-conditioning, but for a more open-minded policy toward a viable alternative. There is talk of following the lead in Holland and Switzerland and implementing legislation to limit the use of air-conditioning. Legislation is not the answer. Building design, economics and user demands will force the issue.

One can point to a number of outstanding buildings that reveal an ambitious architectural and engineering concept in the bold new tradition of the environmental aesthetic. They show by example that it is possible to address the social issues of the day in a balanced and harmonious manner. At one extreme there is the zero energy concept and environmental evangelism of Brenda and Robert Vale's house, and at the other the simple energy reduction targets of the earlier Basildon housing to provide affordable heating for the socially deprived. Both are equally valid.

What is apparent in the design sense is how architecture and engineering come closer together than in the conventional design situation. The sequential process of architecture, then engineering, has no place; the simultaneous solution of ideas is the new order. Low energy and environmental awareness are mutually compatible. It is difficult, though not impossible, to envisage an environmental design which pays little concern to energy consumption.

It is often said that designing passive buildings is more difficult and demanding than designing traditional active systems. To a large degree this is true, and most environmental designs can be thought of as active design with a passive outcome, whereas the traditional approach often has the roles reversed with an active outcome from a woefully passive design input. As the passive components form the architecture, the engineer enjoys a more prominent role in the overall design than is usual, and is involved from the outset. This is an important step, not only from the view of recognition, but to be genuinely able to contribute to the building as a whole. Without doubt the façade and building envelope play a leading part in the green building concept. It is here that the natural forces of sun, wind, and light are moderated and filtered before entering the occupied rooms. It is the thermal and visual barrier between the calm interior and the natural and sometimes hostile exterior. The most significant contribution to the performance of the building as a whole occurs at this interface, and many buildings fail as a result of inadequate façade design – more often than not the air-conditioned variety. How often do we see the glazed mirror glass stump appear with scant regard for its surroundings? Is it any wonder that the internal systems cannot cope, that even air conditioning has its limits? Thermal comfort is absolute and cannot distinguish between the niceties of system choice. Why is the simple engineering fact that the façade is an integral part of the overall system so often ignored or forgotten? A passive design does not have the luxury of a powerful climate control system to fall

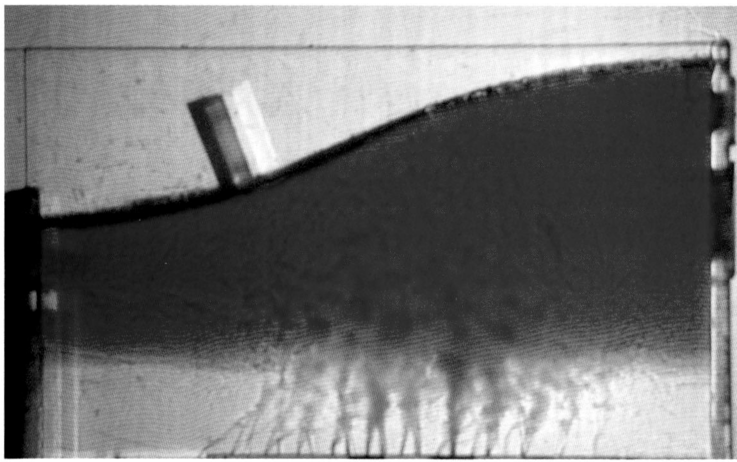

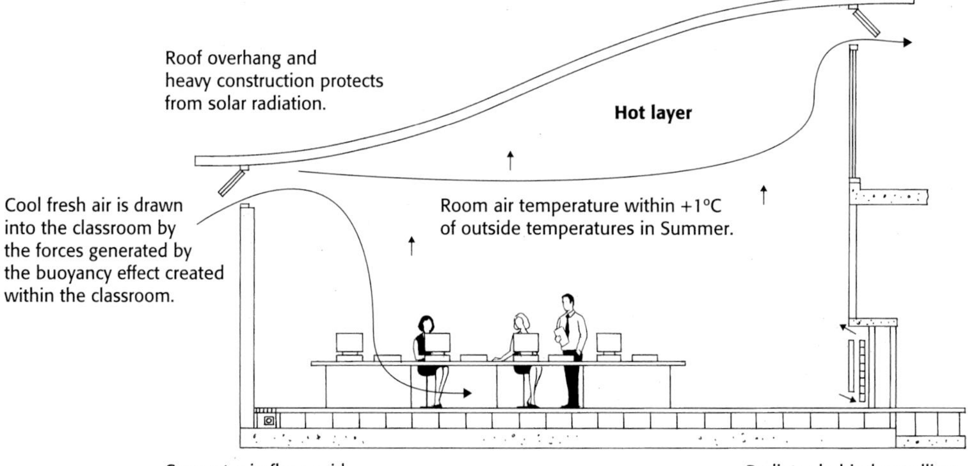
Hot layer forms at high level to be exhausted through openable glazing, driven by the buoyancy effect or cross ventilation forces.
Roof overhang and heavy construction protects from solar radiation.
Hot layer
Cool fresh air is drawn into the classroom by the forces generated by the buoyancy effect created within the classroom.
Room air temperature within +1°C of outside temperatures in Summer.
Convector in floor void beneath glazing.
Radiator behind panelling adjacent to corridor.

Cable & Wireless Training College, Coventry.
(Above centre)
Saline model test showing temperature gradient and air flow (University of Cambridge).

(Right) Concept diagram of typical teaching classroom showing principal of natural ventilation.

Cable & Wireless Training College, Coventry. East elevation of teaching wings.

back on, therefore it tends to take a more rigorous line of thinking about the whole building, not just a small part. While the façade represents the first line of defence, after external landscaping of course, the interior also has a role to play: mass to absorb excessive temperature fluctuations and store night cooling; the capacitive effect, or volume for enhanced ventilation efficiency; surfaces that reflect light in a particular way or are warmed by the sun to induce ventilation; inlets and outlets oriented and designed to catch the wind. All are important considerations. The preoccupation with the desire to utilise passive elements wherever possible may lead to interesting architectural/engineering designs but would be doomed if they failed to address the needs of the occupant. Fortunately the green attitude of dealing with the whole rather than its constituents ensures that the needs of the users become an integral part of the design process – a truly holistic design. In fact far more attention is likely to be paid to comfort than is normal. The view that an ideal environment has an air temperature of 22°C and a lighting level of 350–500 lux is suspect. Fanger's thermal comfort equations, on which most assessments of comfort are made, does not support such a simple hypothesis, and a much wider latitude is permissible. But what is important is the realisation that Fanger's model fails to take account of expectation and that the adaptive model is probably more appropriate to natural or free-running buildings. In simple terms, current methods underestimate people's tolerance to adapt to extremes when they have a degree of control over their environment; this augers well for passive buildings. Of course some external environments are hostile and a sealed building therefore is essential. Even so, the green attitude of minimising dependence on complex and sophisticated environmental control systems is still valid. Installed capacity can be halved and new system opportunities explored which are far removed from the conventional low-temperature solutions traditionally adopted.

So where does this leave us? The environmental lobby is well-established and will not go away! Unease with the current dichotomy of plentiful energy, low cost and environmental damage is apparent. Society demands that we should do better, and we can do better as the buildings here testify. It is crossing the boundary between engineering and architecture to produce a lasting environmental aesthetic which holds the key to the future – Architectural Engineering in the truest sense.

Richard Cowell

Early acoustic design tended towards witchcraft. Latterly, prediction of acoustic performance has improved due to more advanced techniques of simulation. The use of scale and computer models raises new problems associated with the power of the aural environment and the ability to listen to real and not recorded sound.

Simulating the acoustic environment

Based on well-developed knowledge of the physics of sound and the early work of enthusiasts such as Helmholtz, Sabine, Olson, and the BBC, by 1946 development in practical building acoustics was already substantial. Some fundamentals of good practice were available for the design of post-war building developments, but in the young discipline the principles were not as advanced as in the application of the physics of heat and light to building design. The ability to predict the acoustic response of rooms was basic, centred on Sabine's reverberation time theory and the assumed sound absorption of materials. The most powerful elements available for simulation of an acoustic environment were the gramophone record, amplifier, and loudspeaker. Re-creation of sounds was the order of the day. Emphasis was placed on accuracy and developments in the details of sound measurement. Speed of measurement and analysis, with the consequent benefits for iterative design, were not a priority. How a room would perform acoustically was another matter still then considered dangerously close to witchcraft and folklore. Errors made since then have prolonged this mythology, with the result that confidence has only recently returned to auditorium design. Currently, a powerful armoury of prediction methodology is available to be directed at advanced simulation of the aural environment.

We have reached a stage where simulation may now offer us more than we are ready for. As with most advances, there are substantial benefits and significant dangers. On the whole, when put in context by good consultancy, simulation techniques offer very powerful design tools. There have been some useful lessons learned as they have developed, and underlying this has been the need to understand the tools and their limitations.

For the 1951 Festival of Britain, the building of the Royal Festival Hall in London provided the opportunity for a major step forward in the prediction of room acoustic response. Careful and well-recorded analysis was carried out; predictions were made. The results were broadly successful at the time, except for a substantial miscalculation of the bass response of the hall. Reduction of volume (a ceiling lower than originally proposed), and substantial absorption by the ceiling itself, left the hall 'dry' rather than 'warm'. This provided a unique opportunity for Parkin's work in the development of an artificial reverberation system. Using microphones set in tuned resonators and an array of loudspeakers in the ceiling, the response of the hall was modified electronically to add to reverberation. By approaching 'feedback', the decay of sound was extended. Since then, the advance of electroacoustic component technology has brought newer and more stable options to reverberation simulation.

In the 1960s the spectacular failure of the Lincoln Centre auditorium in New York overshadowed pioneering research in auditoria by Leo Beranek. In the UK, 1:8 scale models were used as research tools in association with the acoustic designers for the Barbican Concert Hall and the Olivier Theatre. These were used later for corrective works for both. The models were used for prediction of changes arising from removal (Barbican) or addition (Olivier Theatre) of absorptive materials. Long path reflections and focusing effects in the Olivier Theatre were also removed, first in the model and finally on site. It was clear that such

Acoustic scale model of proposals for Canary Wharf underground railway station (London Underground Jubilee Line). The model includes simulated model loudspeakers and has been used to pursue the optimum balance of investment.

Testing and commissioning of the Glyndebourne Opera House using 1:50 scale model (above) and site tests (right).

models performed well in the examination of room geometry, but it was less clear that sound absorption could be well matched at model scale. With time, and use of model scale reverberation chambers, accuracy of modelling of surfaces and room contents has improved, particularly for mid- and high-frequency sound.

At full scale, the development of commercial acoustic testing laboratories extended the available data on sound absorption, sound insulation, and impact noise isolation for a wide variety of building components. Better field and laboratory data became available and the scope for more accurate calculation grew. Acoustic consultancy grew into a recognised and regularly-used discipline in building design. The opportunities for more accurate and convincing simulation techniques grew with the practice of acoustic design. In the late 1960s and early 1970s, research into listening took individuals into unnatural environments, for the study of artificially produced reflections. In anechoic conditions, arrays of loudspeakers provided sound signals with a variety of timing, level, and spectra. This formed not only the basis for subjective and objective criteria for individual listening parameters but also the building blocks for simulation of an acoustic. Conferences on architectural acoustics were accompanied by demonstrations of aural effects recorded on dummy heads in anechoic chambers.

A wide variety of parameters were invented in an attempt to replicate qualities important to the listener. For building design, perhaps the most influential step at this stage was the development of an understanding that early lateral reflections provide a sense of spaciousness of sound. The importance of this to good auditorium acoustics helped to explain why certain types of rooms (eg the 'shoebox' form, the vineyard form) were so successful. Fundamentals like adequate loudness were also better appreciated.

Physical scale models of buildings, particularly auditoria, had already been used for many years. During the 1970s, the improvement of instrumentation brought the opportunity for operation to higher frequencies, and thereby the capacity to reduce the size of the models. The loss of sound energy in the air itself could be avoided by replacing air with nitrogen. Analysis programs can now correct for absorption in air.

The small-scale physical model has been attractive, not least because it is cost-effective. It also provides another means for the development of the design to be shared with the design team. The visual impact of the model can play an important part in the involvement of the client, architect and engineer in the integration of acoustic requirements into proposals. Simplified acoustic models can also be used to develop concepts, using either laser light and mirror card studies (for high frequency reflections), or simple-scaled acoustic model measurement in very simple constructions. Scale model techniques are applicable over a wide range of building types and have been used not only for auditorium design, but also for the design of atria, meeting rooms, underground railway tunnels, and ticket halls. In the last case, model loudspeakers have been used successfully within scale models as a basis for evaluation of speech intelligibility for a variety of room treatments and loudspeaker layouts.

Over the last 20 years, the development of computer software has opened up the most exciting scope for aural simulation. Rooms can be built within computers and simulation of sound fed into them. The response of the room can then be assessed, assuming the appropriate source, boundary, and receiver positions are identified. Ray tracing has been the primary means for analysis. The complexity of the diffraction of sound at the edges of surfaces or local to sound-absorbent materials has been more difficult to build into the computer models. Nevertheless, as the power of the computers increases, these complexities are gradually being introduced. Ray tracing and image source models are combined to advantage. A special contribution has been made in the field of electroacoustics. As a result of quality improvements in recorded and live show sound and large public address systems, prediction and simulation software has been developed in the industry for loudspeaker coverage. However, for too long, electroacoustic system designs were specified with a degree of emphasis on technical specification of advanced components

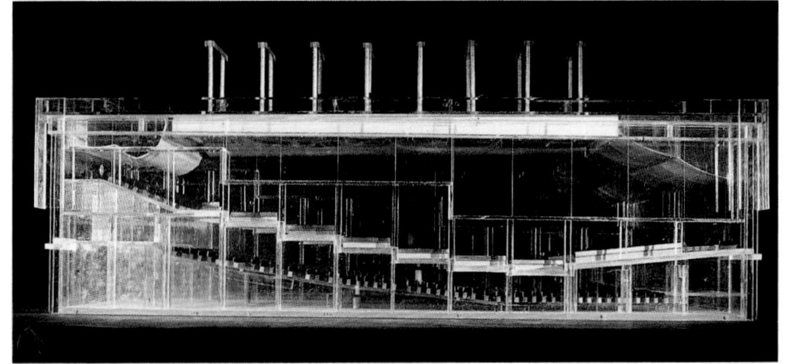

1:50 scale acoustic models of the Manchester Concert Hall (above) and the Lingotto Concert Hall/Congress Hall (below) for Fiat. Both models have been used for diagnostic checks on the performance of the interior geometry and acoustic treatment for optimised performance.

Glyndebourne Opera House (overleaf). The auditorium.

Use of computer simulation of soundscape by sound reinforcement system as a basis for optimised design.

which contrasted strongly with the attention to the response of the room into which the sound was to be fed. More recently, the interaction of loud-speaker output with room acoustic response has been focused upon – a highly significant development.

The growing attractions of computer simulation are manifold. It is now usual for buildings to be designed by computer-aided design (CAD). Developments from CAD have allowed buildings to be visualised by moving through perspective simulations. With further careful integration of software, proposals developed by designers can be subjected to acoustic testing. Most recently, the auralisation of sound in computer and physical models has become available. Here we are now with the dangerous capacity to mislead the client's ears in the same way that designers have used computer-aided design and visualisation to develop the theme of the architect's beguiling perspectives.

The development of broadcast and recorded sound (the tape, the CD, the Digital Audio Tape, the mini-disc, etc.) has been accompanied by dramatic improvement in sound quality. Expectation of sound quality is now so high that this puts pressure on the 'real thing' to be of the very highest quality. There are those who find live performances disturbing to the process of listening. Audience or other ambient noise can prove a major distraction. The pleasure of live performance clearly encompasses a wide range of non-acoustic factors. Because listening is only part of a much wider human experience, it is difficult to attribute too much accuracy to the process of simulation of sound only. The importance of integration in building design and simulation procedures is self-evident, although often difficult to achieve in practice.

It is interesting to speculate on the real commitment which society has to a pleasing aural environment. The battle against noise is not going well in urban environments. The feed of popular music directly into the ears seems like a substitute aural environment rather than an improvement. Is good sound quality reserved only for music, theatrical sound effects, cinema, and video? Some of the experiments of the early and mid-1970s on the use of background masking sound, usually for office environments, made one wonder what sound we would enjoy living with.

It was interesting to find that the preferred spectrum shape for simulated background sound was similar to the spectrum for natural sounds, eg a distant seashore, wind in the trees, waterfalls, and human speech. At the same time the criteria most often used are set as limits on noise, ie controlling the negative aspects of our aural environment, not building the positive. The bottom line may still be that not many people listen, they only hear. We can ask them to listen, using the techniques of simulation as a start. We can ask why we cannot have better quality sound in railway stations, shopping malls, in streets and squares, at home, in court, in hotels, and on board ships. Here is an opportunity to get people interested in reversing the loss of a pleasing aural environment.

Perhaps our skills in simulating an acoustic are developing faster than the understanding of our objectives. Criteria for acoustic design have all too often developed from what we would rather not hear. It is more important and often more challenging to establish the positive features. It is tempting to believe that education has not explored the opportunities which our aural environment presents for us. Many who design or use buildings today have not learned to listen. Yet the new generation have been listening to high quality sound reproduction, and are exploring the experiences of virtual reality. We are approaching a new century at a time when it is becoming easier and easier to engineer aural experience.

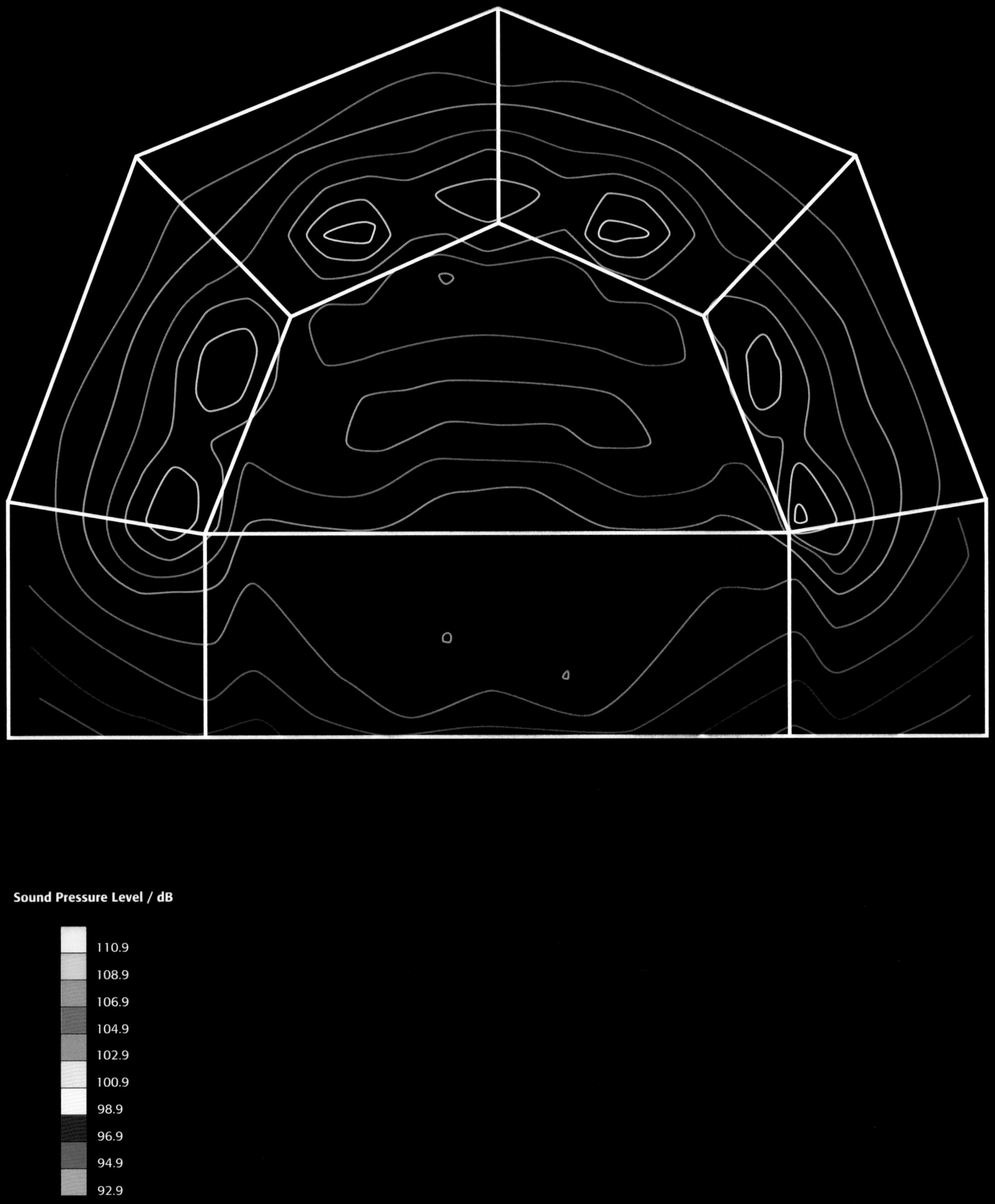

Sound Pressure Level / dB

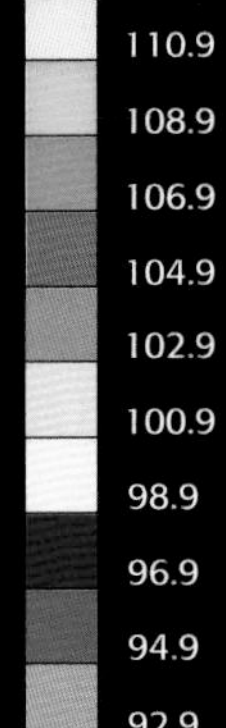

Arups' expertise arose from new construction. With existing structures, the problems the engineer faces are the reverse of those in new build. In order to conserve, the precise way in which the existing has worked well must be understood, as must the defects which time and nature have brought. Arguments about the authenticity of construction must be set against the possibilities of using current techniques without damaging the historic value of the buildings.

Peter Ross

Work to existing structures

The firm's original aim was, of course, to design new structures, but it was perhaps inevitable that, as our expertise grew, we were asked to look at structures which already existed. These 'structures' have ranged, over the years, from a statue of Lord Hill in Coade stone to the 19th century frigate *Unicorn*, but in the main they were buildings which were in need of repair due to a structural defect or accidental damage, or which had outlived their original purpose and were to be adapted for an alternative use.

Understanding the built structure

When we design new buildings, we generally start with an overall layout, and then determine the member sizes, amending them if necessary, using current codes and standards, so that the structure is safe and serviceable. We are in control, and can modify the design as appropriate.

In the appraisal of an existing building, this process is in effect reversed. The structure exists, and is, or is not, performing, as the case may be. The engineer has to discover the way in which it has been built, and how it is behaving, before he can make any meaningful recommendations for repair work. It is of little use, for instance, pointing up a crack in a brick wall if the movement is on-going.

The first task is to carry out a survey. Since most structures are in part concealed, investigative work may be needed, although we may sometimes have the benefit of documentary information such as original drawings. There is then a need to model the structure, either for qualitative assessment or more formal analysis, and come to some view of the material strengths.

This need to understand the structure and how it is behaving applies to the simplest projects. The aisle roof of Cranbrook Church in Kent is in timber, with purlins supported on what appear to be trusses. However, investigation shows that there is only a nominal connection between the rafters and the bottom tie – this means that truss action is not possible, and that the bottom 'tie' is really a beam. A repair to this member will have to be designed for a significant bending moment, rather than the simple tension which is normal in a tie member.

Traditional forms of construction

Cranbrook Church itself dates from the 15th century, and it received its 'new' roof over 100 years ago. In fact most of the buildings which we have investigated were built in the 19th century or earlier, and in traditional forms of construction, which differ markedly from those of today.

The modern cavity wall, for instance, with a thin outer leaf of brick with a cementitious mortar, would eventually crack if movement joints were not included. In contrast, traditional walls are of solid construction, and use a relatively weak lime mortar. Their more substantial mass is able to contain moisture and temperature-induced movements, and they were built generally without movement joints. In addition, water penetration is mitigated by the additional thickness, and by the use of protective details such as cornices or overhanging eaves.

These differences between traditional and modern construction occur in practically all the elements of fabric, and we have learned not to look at the traditional forms of construction through 20th century spectacles. Moreover, we have used the lessons of history in some recent projects, such as the new Glyndebourne Opera House, where solid walls built

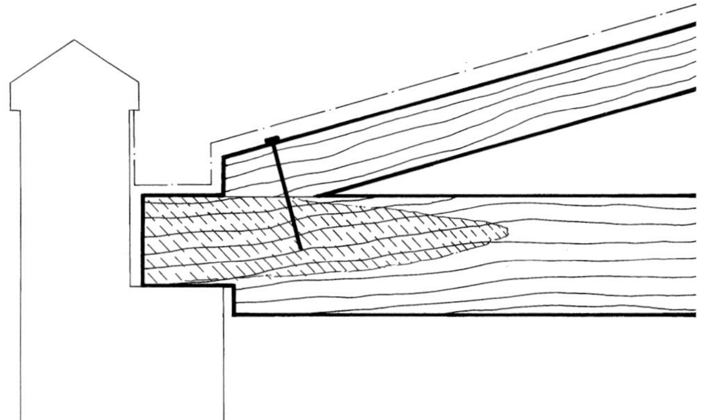

St Dunstan's Church, Cranbrook, Kent
(left) Eaves detail, showing the areas of rot, and the single bolt connection between the 'rafter' and the 'tie'. This is quite inadequate to develop truss action, and the bottom member carries the load as a principal beam.

with lime mortar have enabled us to dispense with the conventional jointing pattern.

The interaction between soil and structure

By and large the traditional wall performs well, and any cracking or distortion which develops is more likely to have been caused by settlements or other movements of the foundations. In these cases some form of site investigation is almost always required, and the aim is to explain the observed movements in terms of the weight and stiffness of the structure, the foundation design, and the soil characteristics. The most usual reasons for building movement are that the designer had either been unaware of variations in the soil properties over the site, or that his assumptions of bearing capacity have simply been too optimistic.

In late 1960s an inspection of the fabric of York Minster revealed cracking and distortion of the masonry around the central tower. A level survey of critical features, such as mouldings and string courses, showed that the total settlement relative to the building was of the order of 300mm, and was still progressing, albeit at a very slow rate.

Exploratory pits determined the general outline of the foundations, and showed that the four main piers, each carrying a load of around 4,500 tonnes, stood on remarkably small footings. The resultant bearing pressures, of the order of $750kN/m^2$, were roughly three times the maximum that we would today consider appropriate on the clay subsoil.

The principle of the remedial work was self-evident – the effective bearing area of the footings had to be increased in proportion to the degree of overload. In view of the

(Left) The roof trusses in the north aisle, temporarily propped on scaffolding spanning the organ.

(Left) The repair joint: a vertical splayed scarf, joined by means of bolt groups with 'split ring' connectors.

York Minster tower foundation: (Below) The additional footings around one of the four centre piers: (a) a concrete collar, designed to encase the badly fractured remnants of the Norman walls. (b) the upper foundation, post-tensioned to the existing footings by post-tensioned stainless steel rods. (c) the lower foundation. (d) flat-jacks inflated to transfer part of the pier load to the underlying clay.

(Right) The East End of the Minster was discovered to have an outward lean of over half a metre. With temporary shoring in position, the foundations were rebuilt and extended, in short lengths.

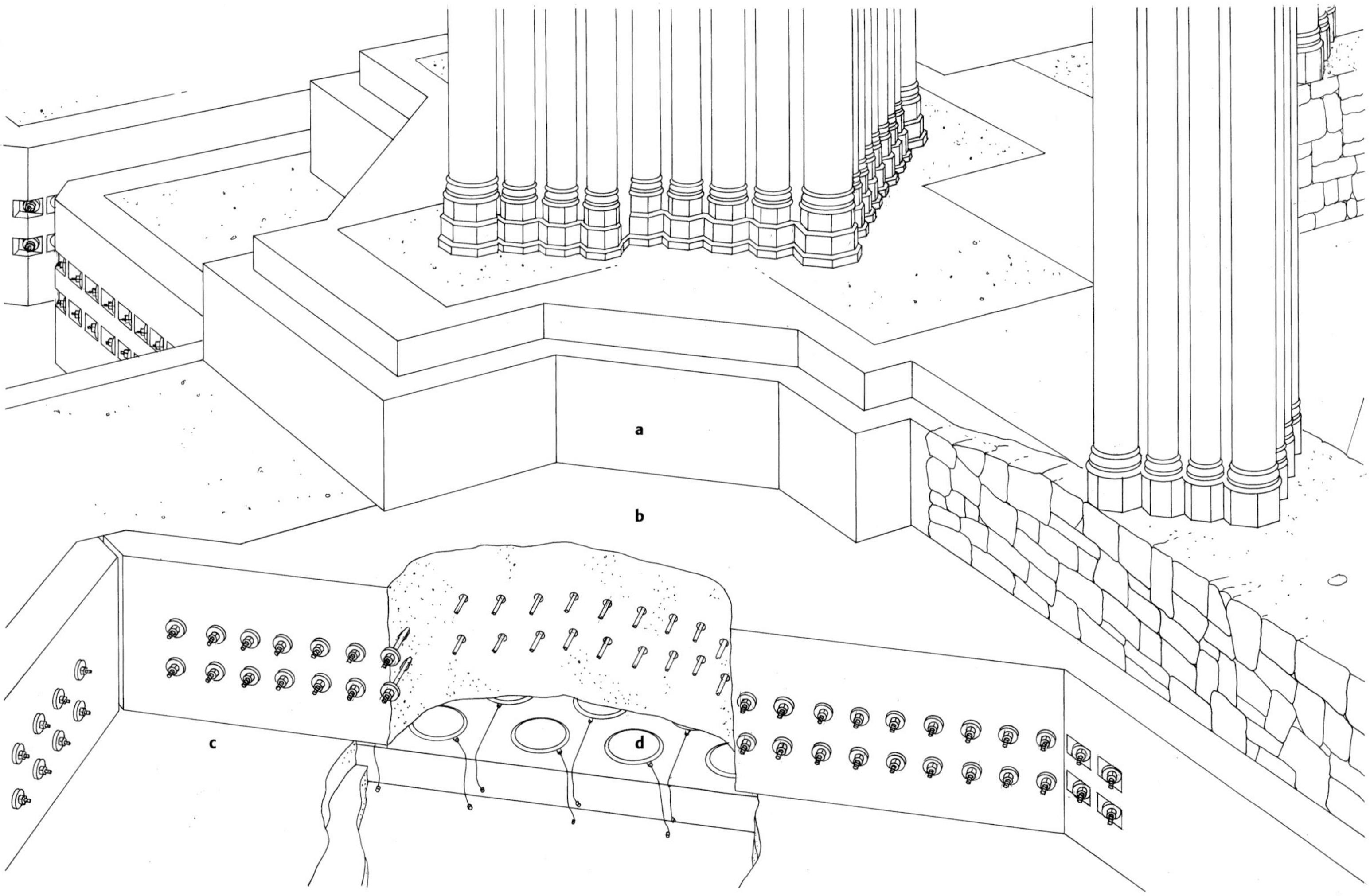

Tyne Theatre, Newcastle. The trusses over the stage, after the fire (left). The new trusses, replicas of the originals, under construction (below, left). The rebuilt fly tower, with machinery (below).

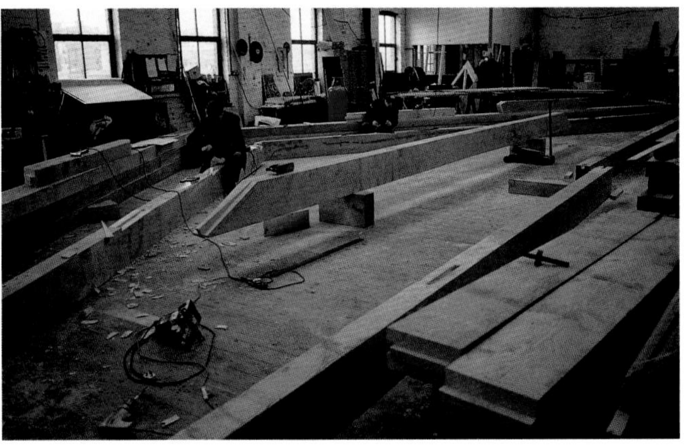

magnitude of the loads, underpinning was considered to be too great a risk. Additional foundations were cast around the original footings, in two layers. The upper layer was attached to the original footing by through-drilling and post-tensioning stainless steel rods. The lower layer could then be made 'active', and prevent further settlement, by inflating flat-jacks set into the gap between the two layers.

During the early stage of the design the idea emerged of creating a new undercroft, which capitalised on the excavated space and allowed much work of Roman times to be seen. Thus, unusually, the work of the foundation engineer is on permanent view to the public.

The aims of conservation

Major repairs of this kind prompt us to consider the nature of historic buildings, and the aims of conservation. Historic buildings may be defined as those which have some form of cultural significance for society. The significance could take various forms – aesthetic, religious, technical, or even social ('Charles Dickens lived here'). The hangers at Duxford Air Museum, for instance, of 1917, are roofed with Belfast trusses – a brilliant technical solution to the problem of providing a 30m span, using only short planks and a large bag of nails. Our aim in conservation is to identify the significance of the structure, and to select the form of repair which, as far as possible, maintains that significance. Two projects serve to illustrate this process – both timber roofs, damaged by fire, on masonry walls.

In 1985 the Tyne Theatre, Newcastle, suffered a major fire. It dates from 1860, and has typical 19th century timber queen post trusses on brick walls. The trusses over the auditorium suffered only minor damage, and could be repaired. However, the four trusses over the stage itself were on the point of collapse, and replacements were needed. The theatre has a Grade 1 listing, primarily for its mid-19th century stage machinery, thought to be the most complete set in England. The trusses, while supporting the roof, contain the fly tower, with all the apparatus for 'magical' effects and scene changing. The new trusses were therefore constructed as replicas, using much of the metalwork salvaged from the fire and copying the profile of the auditorium trusses. The various stage mechanisms were also rebuilt to an operational standard, mainly from the detailed knowledge of the staff, and thus the main cultural significance of the theatre, the stage machinery, was retained.

Some six months earlier, the roof of the South Transept of York Minster was totally destroyed by fire. Daylight found the smoking transept open to the sky, its roof removed with an almost surgical precision. The Minster has a long and complex building history: the transepts themselves date from the 13th century, only receiving their timber vaults in the 15th century. The roof structure itself dated originally from the 18th century, but was inadequately designed, and major strengthening members were added in 1846. Thus, although the roof was of archaeological significance in terms of the record which it gave of the work of various periods, it had in fact been much altered and strengthened with time as was found necessary.

Ideas were offered in abundance for the rebuilding. Why not leave the transept roofless, or provide a glazed roof? Or build new vaults in concrete or glass-fibre? Against the propos-

York Minster
South Transept roof.

(Above) The transept after the fire, the roof and timber vault completely destroyed.

(Above right) The new vault under assembly, with ribs and bosses of glued laminated oak.

(Right) Individual ribs, some 6m long, and each made from profiled kiln dried laminations, glued and bolted together. The originals, of course, would have been single pieces, but the task of finding curved trees of the right radii today was agreed to be impossible!

als of the modernists, the traditionalists demanded an exact replica of what was there before the fire – or even a 'day one' approach, rebuilding a vaultless 13th century roof, although the details were totally unknown. An exact replica, however, would give the same performance in a future fire, and there were arguments for increasing the fire resistance of the vault, and the survival time of the structure above.

In the event, sanity prevailed. The vault has been restored to the original lines, with modifications to the web construction to improve the fire resistance. Replicated trusses, however, including all the repair straps, would have been a pointless contrivance, and so we were free to design '20th century' trusses, which were simply a direct response to the brief. They would then contribute to the record of time provided by the Minster roofs – the Nave trusses, replaced in the 18th century, being classical queen posts, and the North Transept roof, of the 19th century, a quintessentially Victorian composite of timber and cast iron.

A typical truss consists of an upper and lower collar with scissors legs applied to each side. The programme of construction dictating that the timbers would have to be used `green', with moisture contents between 80% and 100%. Modern structures are usually made with seasoned timber, but the evidence for the use of green, or at best partly-seasoned timber in medieval structures is overwhelming. It is, however, necessary to do what medieval designers did, which is to take account of the shrinkage which will take place across (but not along) the grain as the timber dries out. So for this reason all the joints are simple laps, connected with bolts which were periodically tightened over a period of about six years until

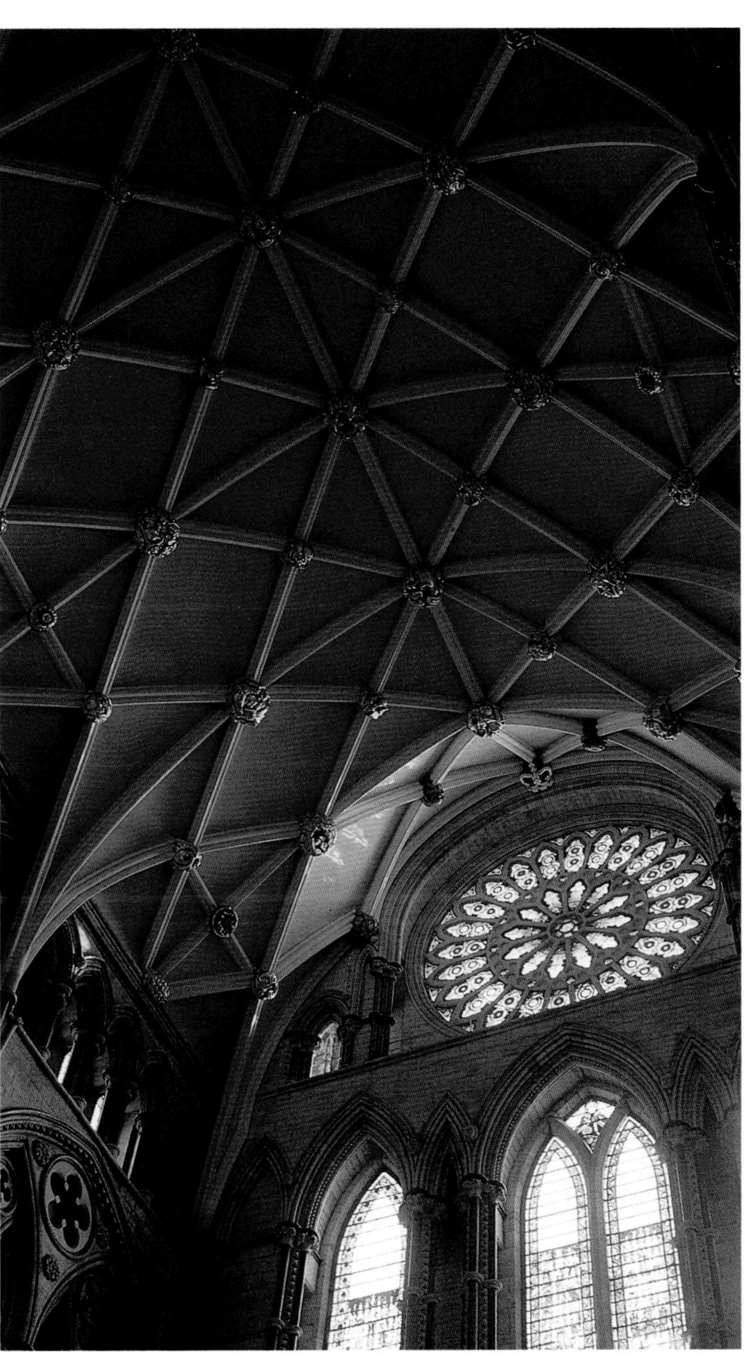

(Top left) Lifting the assembled trusses into place. Each truss weighs about 4 tonnes, and required a crane with this capacity at a maximum of 30m radius.

(Top right) Drilling the members using a purpose made guide. A relatively easy task in green oak.

(Above) Assembled trusses ready for lifting, with the crane prepared to erect the jib.

(Right) The completed vault, shaving the sharp uplift of the ridge in the last bay to over top the Rose window.

Tobacco Dock, London (opposite page) A section through the warehouse, showing the queen post trusses in timber, with their elegant slightly raking posts, supported by tree-like cast iron column assemblies. The groin vaults lead to the column tops, which taper from an octagon to a square. In its sureness of line and use of materials, the structure is close to perfection.

(Left) The same view, with work complete.

the timber had reached an equilibrium moisture content.

The truss is basically triangulated, but a strict application of the principle of intersecting centrelines would make it necessary to connect three members together at most joints. This time we can take a tip from the Victorians, and introduce small offsets between the members. These allow joints to be made between two pieces of timber only, while still keeping the moments and shears within the member capacity. Once again, the past provides a lesson for the present.

Adaptation and re-use

All these buildings continue to serve the purpose for which they were originally built, and so in a broad sense continue to be of use to society. This is a good thing, for the useful building is, overall, more likely to be kept in good repair. Traditional buildings are also surprisingly accommodating of the minor revolutions in heating and lighting which we take as the norm today.

However, there is a great range of buildings which, through no fault of their own, are no longer of direct use; perhaps because they have become superseded, or are simply in the wrong place, such as the country churches which are no longer centres of population. This range of buildings is dominated by a vast legacy of mainly 19th century industrial structures built perhaps for the weaving of cloth or the storage of grain in sacks, and which are now simply redundant as the commercial scene changes.

One such building is the warehouse at Tobacco Dock, designed by Daniel Alexander, Surveyor to the London Dock Company, and built in 1813. It was intended originally for the storage of tobacco and the collection of customs duty. However, this trade moved down river, and so quite early in its life the building was re-used for the storage of furs and skins. It continued in use into the 20th century, suffering minor damage in the War, but finally falling vacant with the closure of the London Docks in 1968. Its condition deteriorated, although it was listed Grade 1 in the mid-1970s.

In 1983 a company was formed with the aim of redeveloping the site, while retaining most of the building. Step one, for us as the consulting engineer, was to carry out a survey, as I said at the outset.

A closer look shows a clearly articulated and elegant structure, with each element made from the most appropriate material. Thus the roof trusses are in timber, resting on rows of unique bifurcated columns of cast iron, a material particularly strong in compression. The main suspended floor is of brick groin vaulting on granite columns, which in turn bear on bases supported by timber piles.

Timber piles were widely used in the 19th century, but with the advent of alternative materials such as steel and concrete they are not allowed under today's regulations. The District Surveyor was initially of the view that new foundations would be required, although the cost would obviously be prohibitive. Exploratory pits were dug, which showed that the piles were of pine, approximately 225mm square, and in good condition. The reason for this is that the cut-off level had been set below the permanent ground water level, and saturated timber, with virtually no oxygen present, is immune to wet and dry rots. Examination and withdrawal of several piles, together with load tests on others, was sufficient to obtain approval for the re-use of the piles.

Overall, then, the structure was in remarkably good condition for its age, repairs mainly being needed only to the ends of some trusses due to leaks from the valley gutter – a recurring theme in historic buildings. However, the basement area was virtually unusable in its current condition. Large openings would be needed through the ground floor vaults to allow light and air in, and to provide access and means of escape in case of fire.

The vaults derive their stability from arch action in two directions, with compressive forces led down to the heads of the columns. At the boundaries of any new opening, reactions must be provided to maintain equilibrium, without obscuring the vault form. If the 'cut' line was at or near the crown of the vault, the thrust line would be sensibly horizontal, and the thrusts could simply be diverted round the opening by means of a ring beam. The beam would have to be very stiff, in order to keep deflections to a minimum. Advantage was taken of the fill zone above the actual vaults to install a beam of the necessary width, which still remained within the overall structural profile.

Work in the field of historic buildings has particular and interesting challenges, but above all it develops a respect for and appreciation of the achievements of our forbears, who worked within the range of traditional materials, and with little or no mechanical assistance. Inevitably, we focus primarily on the achievements of the 20th century, but the best of the past is a difficult act to follow.

Arguing that engineering has always been concerned with the communication of information, this paper looks at the relations between communications and states, cities, the workplace, and the home. The use of different media, radio waves, cables and lasers, is argued to depend upon technological changes which may bring back into use previous information infrastructures which had been discounted.

Bill Southwood

The world wired up

In sending the first commercial inter-city telegraph message (between Washington and Baltimore) in May 1844, Samuel Morse signalled: 'What hath God wrought?' He could have had little idea how that question might be answered a century and a half later. The telecommunication system which his innovation spawned:

- is the largest single machine ever created, connecting home, factory, and office in every country of the world
- provides a seamless web of voice, text, image, and data links
- affects everybody in their work and leisure
- is economically one of the largest industries
- continues to inspire both scientific and cultural change.

Growth in numbers, increase in the capacity of communications links, the power of the devices connected to them, and the uses to which they are put, continue to increase exponentially; the cost of each decreases by the same law. The only limits to this growth appear to be the population of the world, the velocity of light, and the size of the molecule. Whether this be attributed to God or mere mortals, what hath been wrought is spectacular indeed. The figure opposite shows just one measure – the traffic-carrying capacity of cable and radio over a 200-year period.

Where is this taking us and what do we as engineers have to offer? Clearly the machine is a creature of those who design, build and maintain it; what role should those technicians take in the uses to which it is put, the means of accessing it; the availability as a public good? All engineering is about communicating information, so engineering a communications system

Capacity of major telecommunications routes

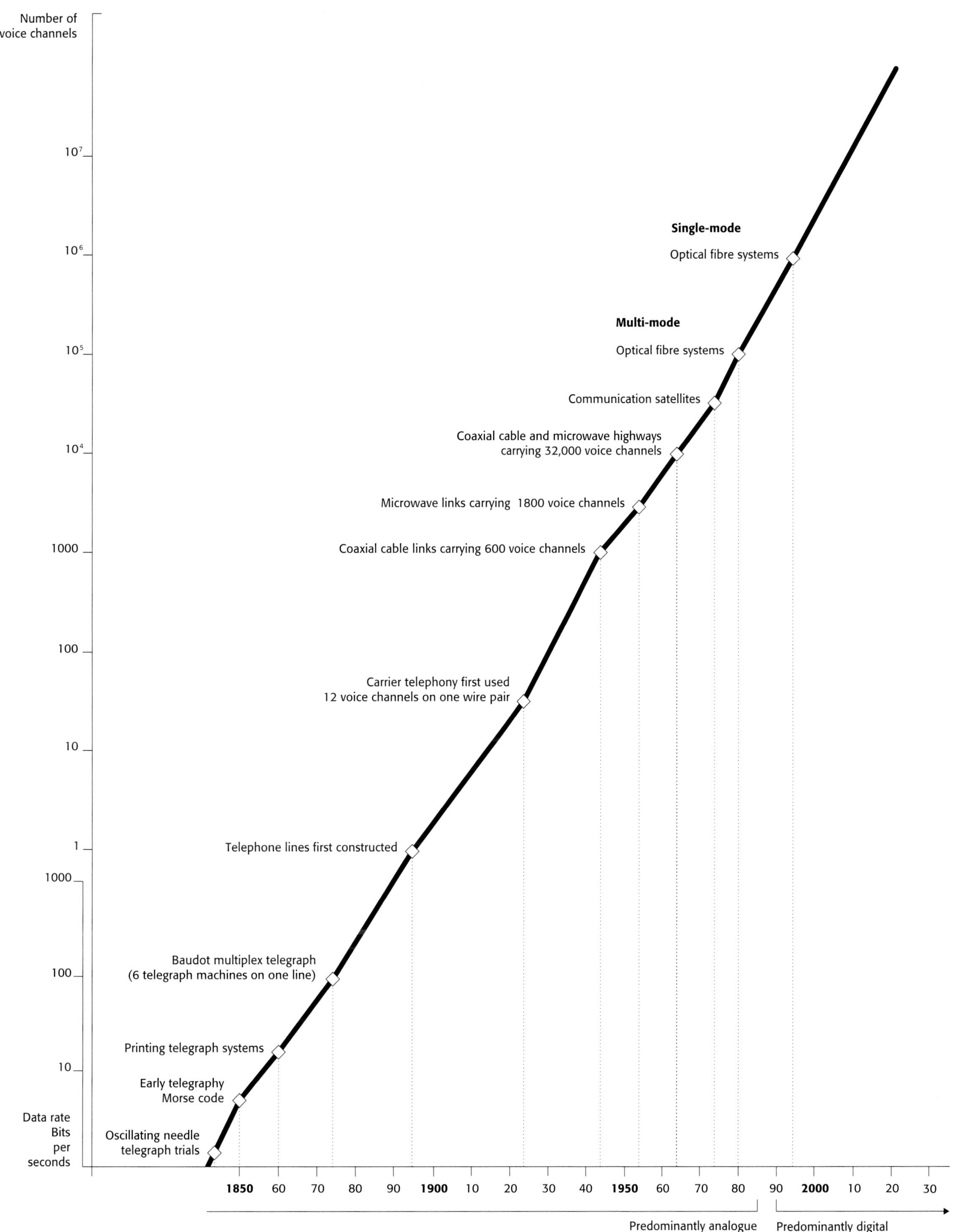

Number of voice channels
10⁷
10⁶
10⁵
10⁴
1000
100
10
1
1000
100
10
Data rate Bits per seconds
Single-mode
Optical fibre systems
Multi-mode
Optical fibre systems
Communication satellites
Coaxial cable and microwave highways carrying 32,000 voice channels
Microwave links carrying 1800 voice channels
Coaxial cable links carrying 600 voice channels
Carrier telephony first used 12 voice channels on one wire pair
Telephone lines first constructed
Baudot multiplex telegraph (6 telegraph machines on one line)
Printing telegraph systems
Early telegraphy Morse code
Oscillating needle telegraph trials
1850
60
70
80
90
1900
10
20
30
40
1950
60
70
80
90
2000
10
20
30
Predominantly analogue
Predominantly digital

(Left) The Lighthouse at Alexandria. Commissioned by Ptolemy II in 280 BC, the lighthouse communicated its message – the location of a major port – to mariners for more than 1000 years. It was reported that in the last few centuries of its life, the upper levels accomodated a small mosque, an early example of a structure adapting to the changing priorities for communication.

(Right) The Herald, Glasgow. A fine example of the work of Charles Rennie Mackintosh, the *Glasgow Herald's* dominant feature is a pigeon loft. The contemporary equivalent of the pigeon would be a small microwave dish serving a metropolitan area network.

should be the prime example of the craft. The way communications has affected, and been affected by, the state, city and building is a history of the engineers who communicated to construct the edifices and their communications systems.

Communications and the State

The International Telecommunications Union (ITU), the Geneva-based agency of the United Nations which sets the standards for the sector – has always had more members than the UN itself. This is not surprising given the need for parties to communicate for commercial or social reasons, whether or not they were on speaking terms, or even at war. Setting standards is essential in an industry which is defined by my hearing what you have said.

From the earliest days of the electric telegraph, which in the mid-19th century became Britain's first nationalised industry, the State has played a large part in communications. Many countries' telephone companies are operated as state monopolies, allowing the state control of the resource and its treasury with a highly profitable business.

Even where telecommunications was in private hands, the power of the companies often exceeded by far those of the states they served. The telegraph operator in a distant principality could know the contents of every un-coded signal into and out of the country before the recipient, and most employees of such companies were under stern orders neither to steal the mail nor reveal the contents of telegrams.

The participation of the US company ITT in the downfall of President Allende of Chile in 1973 is well documented. And nowhere was the influence of communications companies better articulated than by the late Robert Maxwell who, in referring to difficulties experienced in obtaining cable television franchise rights in France, said: 'If the politicians will not give us the decisions we need, we will get the politicians who will'.

The city state of Singapore, justifiably proud of its well-deserved reputation as the world's most wired city was, in 1995, faced with a delicious contradiction: how to allow its citizens to make use of the information available on the Internet without making a mockery of its own censorship laws.

Communications and the city

Long before telegraphy, communications fashioned the city. The importance of Lyon to the Romans owed much to its position where the Rhône met the Saône. Today it is the hub of an extensive network of road, rail and air services. Samarkand on the Silk Route from Cadiz to Peking and, more recently, Hong Kong as a gateway to China both owed their early fortunes to land communication.

Alexandria, as one of many significant Mediterranean foci, boasted two kinds of communications facility as its contribution to the Seven Wonders of the World. Initially constructed by Ptolemy II in 280 BC, the beacon at the top of the 135m tall lighthouse of Pharos was converted to accommodate a mosque during the Middle Ages.

The contemporary city is connected with its near and

CN Tower, Toronto, (left).

Torre de Collserolla, Barcelona, (right). In presenting the competition entry to the judges, the design team led by architect Norman Foster emphasised that the tower's Wankel shape represented the three drivers for the project: communications as the 'engine', engineering as the private face, and architecture as the public face.

Monte Pedroso model, Santiago de Compostela, (following page). The changing nature of technology and business is calling for a new kind of adaptability in telecommunications facilities, illustrated in the evolution of tower design from Barcelona to Galicia. To the theme of adaptability was added a clear need for the project to be bankable, issues addressed in the Foster/Arup design.

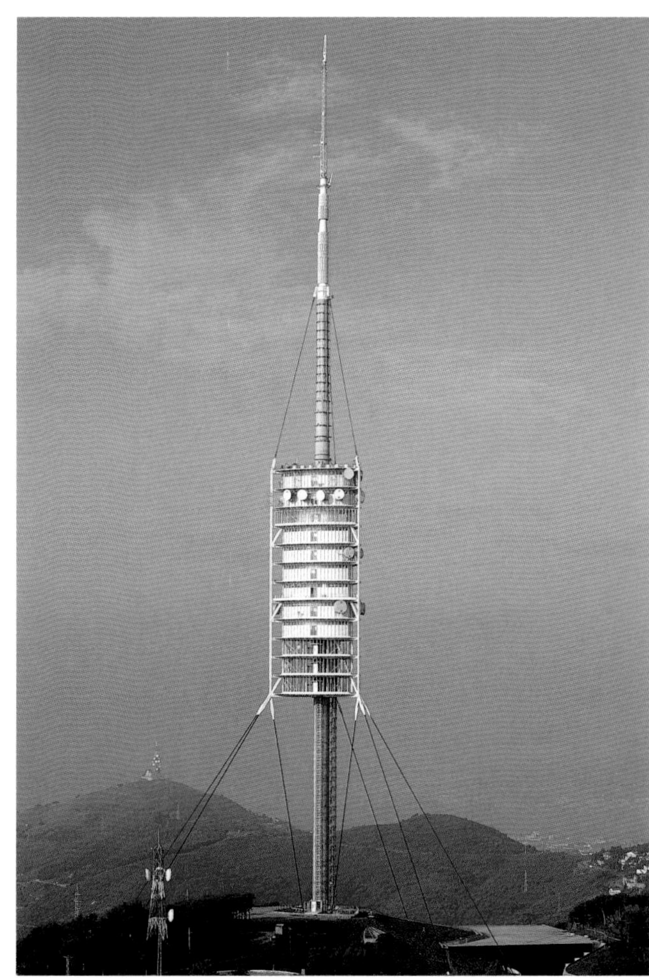

distant neighbours by a mesh of copper cable, optical fibre, microwave links, and satellite circuits. Much of the prosperity of the cities of London, New York, and Zurich comes from the communication between their financial markets. Looked at differently, a good indicator of demand for communications services has always been the volume of trade between two places, whether they be western capitals or Melanesian trading posts.

Within the city, the communications networks complement other utilities and compete with them for scarce space in underground chambers. In dense city centres, technology has come to the rescue, with optical fibre able to carry information at the rate of encyclopedias per second in a fraction of the space taken up by copper cable. Fibre will soon compete economically with copper as the preferred method of connecting individuals in the home or at the desk into the network.

Ownership and control of the physical infrastructure of ducts and chambers is often the key to success. The pipes laid by the London Hydraulic Company to distribute high pressure water to power the capital's lifts had been unused since the Second World War. Buying the company allowed Mercury Communications Limited, established to compete with British Telecom in the early 1980s, to make a rapid start in laying optical fibre in the City.

Stockley Park, constructed on a reclaimed refuse tip near Heathrow in the same period, was one of the first UK developments in which telecommunications was made a central theme. Politically it was important that the business park have a London telephone number; logistically it was essential that the owner have control of the duct infrastructure to allow tenants to adapt their communications to their use of space on the park. Duplicated communications access could be provided by either of the telephone companies as well as the Windsor cable television franchise. A ring of ducts, co-ordinated with the ring road, distributed information to each building, linking them with the UK and international networks.

Communications and the communications building

Information and communications systems have produced a few excellent buildings of their own. The Charles Rennie Mackintosh *Glasgow Herald* building, designed in 1897, has as its dominant feature a pigeon loft. In the days before multimedia and palmtop computers, essential tools for the reporter covering a strike or a football match were a portable typewriter, a supply of rice paper and a couple of pigeons in a cage! It would be pleasing to be able to report that the loft now houses a mobile radio base station but sadly, at the time of writing, the building is vacant and semi-derelict. This is both an omen to those who do not keep pace with technology, and a reflection on changes in the inner city.

Communications towers became dominant expressions of a city's or a company's image. The Eiffel Tower, Toronto's CN Tower, and Barcelona's Torre de Collserola are potent examples. It is interesting that technology is now re-defining the form of communications facilities. The form of the new facility proposed for Santiago de Compostela in Spain is determined by three

communications needs:

- for internal space to accommodate many small companies each providing mobile or portable communications services:
- for roof space to accommodate their satellite terrestrial antennas;
- for the shortest possible cable lengths between broadcast transmitters and their antennas.

Far from describing a conventional tall tower, this technical brief describes a wing, accepted enthusiastically by the Concello and communications companies.

Communications and the office building

Until the late 1960s, telecommunications and the office lived in relative harmony with one another. The modest demands made by cables and computer rooms were fairly readily met by conventional design, slotting in a riser for 'The Post Office' somewhere between the electrical closet and the toilet stacks.

Intelligent building components - Indicative lifetimes

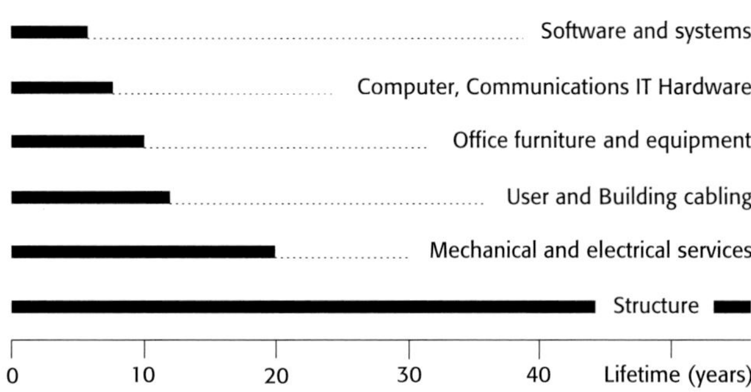

But, some time in the 1970s, things changed. The mainframe computer became an acceptable piece of equipment for most corporations; dumb terminals appeared on desks in the office; a tangle of cables competed with an overheating office for the attention of the building manager. The buildings of the 1960s, designed to a specification which included no allowances for such demands, could not and did not cope. Ironically it was the earlier buildings with their higher ceilings and narrower depths which proved more adaptable.

The science or craft of information technology and the office building was explored in a study with Frank Duffy. Published in 1983, the first ORBIT Report set out to define the interaction in terms of lifetimes of each component of the building.

Today the design of an office recognises both the need for timely information to define the structure and servicing of a building; it recognises the need for the building to adapt to a future which we cannot today imagine. With the cost of information systems dropping rapidly, it would be irresponsible to force the client to make decisions on information systems before it was absolutely necessary. The art is in allowing design of long-lived elements to proceed with each decision taken at 'the last responsible moment'. This is to be distinguished from the more exciting 'last possible moment'!

Communications and the factory

Whilst automation of the office took until the mid-20th century to affect the workplace, the factory was far ahead. Information systems allowing components to be delivered only when they were needed cut manufacturers' inventories and reduced warehouse space. 'Just in time' became a way of working.

The buildings themselves changed little as the amount of information being circulated around them increased. Factories are designed to accommodate far more demanding occupants than a few data cables, and the Manufacturing Automation Protocol (MAP) developed by General Motors in the 1970s used robust, low tech 'analogue' coaxial cable methods developed for the cable television industry.

An interesting marriage of this with the building was achieved in the Toyota car manufacturing plant in Derby, England, where the building control systems and the MAP network used the same infrastructure of cables to move information over the vast site.

Communications and the home

To appreciate that the home has undergone a revolution every bit as intense as the factory or workplace, one has only to count the number of microprocessors and electric motors in a fairly large house. From the door chimes, via the sound system, through the oven timer and boiler to the alarm clock in the bedroom, it is not uncommon to find 30 'computers' and a similar number of motors.

There is another revolution overtaking us in the way we live and work. The telephone, once firmly wired to either the office desk or hall table, has become peripatetic. The analogue cellular radio revolution of the 1980s and its progression into its digital successors a decade later meant that, increasingly, the humble phone could be associated with an individual rather than a place.

The demand this places on radio bandwidth gives rise to one of those about-turns in the use of technology. Radio, once used almost exclusively for delivering entertainment, is now more and more being used for telecommunications; conversely, the dozens of channels of undifferentiated rubbish which pass for a television service in many countries are increasingly being delivered over cable. The radio spectrum, like clean air and drinkable water, is a finite and precious resource – we would be well advised to husband it carefully.

The future of technology and buildings

We now design our buildings to outlast many generations of the technology they accommodate, using the techniques described earlier. It is not difficult to predict the technologies we will be using in our homes and workplaces in 10 years' time: they exist today just as those we are using today existed 10 years ago. Infra-red controls for our video recorders, and holograms on our credit cards are domestic examples of technologies common today which were in the laboratory a decade ago.

The same cannot be said of a 20-year period or longer. Whilst 1996 might have been foreseen in 1986, predictions made in 1976 look naive 20 years on. What hope for our buildings with their indefinitely-lived structures and 20–year services?

One crumb of comfort for the futurologist comes from the cyclical nature of discovery, with every good idea being re-invented about once a generation. Intercontinental telephone calls were originally carried over noisy short wave radio circuits at great expense, with long waiting times and dreadful quality. With the arrival of submarine telephone cables, everything improved to the point where international direct dialling became common. The communications satellite, in the late 1960s, looked like spelling the end to submarine cable production by providing even cheaper calls despite requiring a bizarrely-situated repeater station 36,000km above the equator.

But such was the demand generated by satellite, and such the technical advance of optical fibre, that the submarine cable industry was re-born. Without the need for a rocket launch, and without the half second delay in getting a reply inherent in a satellite call, cable was once again the preferred means of international communication.

The future and the engineer's place in forging it

Our cities are having to adapt to changes in patterns of work and leisure. Tele-presence – the result of marrying virtual reality and video-conferencing – promises to replace many routine face-to-face encounters. This can release people from unwanted travel, and the environment from its undesirable effects; it can also increase alienation, reduce the cultural richness, and remove some of the incentives to working internationally.

These are social issues brought about by technical change. The cities and buildings within which the change takes place must adapt to these social, economic, and political forces. It will always be a mistake to seek technical answers to human problems, even when these problems have been brought about by technical advances. Buildings, as stated earlier, will outlast any of the technologies which are imposed on them. Some of these technologies – computers and the heat they generate for example – have a negative effect on the occupants. Others, such as optical fibre and its very small size, make management of a building easier.

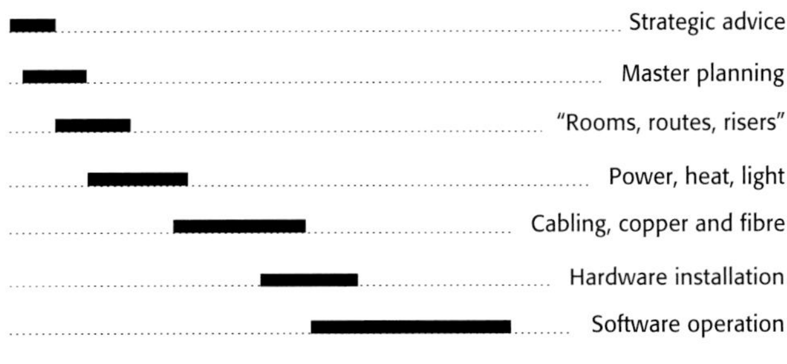

The ebb and flow of technology, often looking like 'fashion engineering', sees cable, overtaken by satellite for international phone calls, bouncing back.

The cable-less revolution which is affecting our lives will indeed affect our buildings; it will become easier to connect our phone or computer by an infra-red link; it will be economical to design such capacity into our buildings and cities. But there will still be the need for cable to deliver energy and information.

Within this changing environment, the engineer must be familiar with technology, buildings, the city, history and, most importantly, people. It is an exciting challenge, and one which can only get more demanding.

Recently society has discovered to its cost that the earth is not an infinite resource, and that the damage we do to it may be irreversible. In England and Wales 100,000 ha are conservatively estimated to be unusable. How to improve such land, and how to prevent further damage, is the subject of Walker's essay. She describes the growth of technologies using natural forces and micro-organisms in the engineering of waste, and its by-products.

Lorna Walker

Environmentalism and efficient engineering

(Above) Northumbrian Water's long sea outfall at Seaton Carew – a section of the outfall about to be pulled out to sea.

(Right) Landfill site, West New Territories, Hong Kong.

The earth is an incredibly complex yet fragile organism. It has been in existence for thousands of millions of years. However, in the last 50 years or so it has started to show signs of stress and fracture. In simple terms, the world is comprised of energy and matter. When this is in balance all is well. However, when this balance is disturbed, stresses begin to occur, and finally there is a fracture or a complete failure. All components of this planet are inextricably linked and totally interdependent. The human race is a part of this delicate balance.

As a species we have been incredibly successful in dominating our environment. However, this success has not been without cost. Our numbers have increased to such an extent that we are beginning to affect the balance. We require even greater amounts of food and energy and produce more and more waste. The resources of the world are finite, and we are using them at a greater rate that any other time in the history of the world. Our activities are causing the pollution of land and water, and the balance of the atmosphere has been so changed that we are in danger of destroying that fragile mantle that

(Above) Mmabatho sewage treatment works, Africa.

protects us, the ozone layer. In order to survive and to leave a world fit for our children's children, we need to treat it with respect.

What has this to do with engineers? Engineers have had a definitive role for many years. Those magnificent men of the Victorian engineering fraternity changed the way we lived in Britain with vision and arrogance. The causal relationship between polluted water and disease was identified. The polluted water running in the drains along the roads was collected and channelled to the sea. Clean wholesome water was provided for drinking and hygiene uses. This probably did more to eradicate the killer diseases such as typhoid and cholera than the medical profession.

After controlling the flow of effluent, they then set about treating the effluent. The early biological treatment processes were 'invented'. In the early days the treatment mainly consisted of removing the gross solids from the flow. Later it was found that if the bacteria naturally present in the flow were cultured, they 'purified' the water. This was one of the earliest examples of man using the power of the microbial world. This process has remained essentially the same to this day. The activated sludge process which is widely used today was first described in the literature in the 1920s. It is still being used and improved today.

These processes rely on the activities of micro-organisms that are invisible to the naked eye. They are living organisms that need to be coaxed to do our work for us. Although a sewage treatment plant is seen to be a major civil engineering construction project, it is in fact a luxury hotel for bacteria. We describe the design of this hotel in words such as anaerobic, anoxic, and aerobic. We label the bacteria as chemical autotrophs, photosynthetic heterotrophs, nitrobacter, etc. We label the bacteria that cause us harm as pathogenic, with names such as *escherichia coliform* and *faecal streptococii*.

The main factor in design is to create the conditions for the bacteria to live with enough water, oxygen, food, and time to be of use to us, that is to consume and metamorphise our waste into a less noxious material. We give them food,

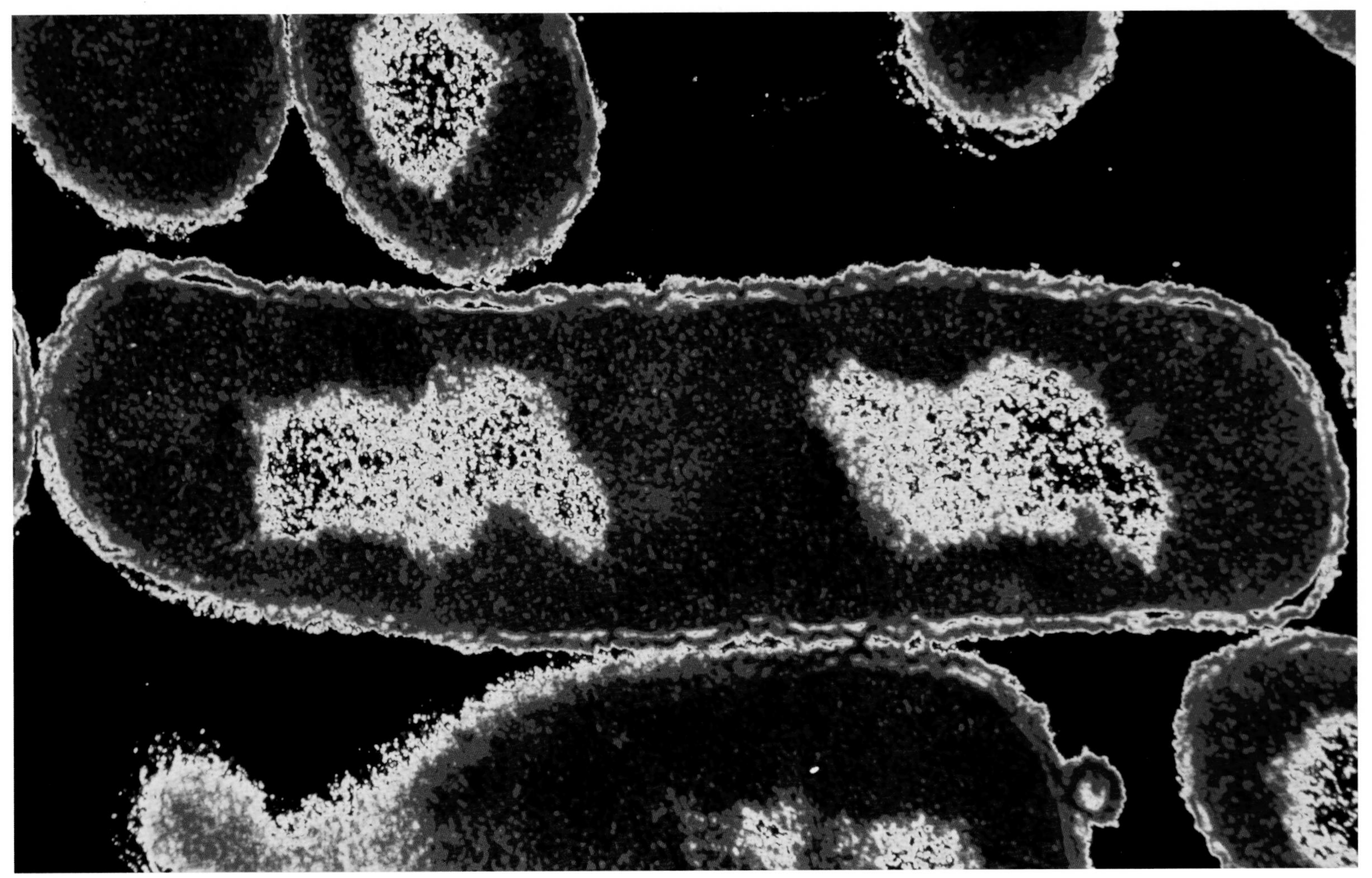

(Above) *Escherichia coli*. Bacteria relates to the sewage treatment process.

shelter, and oxygen. When there are too many bacteria, or a higher proportion of dead bacteria than required, we remove them from the system and they become sludge. If properly managed, this sludge can become fertiliser for our crops or fuel to provide energy. In this way our waste can be reused.

The treatment of the solid waste we produce requires a slightly different approach. In earlier days this comprised a greater proportion of biodegradable material – food and farming wastes. These were normally thrown onto a nearby pile or in a convenient hole in the ground. These early waste disposal sites were in effect anaerobic digesters. In these bacterial communities a different population of bacteria treat our waste. In the absence of oxygen, these bacteria consume this material and transform it into gases such as hydrogen, carbon dioxide and methane.

The modern well-engineered waste management facility, previously known as a tip, is merely a better designed digester where the intermediary products, which can be more noxious than the original substrate, are better contained and controlled. With different design parameters we ask the bacteria to produce methane earlier and more efficiently so that the methane can be used as energy. The unwanted leachate is contained until it has undergone sufficient degradation to render it harmless. The gas is prevented from leaking to places we do not want it because of risk of explosion and asphyxiation, or channelled to where it can be contained and used.

The remediation of water that has been so overloaded by waste that the oxygen has been exhausted and respiring organisms such as fish and invertebrates can no longer survive uses similar principles. At a site near Manchester, 11ha of dock had been reduced to a foul-smelling, lifeless stretch of water. Many options were investigated; some included the addition of chemicals to kill the bacteria which were not deemed desirable. This course of action was abandoned as it was considered that this would only add another stress and the system would still be metastable.

The final design incorporated the

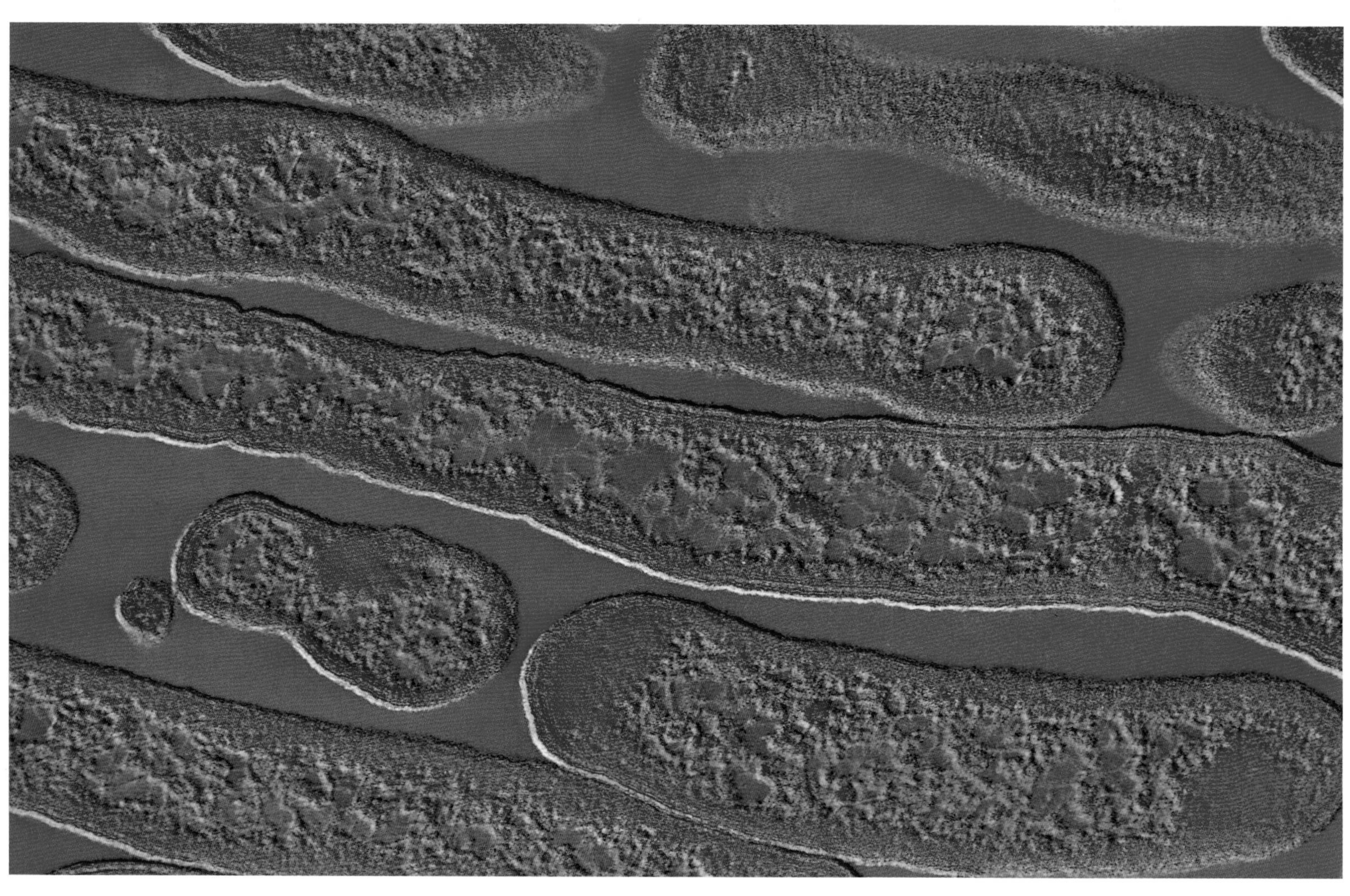

(Above)
Flavobacterium meningosepticum.

control of the polluted source water, and a simple system of mixing of the water body. This mixing achieved two main effects. The water column was destratified and the oxygen-depleted water was brought to the surface. Once at the surface, the rolling movement of the water body allowed the diffusion of oxygen from the air into the water. However, the deciding ingredient was time. Given the right conditions the system changed from one where there were a few dominant species of algae and worms that were known to live in polluted water, to a richly diverse and balanced system where finally fish could survive. One of the early indicators of pollution, a stressed system, is a decline in diversity of species. In this case the process was reversed.

The treatment of land which is contaminated is a fairly new procedure. Until recently land was seen as an infinite resource and when it was damaged to such an extent that it could not be used, people moved onto greener fields. We have a legacy of land from the industrial revolution that is so damaged that it can no longer be used. In England and Wales alone the conservative estimates are over 100,000ha or 100,000 sites. We can no longer move on; the land has to be reclaimed.

New techniques are now being tried to restore land. At first the method was to remove the problem – vast earth-moving projects were undertaken to transport the poisoned material to a landfill site and import clean material. But this merely moves the problem to another site to be dealt with later. Although this is still the main method of land remediation in many countries, other techniques are being employed.

These include washing, by steam or water, burning to destroy organic materials, or encapsulation to neutralise the contaminants. Some designers employ the help of friendly bacteria to consume the unwanted chemicals. These are specially cultured so that a population can be introduced which considers the contaminants to be their favourite food. Now even the lowly white rot fungi are being harnessed to destroy intractable substances such as lignins and complex hydrocarbons.

Many of these techniques are in their infancy and need to be developed. Systems such as washing do not destroy the contaminants, they merely remove them to another medium where they still require treatment. Soil incineration is exorbitantly expensive and renders the soil sterile so that it cannot support life unless nutrients are added. One of the most promising methods seems to be the use of bacteria – this, however, requires time (and patience) and we have yet to discover bacteria which can use these cocktails of chemicals as their food source.

What of the future? We are beginning to believe that we can deal with the legacy of pollution from our forefathers and we can, in some cases, treat the waste which we currently produce. However, our numbers are growing; we need more food, energy and shelter, and produce more and more waste. We need to recognise that the resources of the world are finite. They must be used in such a way that they can either regenerate or be used at a rate where they will not be exhausted.

The role of engineers is vital. They have the skills to acquire and utilise knowledge of our environment to manage and protect our finite resources. The responsibility is daunting, but not new. The challenge is ongoing.

(Above) Stockley Park; treatment of contaminated land.

(Overleaf, left) Salford Quays, Manchester. Dock regeneration.

(Overleaf, right) *Escherichia coli.*

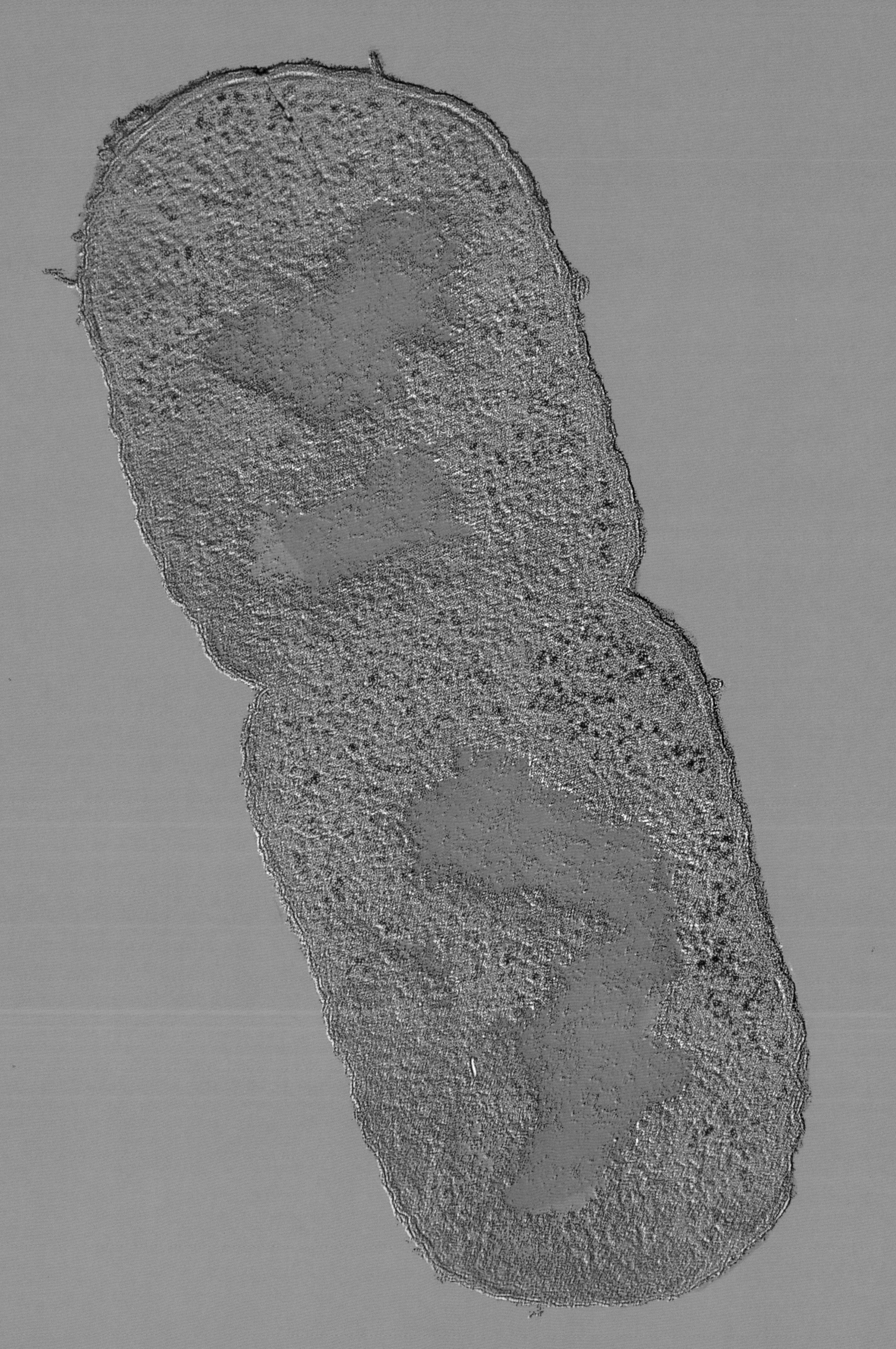

Designing

Bridge design is an intimate mixture of design and construction resolving a 'chaos of facts and circumstances into a unique and beautiful unified whole'. Using the examples of bridges designed throughout the world, the authors make the case that while a bridge engineer will seek unity, consistency, scale and proportion, and an eye for detail, the overriding factor in bridge design is the importance of the site, its precise shape, and the views from it.

Bill Smyth and Jørgen Nissen

Total design of bridges

'This consideration of every aspect of the design and fitting the bits together to produce the best possible total solution is the real art of designing. And as in all art, there is more to it than meets the eye. It is not only a question of meeting all the different requirements in the fullest possible way at the least cost, it depends also on the simplicity or elegance of the solution, the felicity of design which has the power to inspire those who comprehend it, and which is the reward the designers hope for. It can rarely be produced without taking great pains and it cannot be defined or measured.'

Ove Arup: *The Arup Journal*, September 1972

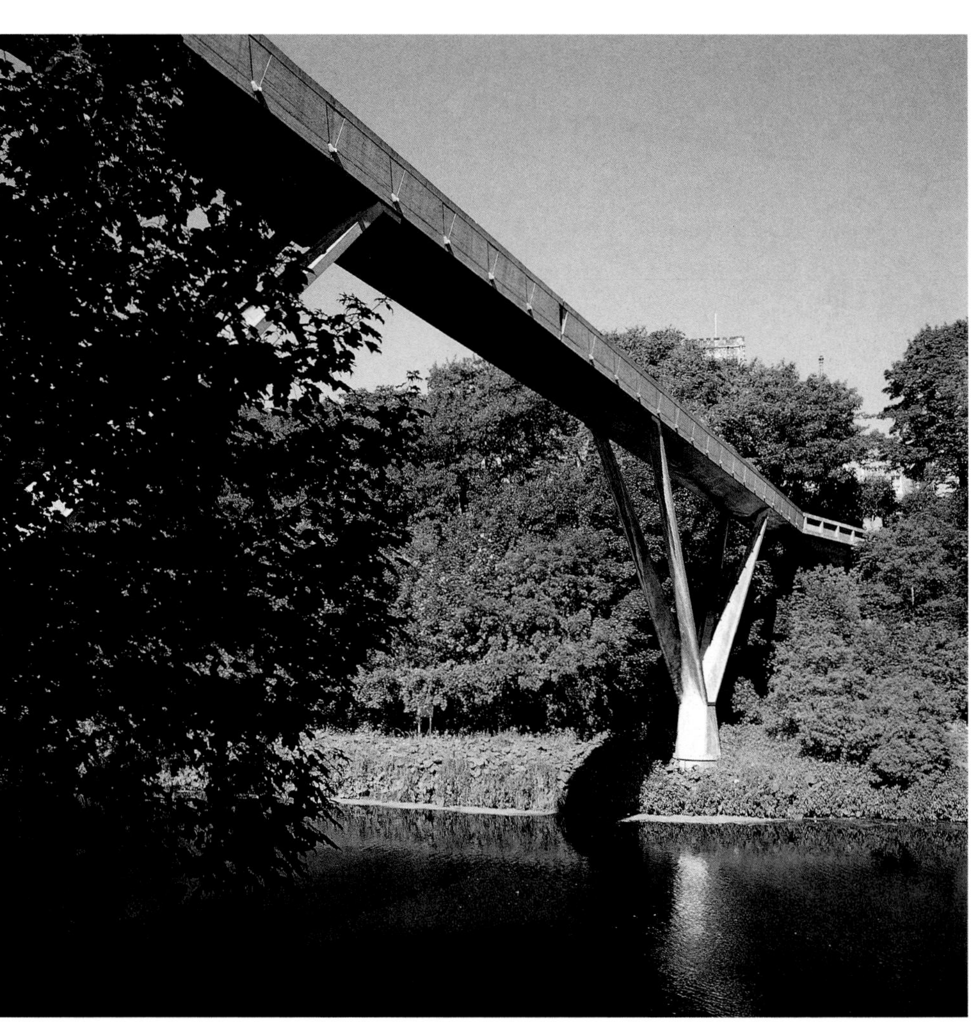

(Above) The rhythm created by the recurring patterning from the timber framework can be clearly seen on the parapet of the Kingsgate Footbridge at Durham. This is reinforced by the gargoyles which can be seen from a greater distance.

There is almost always something by Ove which says what one would like to say much better than one could oneself, particularly about design. In the field of bridges he also gave us the well-known footbridge at Durham in which function, structure, construction, and appearance come together with beautiful detail to create a superb example of Total Design.

On Durham Ove had the architect Yuzo Mikami working with him. According to Yuzo, Ove made lots of sketches, some of which Yuzo had to work up into drawings which established proportions and dimensions which were in turn then changed because often they were not quite what Ove wanted. Ove had very clear ideas about what he wanted and he thought about every aspect and every detail of the bridge, how it would be built and how it would work and how it would look. His collaborators no doubt contributed to it, but it is very much Ove's bridge in conception and detail. Ove took an interest and more than an interest in some subsequent Arup bridges, but probably in no other did he have such complete control over the design.

The Salginatobel Bridge, designed by another great engineer, Robert Maillart, is a very different kind of bridge but nevertheless has affinities with the Durham bridge. The most obvious one is the rhythmic repetition in the parapet walls of semicircular scuppers at Salginatobel and gargoyles at Durham. Did Ove get the idea from Maillart? Designers often copy ideas which they like and sometimes improve on them, and there is no doubt that spouts are a better way of shedding water than simple holes, although more complicated in construction and appearance. Maillart himself must have got the idea from older masonry bridges in Switzerland which have almost identical semicircular scuppers. The scuppers/spouts in both bridges and the older ones are almost certainly closer together than is necessary just to drain the decks, but they are are at the right sort of spacing to give visual rhythm and pattern to the parapets.

Again both designers were also constructors; both were contractors before they became consultants. The form of the Durham bridge is intimately linked with the method of construction, each half built on a bank and swivelled to meet the other, thus avoiding falsework in the river. The deck dimensions were also based on a module the length of a sheet of shuttering plywood. The Salginatobel Bridge was also designed to keep falsework costs to a minimum, the thin bottom slab of the arch being cast first and then helping to support the upstands which then helped to support the deck construction. Both bridges relate superbly to their sites: Salginatobel with its arch across the gorge completely integrated in appearance with the short approach viaduct by means of the deck and supporting walls; Durham using the height above the banks to create its unique rotating structure.

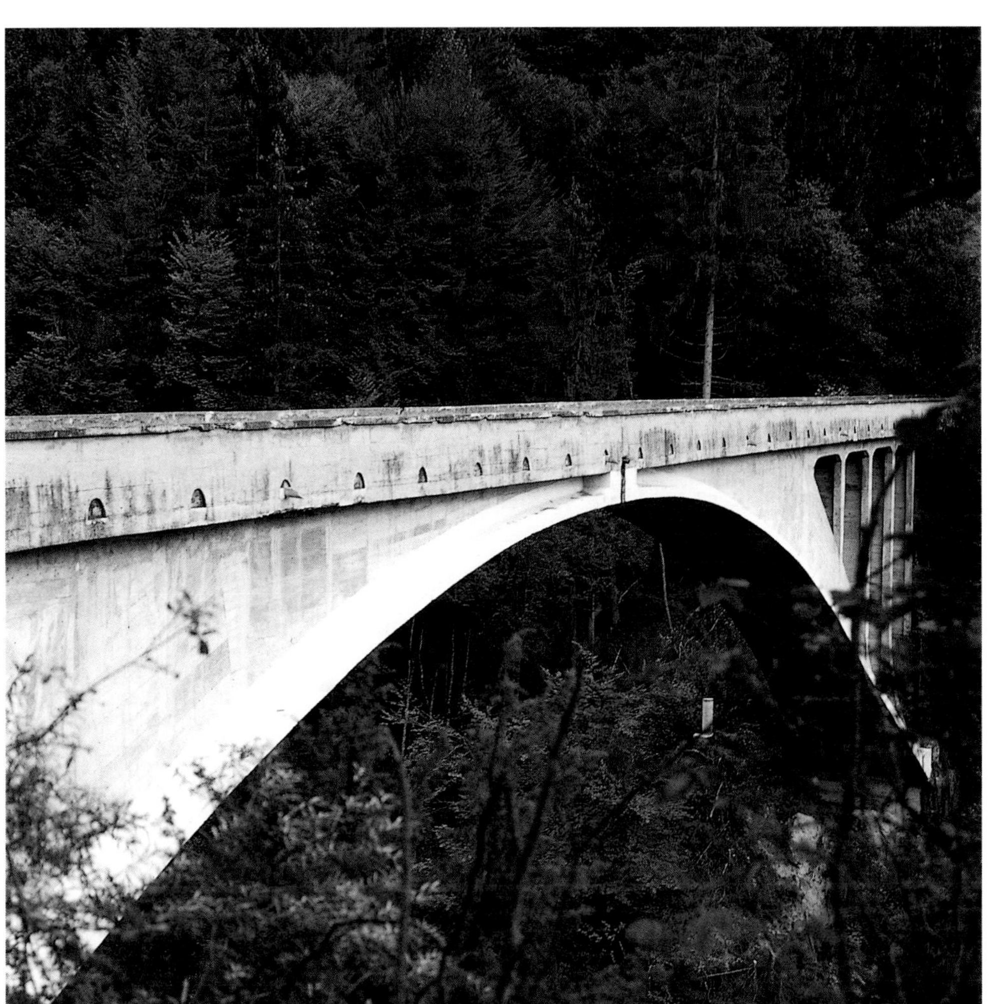

(Above) Maillart's bridge over the Salgina Gorge carries a single track road in a remote mountain area, yet it is one of the most famous bridges of all time, among architects (who were the first to appreciate it) and engineers. Since it was completed in 1930 trees have grown up around it so that the full view of the bridge can no longer be seen.

Each of these two is in its own way an example of Total Design, the resolution of a chaos of facts and circumstances into a unique and beautiful unified whole. Total Design means a design in which all the important factors – the relation to the site, structure, construction, appearance, economy – are taken into account right from the start. It means considering every aspect, non-technical as well as technical, before the basic design decisions are taken. It means not taking anything for granted. (Incidentally, both seem to be more difficult for professionals than amateurs; however, a certain professionalism is also required). It also means a lot of hard work, where a professional might consider that he knows the answer already.

Unity, consistency, scale and proportion, attention to detail: these words should be part of a bridge designer's vocabulary, alongside structure, materials, construction, durability, geotechnics, and of course environment and economy.

The design of a bridge (or anything else) has to start with the facts, and usually the most significant fact (or aggregation of a large number of facts)

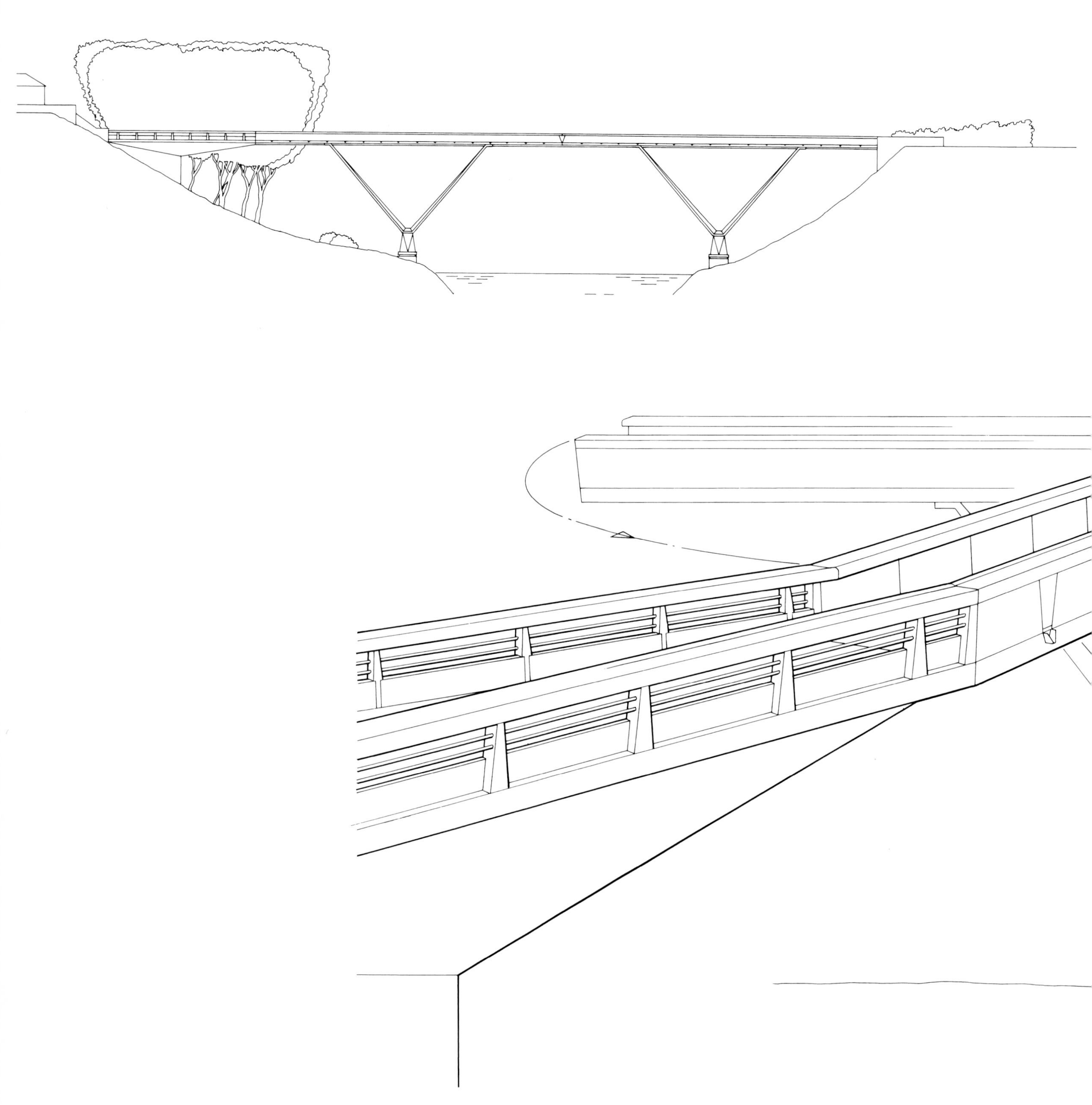

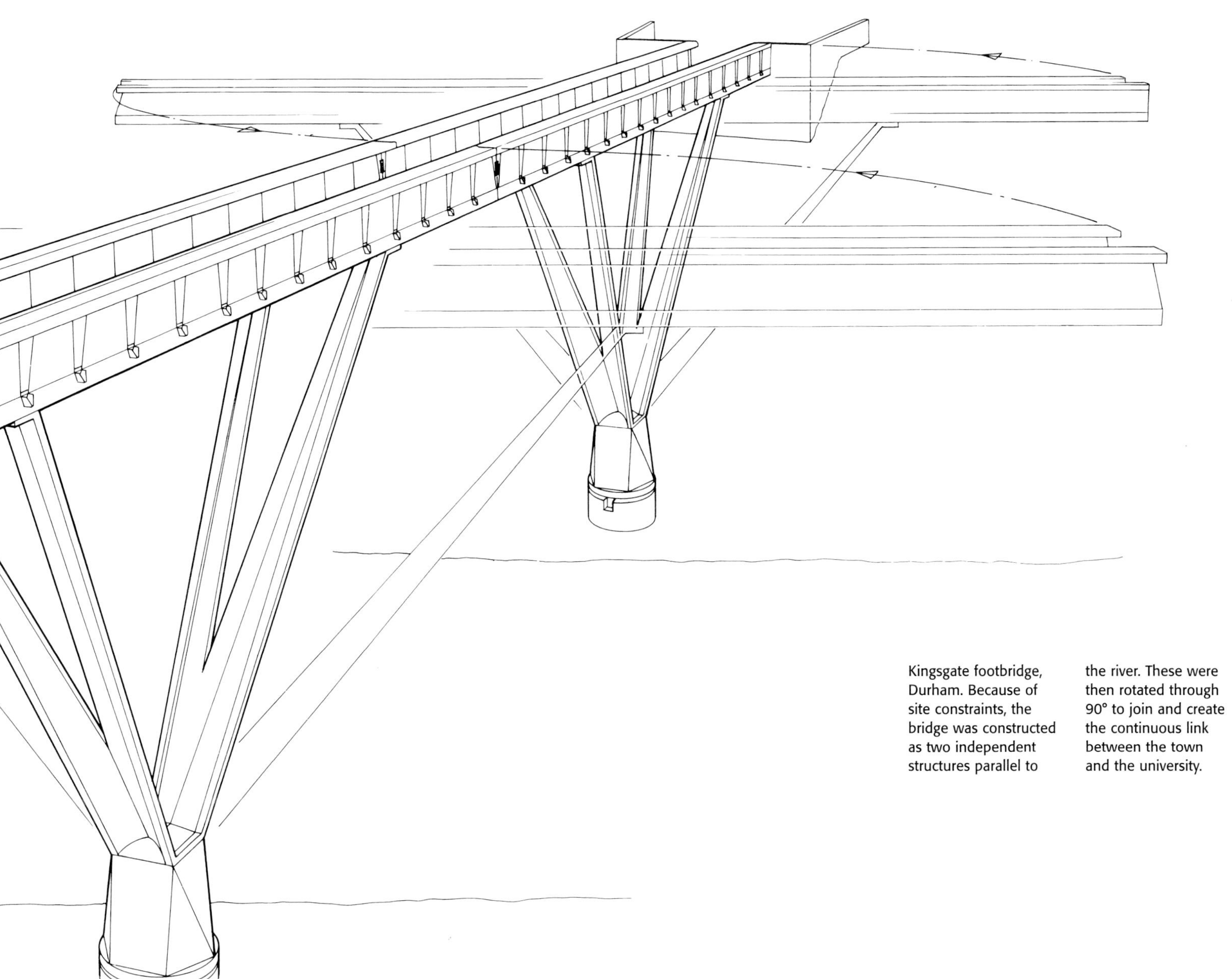

Kingsgate footbridge, Durham. Because of site constraints, the bridge was constructed as two independent structures parallel to the river. These were then rotated through 90° to join and create the continuous link between the town and the university.

is the site. It is often the relation of a bridge to its site which gives it a unique character. Two major factors in the site of the North Seaton Bridge were the shape of the buried valley underlying the estuary which it crosses, and the nearness of the North Sea. The first influenced the arrangement of piers, and the second was the reason for the footways which are at a lower level than usual in relation to the road, so that people in cars crossing the bridge can see the sea. The structure is a very economical one with two prestressed concrete T-beams. These are integral with the deck structurally, and this is expressed visually by the way the forms flow into each other, which also hides the fact that the beams are quite deep in relation to their spans. The ribbing at the outer edges is there to improve the weathering of the surfaces and is arranged so that it gives a rhythm to the edge.

The Kylesku Bridge in Scotland has a most dramatic site. The horizontal curvature arises from fitting the approach roads as closely as possible to the contours of the granite landscape. The bridge has to be high above the sea and this has been used, as at Durham, to create an understructure to reduce the spans of the deck so that the deck itself is relatively shallow and economical. In a situation of very high winds the structure is very stable and expresses that stability.

One of the decisions which is an integral part of the design is how the bridge is going to be built. Many bridges are built on sites where construction and access for construction are difficult and expensive. Even in cases where this is not so, the construction method may be crucial. The Berry Lane Viaduct was built across an attractive wooded valley in Hertfordshire, where one of the important factors was to minimise damage to the woods during construction. This was done by using a deck constructed from precast trough-shaped beams which could be placed from above using a launching girder. These spanned between concrete table-tops which could easily be built working within the band of trees which had to be removed in any case. In order to avoid massive piers, which would have been totally out of scale with the trees, the table-tops were carried by groups of columns in a carefully considered relationship.

There are always two aspects to the form of a bridge, the solid structure itself and the spaces between and around the bridge. Western Bank Bridge shows a case where the space flowing under the bridge is more important than the bridge itself. Sheffield University had the vision of replacing an embankment carrying a busy dual-carriageway road by a bridge, and forming a concourse, a space which would unite a number of its buildings and allow easy movement between them. The maximum opening possible was taken, and various bridge solutions were tried to see which one created the best effect of uniting the spaces either side of the road. The bridge deck is underplayed and the rather sculptural central supports help to lead the eye through and create a feeling of space flowing through from one side to the other.

Where a bridge is long and the site changes character, it is usually unsatisfactory to take different types of bridge

(Right) North Seaton Bridge crossing the estuary of the River Wansbeck. The North Sea, a few hundred metres away, is clearly visible from the deck.

and simply add them end to end. A bridge should have unity of appearance and construction technique as far as possible (and it is often more possible than it seems at first sight). The site of Byker Viaduct at Newcastle, carrying a metropolitan railway, is partly a gently sloping hillside and partly a valley cutting into it. The deck is much higher above the valley floor than it is elsewhere and the economical solution requires much longer spans for the valley than for the hillside. The solution, which took some considerable time and effort to arrive at, and evolved from one of a number of structural ideas, uses similar precast segments for the whole viaduct, and indeed they were all cast in one complex mould. They were all erected using cantilever construction, balanced cantilevers for the valley and continuous cantilevering for the hillside.

(Above) Berry Lane Viaduct. Each of the two parallel bridges carrying the motorway is supported on groups of four columns which relate better to the trees than massive piers would have done.

(Below) The curving deck of Kylesku Bridge allows the road to fit closely to the contours of the rocky landscape.

(Above) The balanced cantilever constuction for the Ouse Burn Valley section of the Byker Viaduct. The other section uses segments of constant depth and was built by continuous cantilevering, using temporary props. The viaduct carries trains of the Tyne and Wear Metro.

The hillside spans are of constant depth. The valley spans are the same depth in the central portions but are haunched towards the supports. The valley piers are also moment stiff to stiffen the spans against the heavy live loads, but they can sway in the span direction to allow for thermal expansion and contraction.

Where the complete design of a bridge is not in the hands of one team, how is it possible to achieve unity? The construction strategy for the Øresund Link between Denmark and Sweden is to have detailed design and construction carried out by different contractors on the various parts of the link, breaking the continuity of the design process. In preparing the preliminary design on which tenders were to be based, the strategy was to create a simple, rational and straightforward form; to express function without unnecessary details, and so to produce a strong and robust unity capable of being divided into smaller parts which could be detailed by different contractors and still produce a harmonious whole.

The Link consists of three very different elements: a tunnel under the sea, an artificial island where the motorway and railway surface, and a long bridge with a large navigation span across the channel over Flintrännan. It is very large and will stand in a seascape without natural forms to set the bridge against. The landscape on either side is gentle and friendly, with small and rolling hills and curved coastlines where the land meets the sea. The Link will often be seen at a distance: from the shores, the sea, and the air. Those on the Link itself will mostly see it at speed.

The sweeping line of the bridge and the way road and rail rise gradually from the sea at the island create a harmonious link between the lands it binds together. The curving road line gives users continuously changing views of the sea, the islands, the coastlines, and the bridge itself. On the bridge the motorway is carried at the upper level and the railway below it. With this arrangement the most economical structural solution is to use steel trusses with diagonals connecting the upper and lower decks. These trusses are uniform throughout the bridge, but are modified at the cable-stayed main spans so that every other diagonal has the same direction as the cables.

All these bridges (except perhaps Salginatobel) were designed by teams in which an architect was included in or alongside the engineering team, and on all of them models were used to study and refine overall proportions, the shaping and relationship of parts, and the relationship between the form of the bridge and its immediate environment. Maillart's bridge shows that some engineers can design beautiful bridges without the help of an architect. Ove, on the other hand, had to have an architect in order to meet his own exacting standards, and Arup bridge designers have mostly followed him in this, and also in the use of models.

(Left) This model of the Øresund Link shows an alignment which has been modified in the final scheme.

(Right and below) Western Bank Bridge and Sheffield University Concourse. The rather sculptural central supports lead the eye and create a feeling of space flowing through from one side of the bridge to the other.

The sheer exhilaration of designing tall buildings, realising that they can have more extreme slenderness ratios following the discovery of vibrational damping, is set against the bad press that high-rise has received. Economies of land use and the advantages of density suggest that the high-rise will continue, not only because of developments in engineering but also because of the symbolic significance of towers.

Tony Fitzpatrick

The balance of merit in designing tall buildings

(Right) Workers on the steel skeleton.

Why do we design tall buildings? Because we are asked to could be, to some, a sufficient reply. But to a firm with the history, traditions, and culture of Arups, with the memory of the man still fresh in our minds, that may be a necessary part of the response (otherwise no-one would pay us and we could not live), but it is not sufficient.

Our work can be 'scored' in three domains: what it does for Society, for the Firm, and for the individual. A positive balance of merit in all three must be achieved: then the higher the total, the more we want it!

The negative aspects of tall buildings in society have been stridently proclaimed. They are perceived as:

- Inhuman in scale, contradicting the idea that buildings are for people: Frank Lloyd Wright's Mile High Tower can be superimposed upon Hyde Park as a mental and graphic exercise but is perceived also as a representation of Orwell's and Huxley's technocratic future.
- Oppressive and aggressive. The wind-swept canyons of mid-town Manhattan do not invite a late-night stroll amongst the seemingly unbroken flanks of skyscrapers. And the visual chaos of Tokyo can also be added to this fear of the city of towers.
- Destroying neighbourhoods and communities: Even though high-rise housing rarely rises above 20 storeys, where tower blocks have been used to 'solve' a housing problem the solution is seen to have produced vandalism and violence, destroying communities and making the housing uninhabitable. Both the film *The Towering Inferno* and the Ronan Point disaster support these fears.

Yet these are not inherent traits of the buildings: rather, the sins of the designer have been laid at the door of the progeny. With care and with thoughtful and intelligent designers, not only can the above be reversed but real (and otherwise unobtainable) benefits can be generated.

Cities are real, and remote working from self-sufficient homesteads via the Internet cannot replace the powerhouse of personal interaction which drives teamwork, creativity, and the consequent improvement in well-being of society and its individuals. Predicted city populations reflect this. The challenge is to enable people to be together in an environment which enhances the quality of life. To live and work and play with good space but low-rise results in massive urban sprawl, with the consequent vicious spiral of distance, travel, waste, and environmental pollution.

Tall but slender is a key, and one to which in our work as structural engineers we hold: to create the needed space at height, yet allow the ground-scape to remain at human scale; to empower planners and architects to engage in the social design of cities at street level, and to employ our skills and technology to create above.

Tall building structures are dominated by wind; resistance to wind requires strength to stand up and stiffness not to shake about. Materials technology has advanced dramatically, and continues to do so. Concrete of twice the strength considered good 20 years ago is now reliable and affordable. Structural steel is today strong, cheap, and reliable. Together with high-powered analytical techniques, there is practically no limit on height from strength.

Stiffness, however, is by comparison in the Dark Ages. The vast majority of tall buildings to date have achieved their

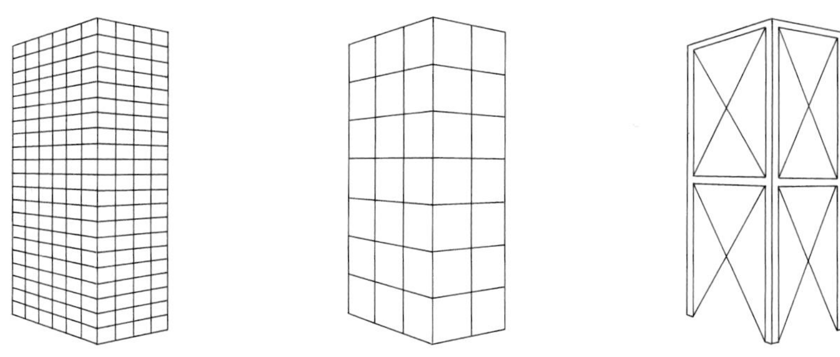

Japan 9/3/87 KS/PN/CB/CW/AJF.

1. CW drg lateral stability

2.

500

tie back

2 floors

check: take out

OAP to check this pin.

total flexibility floors

Preferred System.

OAP check:
- tie-back on struts
- pins @ ends.

3. Floors
- 3.6m Tapered beams, 24m span
- 500 cantilever.
- 150 Y/ut conc slab.

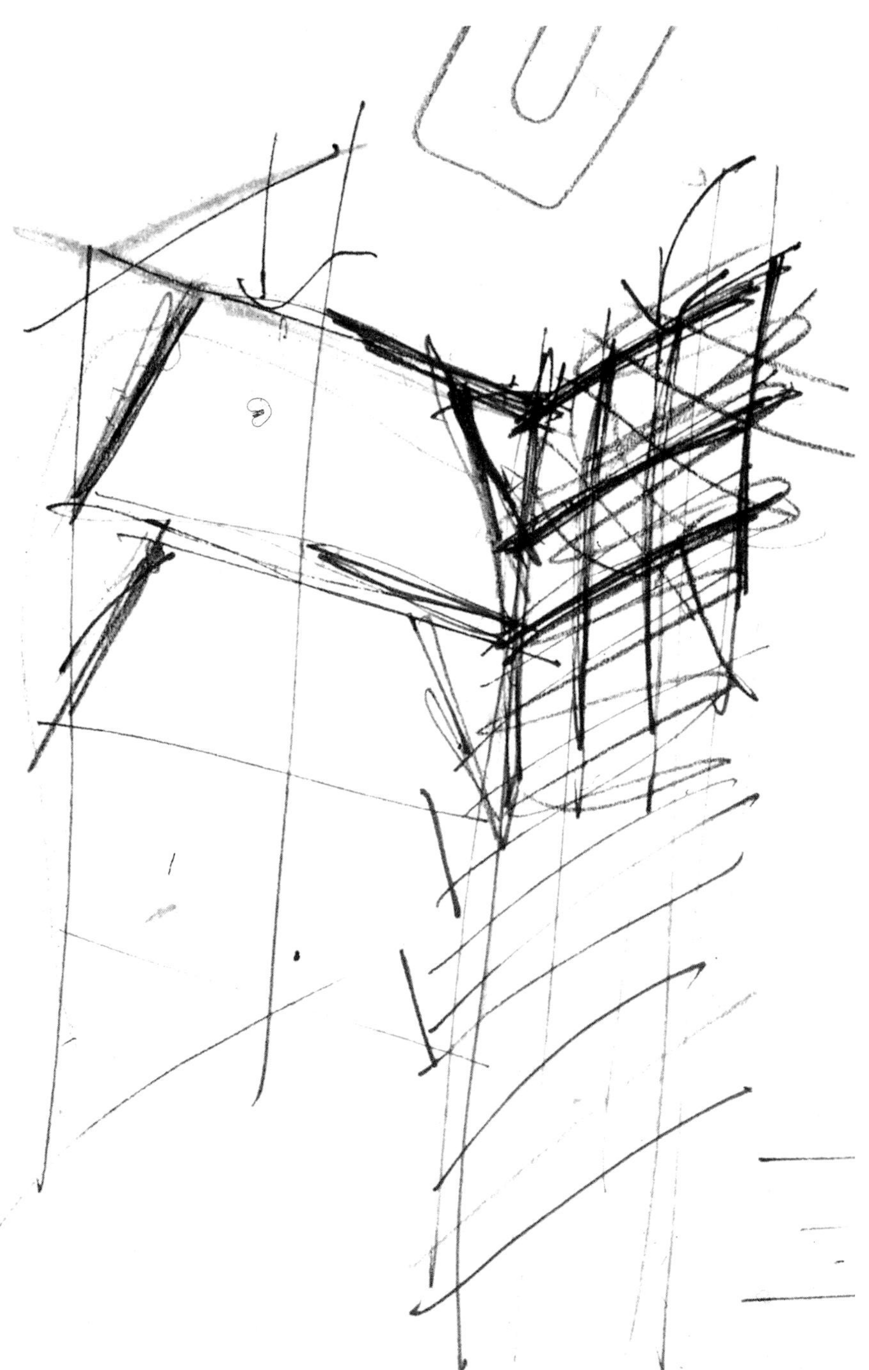

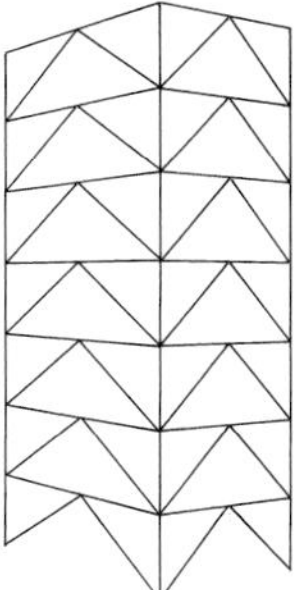

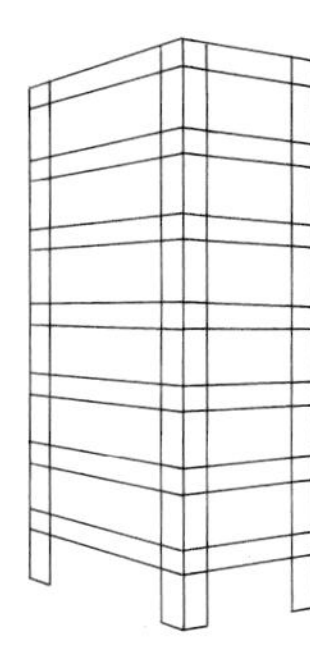

Century Tower, Tokyo. Alternative framing proposals (left), (below left). Completed building (below).

stiffness through enlarging their base plan size. A base of 1/7th of the height is considered small: 50m for a height of 350m. This is what generates the canyons, the walls along the street.

A few inspired and caring groups are breaking this barrier through the use of active damping systems: moving masses at height within the building which absorb the wind energy. Base widths of 1/10th the height are achievable now, as Jean Nouvel's Tour Sans Fins, and 1/15th with only a small extension of the same systems. Japanese research into 'prior-control' systems promises either the same at lower cost or even greater slenderness.

The real gain from slenderness is not simply more height. It is rather the freedom it gives to spatial organisation: the medium high building on a small plot with a friendly street-face can revitalise a decaying centre, bringing people and critical mass back from the flight to the suburbs. Care in the design is fundamental to a good environment. The recent successful work in rehabilitating public housing blocks once thought uninhabitable, and the success of the apartments in the John Hancock building, Chicago, are proof of what can be achieved.

Safety, or rather the sense of being safe, is what the designer must bring about to the public. It is probably more dangerous to cross a farmer's field than live on the 100th floor. Living in an earthquake zone is dangerous; yet who realises that the safest building to be in during an earthquake is a tall one?

Earthquakes put load into a structure through repeated cyclical ground motion. The amount of load depends upon the degree of energy transfer from the ground. The taller a building is, the further away from the peak power spectrum of the earthquake is the structure's natural response and hence the less load it attracts. In fact, a 50 storey building is so far down the earthquake's power curve that the wind loads are almost always greater. Hence, once in existence for a few years, it could well have been 'load-tested' by the wind to a greater load than from the earthquake. Conversely buildings between 5 and 10 storeys are at the peak of the earthquake power curve and will be subjected to many times any previously experienced wind load during the earthquake.

For designers, the balance of merit is more easily positive. Prime amongst the gains is the development of our skillls base. Development and use of materials to create form as well as

basic strength, and use less to improve value; advancement of our understanding and ability to predict and control dynamic behaviour; refinement of construction methods and site organisations to improve quality, speed and reliability of the product. All of these are fed by the needs of tall building design, and the research can be done because the scale of the projects and the need for the skills make it economically viable: there is real added value to be gained. Such skills can then be employed and the technology benefits taken on smaller projects which could not on their own justify such investment in a competitive economic environment.

Tall buildings are international, and designing them both requires and inspires a correspondingly international view. This helps a company to be broad and stable, and to bring that international dimension to domestic clients, friends and projects around the world.

It is not all positive. Tall buildings tend by their nature to be on a grander scale than most and this can easily lead to a remoteness and aloofness on the part of the designers to those very aspects of context and detail which bring them into disrepute: the concept of designing a civic centre and then looking for somewhere to put it would rightly be greeted with scorn, yet how often this is proposed by designers of tall buildings. Tall buildings must be conceived simultaneously at 1:1000 for the site and at 1:20 for the people; to lose sight of either is to fail. To work on a house and a tall building concurrently keeps the mind aware. Arups comprises nothing but people and information. Satisfaction of the individuals is paramount to continuing good design. From the inner satisfaction of personal skill enhancement to the global basking in complimentary magazine articles, tall building design has few competitors in the satisfaction stakes. The perpetual challenge of closing the gap between the desirable, the achievable, and the affordable is at its greatest when compounded by height, with consequently greater joy from success. Time and cancellation are the banes of the projects: large slices off the designer's life can be taken by each project, and their costs are such that cyclical economics often destroy them before reaching site.

The overall balance of merit for designing tall buildings? Given care, skill and tenacity it can be both a pinnacle of achievement and an example of the very best in balance.

(Far left to right)
Central Plaza, Hong Kong;
UOB Plaza, Hong Kong;
Commerzbank, Frankfurt;
La Tour Sans Fin, Paris;
1, O'Connell Street, Sydney.

This diary of the design and construction of a North Sea oil platform begins with a discussion of the virtues of concrete gravity stuctures versus platforms made of steel. The detailed progress of the project for Ravenspurn North is a fascinating and impressive record of a major civil engineering achievement.

John Roberts

Offshore platforms

Construction complete (right).

Significance

Two concrete gravity substructures (CGSs) supporting production decks were installed in the North Sea in 1989. In June the Gullfaks 'C' platform was installed in the Norwegian sector. At towout the structure weighed 850,000 tonnes – reputedly the largest object ever moved by man. At the beginning of August the Ravenspurn North concrete gravity sub-structure, weighing some 28,000 tonnes, was installed 80km off Flamborough Head in block 43/26 of the UK sector.

A CGS was selected for Ravenspurn North because the operator was convinced that the design developed was cheaper than the steel alternative. The main reasons why concrete gravity substructures are now comparatively cheaper than steel jackets stem from design. In the case of the Ravenspurn North CGS, the principal factors were as follows:

- The decks can be installed offshore in a single lift using a semi-submersible crane vessel after emplacement of the CGS on the seabed. Previous CGSs have been mated with their decks inshore

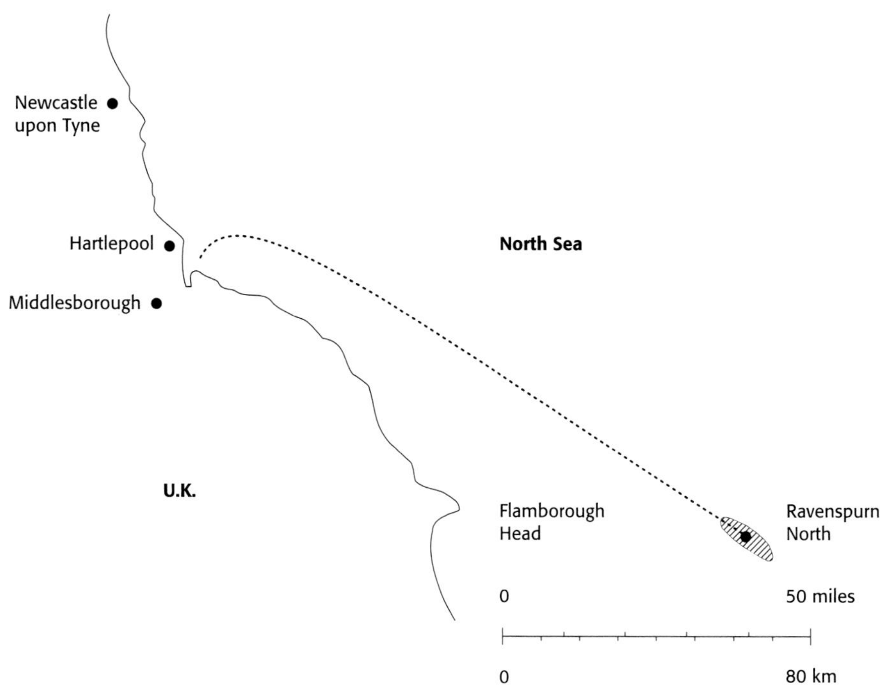

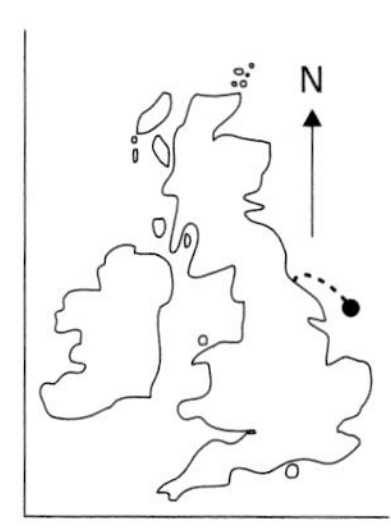

(Above) Location of the Ravenspurn North gas field.

prior to towout, or the topside equipment has been installed offshore in comparatively small lifts.

- The Ravenspurn North CGS will support two separate decks, of which only one will be installed initially, whereas all previous designs supported only a single deck. Thus the cost of a second separate support structure has been avoided.
- The CGS has been built entirely in a dry dock rather than partly as a floating structure – which has been Norwegian practice.
- The structure was designed to be simple, highly repetitive and aimed to disassociate the concrete from the more ephemeral process pipework which it supports.

It comes as a surprise to the majority of civil engineers that the oil industry does not readily appreciate the advantages of adopting concrete as a structural material in a marine environment. In fact concrete is seen as a radical change which therefore involves greater financial and programme risk. Since the cost of the substructure supporting the production equipment on a platform typically represents less than 5% of the total field development cost, any 'innovative' design has to show substantial savings compared with the alternatives, before operators can accept the risks.

Development of the design concept

We were approached by Hamilton Brothers in October 1986 and were surprised to find out that they had already commissioned a number of design studies for the CGS option.

The CGS was required to support the central gas processing platform of the Ravenspurn North field. It will eventually process gas from four remote wellhead platforms imported via 324mm and 356mm diameter infield risers. Processed gas is exported to shore via a 600mm diameter line.

The main production deck was installed in the spring of 1990. The maximum operating weights of the main deck and compression deck are 8,500 tonnes and 4,000 tonnes respectively.

During the course of the feasibility study a number of conclusions were reached about the design features required to realise the most cost-effective CGS:

- By installing the decks offshore using a semi-submersible crane vessel, the size of the base caisson could be significantly reduced compared to a design with the decks installed inshore. One of the chief determinants of the base caisson size is floating stability. For every tonne of structure at the top of the shafts, between 5 and 6 tonnes are required low down in the structure to maintain the same stability characteristics (metacentric height).
- Installing the decks offshore reduced the size of the CGS to the point where the entire CGS could be fabricated in any of the existing UK construction docks. Thus the cost premium associated with completion ofthe CGS while floating at an inshore location was avoided.
- By providing a design which could be constructed entirely within the dry dock, the problem was reduced to one of conventional prestressed concrete construction. The inherent economy of the structure therefore depended upon providing a design which was simple and highly repetitive and could be constructed from readily available materials.
- A number of CGS designs based on these principles, having two, three and

four shafts, were identified during the feasibility study. The concept favoured involved the novel idea of two separate decks, the main production deck supported by two shafts and a compression deck by a single shaft. It was possible to show that the size of the base caisson would be approximately the same, regardless of whether the CGS had two or three shafts. Conventionally, two decks would have been supported by two separate steel jackets. The cost of supporting the compression deck was therefore equivalent to the marginal additional cost of providing the third shaft on the CGS.

The conceptual design work was undertaken in the period from March – July 1987. The detailed design was carried out in parallel with the tender and assessment period from October 1987 to March 1988, and project sanction obtained in May 1988.

The design solution

The component parts of the Ravenspurn North CGS are very simple: connections for the decks, concrete shafts, concrete base caisson and steel foundation skirts. Initially a number of parametric studies were carried out so we could better understand the behaviour of the

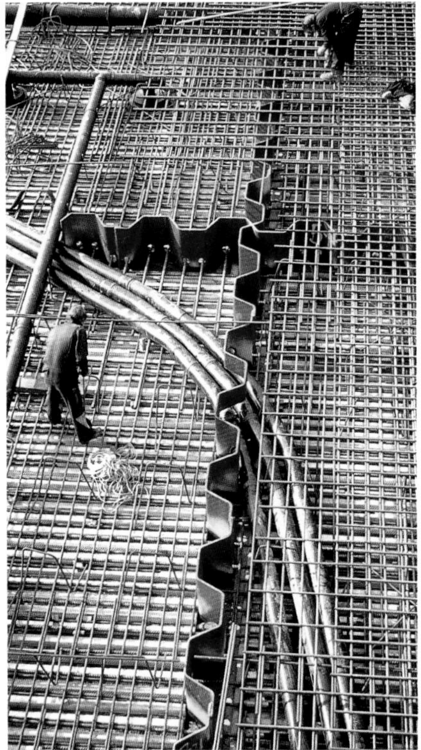

(From left to right) Base slab reinforcement showing embedment of profiled steel skirts, drainage and ballasting pipework, and electrical ducts; Concrete construction nearing completion, deck connections installed and mechanical outfitting under way; Close-up of pipework; Corner detail showing skirts, base slab, and surface-mounted prestressing anchorage.

structure and optimize the size of the structural elements. At the outset of the design the most difficult task was to determine the size of the base caisson. This depends on:

Hydrodynamic loading

Approximately 80% of the horizontal load from wave and current action is generated on the base caisson. The load generated is proportional to the enclosed volume of the caisson, and it is desirable to keep the latter as small as possible. It is also desirable to keep it low since the magnitude of hydrodynamic force reduces exponentially with water depth.

Structural considerations

The caisson must be of sufficient size to transmit the forces from the shafts and those generated on the caisson itself into the seabed soils.

Geotechnical considerations

Sliding of the structure along the seabed rather than bearing pressure is the governing mode of foundation loading. The base area of the caisson has to be such that the shear stress in the soil at the level of the tips of the skirts is less than the soil strength.

On cohesionless seabed soils, such as those at Ravenspurn, the resistance to sliding is therefore a function of the mobilized shear strength of the sand and the submerged weight of the platform.

Naval architecture

The CGS must possess sufficient buoyancy and remain stable at all times while it is still floating. Generally speaking, the larger and taller the base caisson and the greater the diameter and spacing of the shafts, the easier it is to provide adequate stability. It is essential to keep the centre of gravity as low as possible.

Floatout draught

The size of the base caisson must be such that on floatoff, the draft is well within the limit imposed by existing UK docks; the larger the base area, the lower the floatout draught. Some of these requirements are conflicting.

The design solution was to develop the base caisson having the lowest density (weight in air divided by

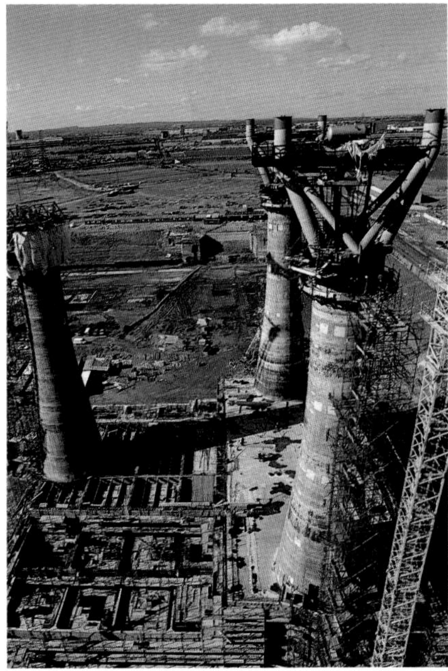

enclosed volume), and to divide the caisson up into open and closed cells.

They also permit the eventual refloating of the structure for abandonment. The open cells improve floatout draught, reduce the centre of gravity of the structure (since the cells have no roofs) and enable the onbottom weight of the CGS to be varied by the addition, after emplacement, of solid ballast. This can be used as a means to improve resistance to sliding.

The size of the cells into which the base caisson is divided was optimized for least caisson density. The optimal size for square cells is 6-7m square. Below this size the concrete cover to the tension reinforcement adds weight without adding to the strength of the

section; above this size, shear rather than bending begins to control wall and slab thickness. Cells 7.5m square were selected, as this dimension was the multiple of the desired support spacing for the main deck.

The geotechnical design problem was to determine the extent and the length of the skirts cast into the underside of the concrete base slab. The determining factors are:

- The need to reach competent soil
- The contributions of passive resistance and the self-weight of the soil within the skirt compartments
- The need to avoid 'piping' during both installation and operation
- The need to guarantee full penetration of the skirts at installation so that substantial contact between the seabed and the underside of the base slab could be achieved.

The last point was important because most previous CGSs had relied on grouting of any spaces remaining between the seabed and the underside of the base after installation. Grouting offshore is extremely expensive and makes subsequent refloating for abandonment technically difficult and costly.

Steel skirts manufactured from 18mm thick Grade 43C steel plate, profiled in a form resembling sheet piling, were adopted. The skirts run under the majority of the base caisson walls.

The design of the shafts is comparatively simple. The principal loadings are from the deck itself and from hydrodynamic loads. In addition it was decided to prestress the shafts vertically to ensure that cracking did not occur (Class II to *BS8110*), and taper them so as to reduce the hydrodynamic forces generated.

Deck connections

The deck connections reflect a problem specific to the Ravenspurn platform. The size and arrangement of the topside facilities are similar to an existing platform. In this deck, the support points were spaced as 12m centres compared to the diameter of the CGS shafts of 6m. Alternative solutions were considered, the majority of which resulted in considerable additional weight at the top

of the shafts. The deck connections or 'antlers' developed are relatively lightweight, being fabricated from Grade 50E plate up to a maximum of 50mm thick, thereby avoiding extensive post-weld heat treatment. Each main strut of the deck connection must support a maximum of approximately 2,500 tonnes in operational conditions. The deck connections are fire-protected with *Chartek*, an intumescent epoxy ablative coating up to 14mm thick.

The deck connections were attached to the top of the shafts by 20 prestressing tendons. High strength neoprene bearings developed with Tico Manufacturing Ltd have been provided between the steel and the concrete to spread the concentrated load from the struts around the perimeter of the top of the concrete shafts.

The CGS supports process pipework consisting of gas risers (which connect the seabed import and export pipelines with the process deck), caissons which allow sea water to be lifted to the platform, cleaning and dumped) and J–tubes (which are the ducts for electrical cabling to the remote wellhead platforms). It was recognised that the sizes and layout of this pipework could easily change both during the design and during the platform's operational life. The design that evolved was to mount all the process pipework externally on the shafts. Circular tubular supports of

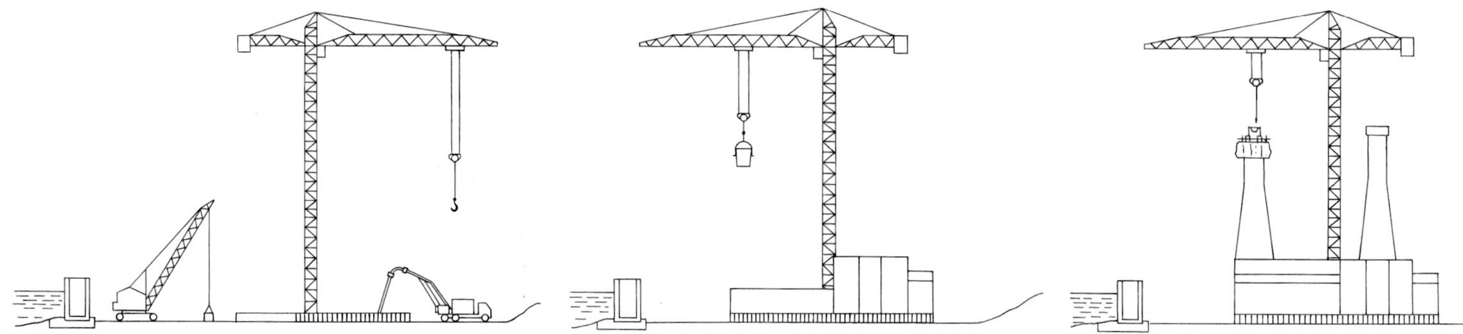

(Above) Casting the base slab to float out.

'cowhorns' have been provided at regular vertical intervals up the shafts, welded to steel embedment plates cast into the concrete. The uppermost cowhorn is attached to the steel deck connections and supports the selfweight of the pipework; the lower cowhorns are guides which allow axial movement only.

The cowhorns were designed to support pipework at a regular spacing but with no particular arrangement in mind. This design decision was vindicated at an early stage. Three months after construction of the CGS began, the Piper Alpha accident occurred. As a consequence it was decided to change the routing of the risers from one shaft to another. The location of a number of caissons was also changed. No alteration to the concrete structure or the cowhorn supports themselves was required and, as a result, there was no impact on the construction programme.

The operating pressures of the risers are up to 400 bar and the routings were modelled for analysis purposes from a point on the seabed some distance from the CGS to the pig launchers (pipe testing devices) and receivers in the deck. Careful detailing was required to avoid longitudinal welds on the pipe.

Weight control is an essential part of designing a CGS. Apart from regularly monitoring the weight of the structure as shown on the drawings (the 'Current' weight) it is essential to establish at the outset a highest expected weight (the 'Contingency' weight). During the design, floating stability and draught calculations were based on the contingency weight and the corresponding position of the centre of gravity of the structure. During construction, weights calculated from the drawings were replaced by data from weighings of individual elements or, in the case of the concrete itself, a weight devised from as-built surveys and densities assessed from 100mm cubes and core samples taken from the structure. The results from over 1,800 cubes and 400 50mm diameter cores were used to calculate the Estimated Final Weight. The weights established were as follows:

Current Weight:	26,938 tonnes
Contingency Weight:	28,875 tonnes
Estimated Final Weight:	28,625 tonnes

Previous CGSs have been installed by keeping the structure very nearly level as it is ballasted down to the seabed (parallel installation). In terms of floating stability the crucial stage is when the water level overtops the base caisson for at that point the water plan area is substantially reduced.

It was realised that if the CGS were deliberately ballasted down to a steep angle, so that the skirts touched the seabed, and was then rotated to a level position, the GM would be higher than for the equivalent parallel descent. The reason is that in the inclined descent with water ballast in the edge cells, Hg is correspondingly lower and Hb correspondingly higher than for the parallel descent. The higher the GM the greater the floating stability and the stiffer the response.

Selection of construction contractor

During conceptual design a list of 35 potential tenderers was drawn up and each firm contacted. It proved hard to reduce the list to the seven firms finally approved. By then each tenderer knew a great deal about the project and the design team also knew, which docks they would use, how they would be made available, their proposed construction methods, programmes, and choice of sub-contractors.

A two-stage tendering process was recommended whereby tenders were sought, based on approximately half of the construction drawings and the full specification. The Conditions of Contract were the Institution of Civil Engineers' 5th Edition modified to suit oil company needs and the peculiarities of the project. First stage bidding was on a bill of quantities.

In the second stage a limited number of tenderers had the opportunity to resubmit prices, based on the rest of the construction information. Rates were as originally tendered. During the final negotiations, the general items and measured works were converted into a lump sum and the bill of quantities was converted into a schedule of rates which was to be applied in evaluating variations.

Construction

Graythorp Dock, on the Seaton Channel 1.5km from the River Tees, was constructed in the early 1970s for building large steel jackets for the Forties A and B and Thistle platforms. Consequently, there were substantial pile foundations in the dock floor.

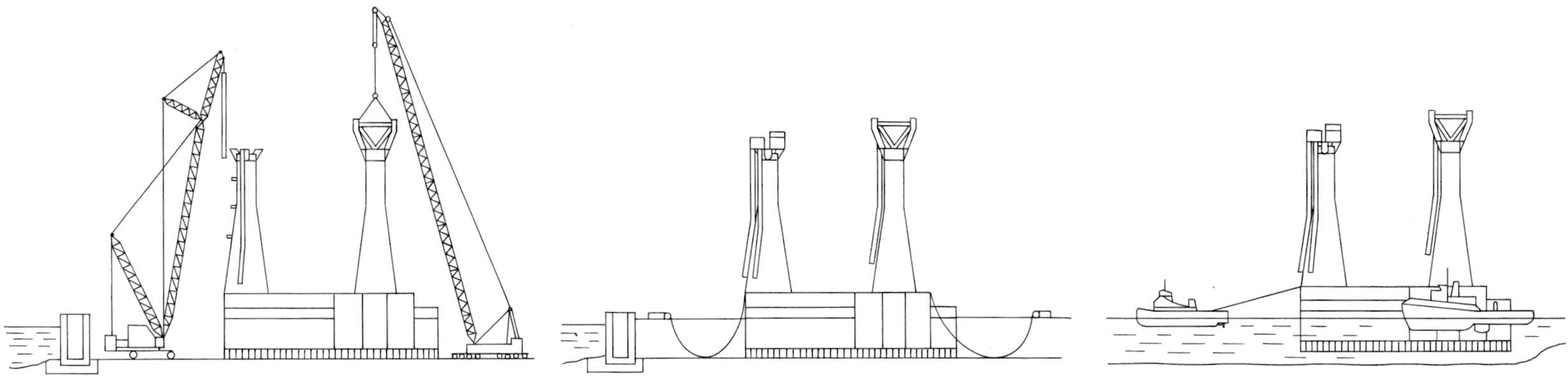

Further piles were provided from the CGS contract to suit the pattern of foundation skirts. After temporary works piling and the construction of ground beams, the steel skirts were erected in sections. No propping to the skirts was required since the cruciform sections provided their own stability.

The concrete base slab was constructed on permanent metal decking supported on shelf angles welded to the skirts. Modified high-density polyethylene (MHDPE) pipework for the under-base drainage and for closed cell ballasting was cast into the concrete.

The skirts were detailed with slots to enable reinforcement to be fed through them at the wall positions. The base slab was cast in five sections with the largest pour being approximately 650m^3.

The base caisson walls were each cast as a series of cruciforms with each leg approximately 3.75m long on plan. Three lifts of up to 4.8m high were required for the full height of the base caisson. The formwork was erected and positioned by holding it on a crane, often for extended periods, as adjustment proved more difficult than Laing had envisaged.

The wet concrete was vibrated using electrically-operated, shutter-mounted vibrators. Their use resulted in high quality dense concrete with few instances of honeycombing. Both horizontal and vertical joints in the caisson walls were keyed and scabbled.

The concrete mix contained 475kg of cement and pfa per m^3 of concrete. The average cube strength achieved was 68N/mm^2 compared to the 50N/mm^2 required by the specification.

As with the base slab, the caisson roof was constructed on permanent metal decking supported in this case on steel beams spanning across the 7.5m cells. The roof is reinforced and prestressed, and also contains pipework which vents air from the closed cells during ballasting.

The shafts within the caisson height were constructed using adjustable steel shuttering to form the circular shape. Above caisson roof level the shafts were slipformed using the Interform system. A variety of steel embedment plates forming attachment points for the process pipework supports were cast in as slipforming progressed.

The deck connections, constructed off site supporting the main deck each weighed over 160 tonnes. They were lifted into place using a 1000 tonne mobile crane and temporarily held in place using steel dowels before being prestressed to the concrete structure with part of the vertical prestressing in the shafts.

With the deck connections in place two activities proceeded in parallel. On the outside of the shafts the 'cowhorns' were welded to embedment plates. On the insides of the shafts, work began to complete the mechanical and electrical systems required for installation. When all the supports on the shafts had been welded into position, individual risers and caissons were lifted onto the dead weight supports at deck connection level and the halving clamps at the other guide positions bolted into place. The infield, export, and methanol risers were routed over the caisson roof and erected on 'goalpost' supports. After completion each riser was hydraulically tested to 1.5 times operating pressure.

The design of the structure is such that, apart from the need to prestress the deck connections to the shafts, all other prestressing operations could be carried out fairly independently of other construction activities.

Moreover, access to the anchorages at both ends was freely available, as they were either surface-mounted on the sides of the caisson walls or accessible from the tops of the shafts or from the skirt compartment below the base slab. The remainder of the construction operations involved the testing and recommissioning of the installation systems, including the hydraulic testing of each open and closed cell. Minor leaks were encountered during testing and repaired. Those construction joints which it was not possible to test hydraulically were visually inspected, tested ultrasonically if their integrity was in doubt, and repaired if necessary.

The installation control cabin and power generators were lifted onto the top of Shaft 3 by crane and hooked up to the onboard electrical and control system. Each component of the ballasting system, including valves, flowmeters, pumps and air pressure bubblers, had been individually tested at the works. The assembled system was pre-commissioned on site.

After pre-commissioning was complete, mooring lines were laid from the CGS to four winches on the sides of the dock and the dock flooded. During

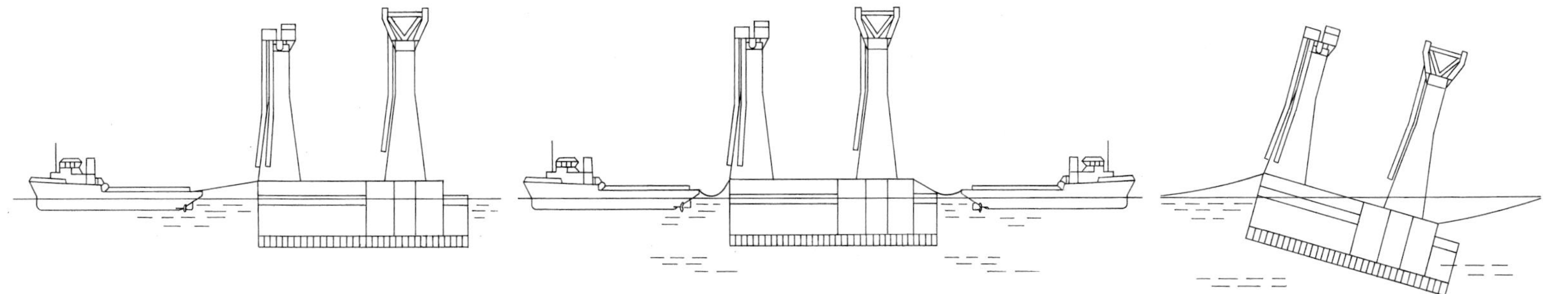

(Above) Tow out.
(Opposite page) Floatout from Graythorp Dock.

flooding, air in the skirt compartment was released via the underbase drainage system and water admitted to the open cells via manually-operated sluice valves to prevent premature floatoff of the CGS. Finally, the cellular concrete dock gates were removed. They were deballasted, floated off the spillway and towed to a temporary foundation area at the rear of the dock.

Installation

The CGS was installed in four stages: flooding the dry dock, towing the structure to the Ravenspurn field, sinking it to the seabed in a controlled ballasting sequence and finally placing ballast in the open cells and scour protection around the entire structure.

The first stage of installation was to manoeuvre the CGS out of the dock by tug and down the Seaton Channel to a location off the mouth of the River Tees. Once the tugs were in position the mooring lines to the dock winches were tensioned.

As the tide fell below mean water level, submersible pumps inside the open cells were started and compressors began to pump air into the skirt compartments below the base. With the open cells empty and a 1.55m air cushion under the base, the CGS floated off on the rising tide.

When the underkeel clearance to the deck floor reached 1m, manoeuvring the CGS out of the dock began using both mooring lines and tugs.

As it reached the dock entrance the first of two ocean-going tugs was attached. The Ravenspurn North CGS then proceeded down the Seaton Channel at a stately 1 knot, reaching Teesmouth three hours after floatoff. The harbour tugs were then demobilised and the second ocean-going tug attached to the bow.

The CGS was then held while the open cells were flooded, the compressed air released and items of floatout equipment (pump, switchgear and positioning systems) removed. Only then could we obtain a definitive measurement of the CGS's final weight of 28,125 tonnes. This compared with the Estimated Final Weight of 28,265 tonnes.

The CGS was then towed using the two ocean going tugs with one of the other tugs as the attendant vessel, the 84 nautical miles to the field taking 27 hours. At the field four anchors had been laid forming a square around the CGS location. Two towing tugs and two further sister vessels were moored to the pre-laid anchors assisted by a fifth anchor-handling tug. Using the winches on the tugs the CGS was manoeuvred into the correct position for emplacement to begin. From that time onwards its position was monitored by surveyors from the wellhead platform 70m from the CGS.

The third stage of installation was sinking the CGS to the seabed. Four men were stationed in each shaft with four others, including the offshore installation manager, in the command position at the top of the shaft which will eventually support the compression deck.

The predefined sequence of ballasting began and the CGS gradually trimmed down by the stern until contact with the seabed was achieved at an angle to the horizontal of 37°. The touchdown position was confirmed as being within acceptable tolerance. Water ballast was then admitted to more of the closed cells to bring the CGS level, with all skirts in contact with the seabed. Finally water was admitted to the remaining closed cells, resulting in full penetration of the skirts into the seabed soils. 10 hours after the ballasting began, the CGS was safety installed on the seabed 2m away from the target position and 1° from its intended orientation. The entire procedure from the start of floatout from the dry dock to completion of installation on the seabed took 55 hours.

The final stage of installation took longer. The open cells were filled with crushed rock, and a scour protection blanket, also of crushed rock, was installed around the base caisson. The priority was to install scour protection around the corners of the CGS where we had predicted scouring of the seabed sand would occur at the extreme ebb and flow of the tide.

The Rocky Giant, a side dumping barge, placed the initial scour protection blanket around the CGS, comprising a 400mm thick filter layer and 200mm thick armour layer. A few days later the Trollness, an 8,000 tonne vessel equipped with a steerable fall pipe, arrived to complete the work.

Altogether 9,600m^3 of ballast was placed in the open cells and over 10,000m^3 of scour protection laid. The three shafts were also flooded, adding about 6,500 tonnes onbottom weight to the CGS. With these activities complete,

(Far left) Offshore installation: touchdown on seabed.

(Left) Offshore installation complete, with skirt having fully penetrated the seabed.

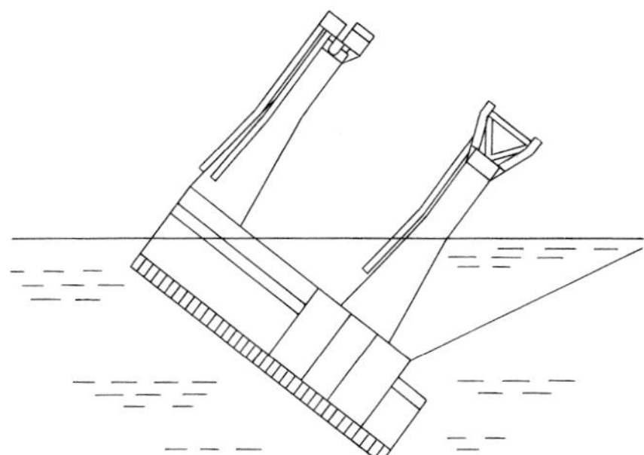

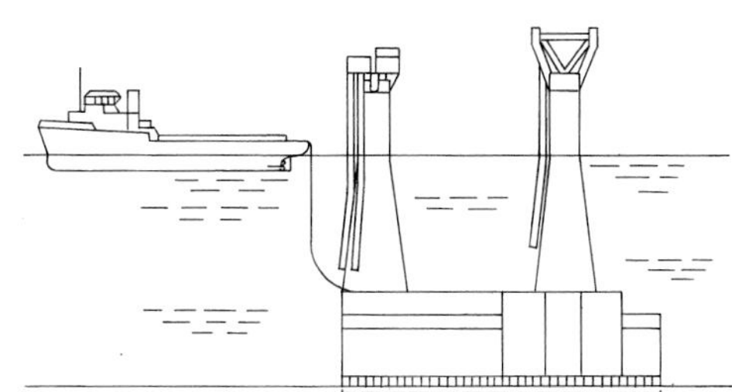

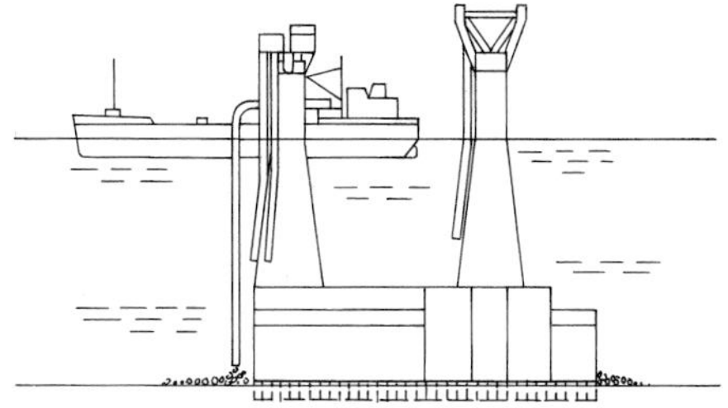

(Above) Solid ballasting and scour protection.

the CGS is able to resist the design storm having a 10-year return period. The weight of the main deck installed next spring will increase further the platform's resistance to sliding, to a level corresponding to the 100-year return period environmental conditions.

Future developments

The Ravenspurn project has led us into a number of field development studies for other oil companies. From these studies we have concluded that for a wide range of applications – water depths from 20m to over 120m – concrete gravity substructures may be more economic than steel jackets. The greatest cost savings are likely to result where:

- Multi-purpose platforms are required (drilling, production and quarters).
- Platforms with relatively heavy topsides in comparatively shallow water are required.
- A phased field development is envisaged such that future topside equipment is difficult to define initially.
- Safety considerations lead to the requirement for a spatial separation between the accommodation and process areas.
- Oil storage (which can be accommodated within the base caisson) combined with tanker export is more economic than laying a new pipeline.
- The presence of rock close to the surface makes piles difficult to drive.

Forms of substructure – Jackets

Piled steel jackets are by far the most common form of support structure for offshore platforms throughout the world. They consist of a tubular steel spaceframe supported on tubular steel piles driven after the jacket is located on a seabed. Today virtually all jackets are transported offshore on purpose built flat barge sand installed either by launching from the end of the barge or by lift installation using a semi-submersible crane vessel.

In operation, hydrodynamic forces generated on the structure are transmitted to the seabed soils through the piles. The tendency for a jacket to overturn is resisted by tension in the piles on the upstream side and this load condition normally governs the design of the piles.

The popularity of piled steel jackets can be traced to the first offshore developments in the Gulf of Mexico. Steel rather than concrete was the material most familiar to the American engineers involved in developing offshore technology. The influence of American oil companies throughout the world is probably the main reason why steel jackets remain predominant.

Gravity-based structures

Gravity-based structures (either steel or concrete) resist overturning in the same manner as a simple pad foundation. To prevent uplift the width of the CGS base must be such as to ensure the resulting force from horizontal hydrodynamic loading, and vertical loads lie within the middle third of the base width. Apart from isolated examples such as the Royal Sovereign Lighthouse, concrete gravity substructures had not been used offshore until the early 1970s when a combination of factors created the need for a new design solution. At that time developments in the North Sea were at the forefront of offshore technology, requiring platforms in a deeper and more hostile environment than had been attempted previously anywhere in the world. In addition the platforms required very heavy topsides. At that time the largest offshore crane vessel could only lift about 1,000 tonnes and therefore a large number of relatively small lifts were required to install the topsides on a conventional steel jacket. Concrete gravity substructures were designed to reduce the cost of offshore hookup by mating the CGS with a single-piece deck inshore prior to towout.

Since the early 1970s concrete gravity substructures have been automatically associated with this sequence of construction and installation which has been adopted for all subsequent CGS designs installed in the Norwegian sector.

The Ravenspurn North CGS is the first concrete structure in the North Sea to break with this tradition.

In Australia the emerging building type of the Exhibition Centre calls upon the engineer to understand the structure as a powerful representation of the Centre which the structure supports. The new role of engineering in providing something akin to the signature or the logo of the function is seen within the context of the emerging national interests of Australia.

Tristram Carfrae

Exhibition centres in Australia

The basic specification for exhibition halls can be summarised as a number of 5000m² spaces with 14m clear headroom, interconnected by enormous operable doors that allow the halls to be used independently or in any combination. The halls have entry foyers for the public on one side and service access on the other, to allow heavy goods vehicles to drive right onto the exhibition floor for fast set up and knock down of exhibits.

This basic specification produces a big shed, which might be much the same as an aircraft hangar or high bay warehouse. Internally, an exhibition hall must not detract or interfere with the exhibits themselves. However, in Australia, all these sheds are in competition with one another for their share of the same market, and therefore all strive for some positive visual identification or public appeal. All these buildings have limited budgets, so the question becomes: how do you make a shed not a shed (or a duck) and still be an economical structure? It is a distinguishing feature of Australian architecture over the past 15 years that many prestigious public buildings have been commissioned via design and construct competitive tender. Immense credit must be given to the participants for producing noteworthy buildings despite the time and budget constraints which they treat as just another design parameter.

The main opportunity for differentiation is with the roof structure and external wall design. The roofs are of more interest to us as structural engineers and are typically of 60m to 80m clear span. Many structural systems are appropriate and cost-effective in this situation. I have no doubt that a system comprising deep trusses, with or without continuity with the wall structure to give portal action, would be the most economical. One look at warehouses or hangars will convince

(Above) Brisbane Exhibition Centre performing its duty as a big, grey, column-free shed to house – but not distract from – the exhibits.

that this is the case. However, for a small increase in structural cost, many other systems may be used. The choice is largely aesthetic but must also take into account the site, its ground conditions and boundary constraints; cable-stayed solutions, for example, are not appropriate if permanent tension anchorages cannot be constructed easily, nor if there is insufficient room on the site for back stays. The actual choice of roof system, therefore, comes from an intimate collaboration between the architect and structural engineer, with each struggling to understand the other's objectives.

While all of the projects to which we have contributed are a specific response to their particular site and requirements, viewed from a historical perspective, they can be seen to take their place in the sequential development of long-span roof structures. This development is normally associated with technological advances, but appears also to have a distinct element of fashion. In the1950s, state-of-the-art was represented by thin concrete shells that were amenable to analysis by mathematical theory for which the calculations, though labourious, were possible without the aid of computers. The arrival of computer analysis unleashed the space frame, or more properly the space truss, which with its distinctive tubes and spherical connections, is strongly identified with architecture of the 1970s.

The Darling Harbour building is a good example of the cable – or more precisely rod-stayed roofs that appeared through the 1980s. More recently the ability to study discrete buckling behaviour using nonlinear computer analysis has created a popularity for steel grid shells. Most examples round the world have been of synclastic curvature, usually domes or

(Top right) Melbourne Exhibition Centre under construction. Simple trusses at 18m centres supporting operable doors can subdivide the space to give maximum operating flexibility.

(Above)
Cairns Convention Centre presents a shell-like appearance with its timber, folded plate roof.

barrel vaults. The Brisbane building breaks new ground, for steel lattice shells, by using the anticlastic form of a hyperbolic paraboloid. The hypar was a common form for concrete shells, so we appear to have come full circle.

Cairns has a curved, folded plate roof executed in a combination of plantation timber and steel as a direct response to the current emphasis on ecologically sustainable development. The Melbourne building does not fit into this evolutionary pattern of roof structures and is the most conventional, using simple one-way steel trusses. The emphasis in Melbourne architecture is much more on the clothes than the bones; the structure remains hidden behind an elegantly crafted façade and a dynamic 'entry statement'. The proposal for Homebush Bay is based on an arch thrusting into an overhead catenary cable which both shares downwards loads and stabilises the . arch against in-plane buckling.

Darling Harbour Exhibition Centre

When commissioned for the Darling Harbour centre in 1985, the roof design was based on a body of work already carried out e.g. Hall Seven of the National Exhibition Centre, Birmingham; Fleetguard, Quimper; Renault, Swindon, and Patscenter, Princeton, New Jersey, about which excellent technical papers had been written. This type of roof, with some elements only capable of carrying tensile forces, requires nonlinear computer analysis similar to that used for membrane structures. Fortunately this technology had been transferred from London to Sydney a few years earlier for the design of the sails at the Yulara Tourist Resort near Ayers Rock.

The cable-stayed roof was wholly appropriate to the site, which was generous in size and surrounded by elevated highways. The masts and cables announced themselves to passers-by in their motor cars and also came down to greet pedestrians in front of the building. The halls were staggered on plan to break up the enormity of the façade and to give each hall a separate identity.

The self-weight of the roof is minimised by using a steel structure clad with lightweight profiled steel sheeting supported by cold-formed steel purlins. Indeed this type of roof construction is all-pervasive in Australia, which manufactures acres of superbly coated steel sheet and does not have an unduly

(Above) Darling Harbour Exhibition uses elegant masts to advertise its presence. Note also the curved eaves to reduce wind uplift.

(Overleaf) A hallmark of the building is the consistent use of forged rod ends and stainless steel pins connecting to oversized plates.

severe environment in which cold bridging is a significant factor. A consequence of this light weight is uplift under wind suction, particularly when combined with the internal pressures generated by the large vehicle access openings.

The basic form of a cable-stayed roof is not as competent under upwards as downwards loads and considerable research effort was put into the eaves detailing to reduce the peak wind suction. However the building is situated alongside an arm of the harbour which forms a natural declivity, and is surrounded by ridges and taller buildings which help reduce the wind uplift. Eventually we were persuaded that the upwards wind loads in service were no greater than the self-weight. By elevating the central roof panel relative to the edge, a shallow catenary was formed under upwards loading, which gives sufficient strength for ultimate wind loads when allowing gross deflections.

The roof structure is carefully, even lavishly, detailed and articulated, not only to give texture and quality to the finished product, but also to assist in erection. As with all large structures, the designers must have a clear idea about how they may be erected. A step-by-step erection method was provided to the potential builders with their tender documents, and was followed almost exactly by the successful company.

The completed building makes a memorable contribution to the Sydney cityscape. It emanates a festival or circus flavour that helps to draw crowds not only to exhibitions, but also to the Darling Harbour precinct in which it is situated.

Brisbane Convention and Exhibition Centre

As mentioned above, the design of the Brisbane Centre's roof not only extends the repertoire of roof forms within Australia but is a distinct step forward for the world as a whole. It builds on a recent trend of using steel lattice shell forms but extends this by using an anticlastic, doubly-curved geometry. Singly-curved or synclastic forms require stiffening against buckling and in the case of a barrel vault, also deliver thrust to their abutments. The hyperbolic paraboloid shell used for Brisbane has little propensity to buckle and is equally competent under both upwards and downwards loading. Brisbane is situated on the edge of the tropical cyclone region and the site is relatively exposed, so wind loads were even more dominant than for Sydney. With its particular arrangement of diagonal truss and

(Right) Brisbane Exhibition Centre. The spine truss also houses the main air duct to give uncluttered roof space.

edge framing, the structural system is entirely self-contained and no forces are delivered to the building below, other than the imposed loads.

The roof form is all the more remarkable when considered in the light of the adopted building procurement method and the constraints thereby applied to the designers. The Queensland Government called for design and construct tenders for the building in late 1992. The site completely fills two city blocks and is bounded by roads on all sides. The resulting building could only be an enormous rectangular box. One of the aims of the roof design was to break up this box into more manageable units. Other aims were to evoke memories of the membrane structures deployed during Expo '88, which had occupied the same site, and most importantly, to be unlike the Sydney building.

The split hypar roof was proposed as a means of economically achieving these aims, and after a brief comparison with a more conventional roof, the whole team adopted it with enthusiasm. Innovation, in this sort of environment, is a major act of faith by all parties; there is little scope for significant contingency sums within the intensely competitive bids. The roof was conceived, through a series of meetings, over a one-week period. Conceptual work of this nature requires engineers to think in terms of three-dimensional form. We must draw at the same time as design. And this design requires immediate and continual analysis. So we enter into an evolutionary loop of form-finding and analysis until we achieve the architectural aims and ambitions with an economical structure.

Fortunately, our unique roof concept helped us to win the bid and proceed to schematic design and eventual project construction. Piling commenced four weeks later and the roof steelwork was sent to tender four weeks after that. During this period: the roof was redesigned using conventional beam and column sections instead of the original circular tubes; the bracing pattern was offset to stabilise the I sections about their weak axis at mid-span; connections were detailed using simple, flexible, bolted end plates. This allowed ease of fabrication and ensured that, despite the complexity of the roof's structural action, the erection procedure could be a simple, step-by-step process, with minimal site survey work and no temporary works.

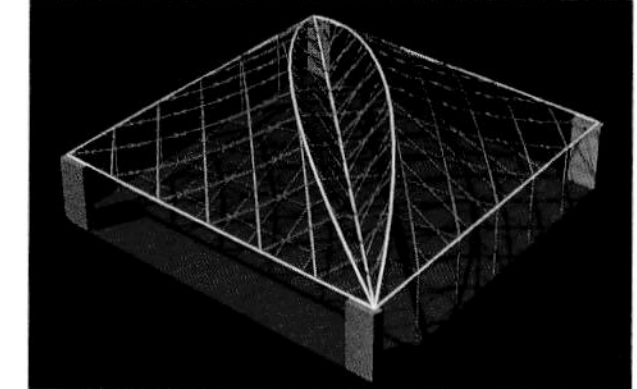

Once all are committed to a certain product, to be built in a given time and for a determined cost, they need to co-operate very closely. Detailed calculation also uncovered the need for more stiffness in the perimeter frame and diagonal brace. In order to provide this at minimal cost, the compression members were filled with concrete and the diagonal tension brace was filled with reinforcing bar running the full 100m from end to end. An independent design check after an intense period of member optimisation, found that no individual member was stressed to less than 90% of its capacity. After the feverish design effort, construction progessed so smoothly as to be almost an anticlimax.

It is undoubtedly the roof form, punctuated by the monumental service shafts which support it, that contributes most to the building's character. The main hall roofs are extended to form the entrance foyers at the front and protect the loading entrances at the rear. They are also repeated over the smaller ballrooms at the rear. They are used as the centre's logo and feature on all publicity material. Viewed across the river from the city, the gentle undulations echo the distant hills and when lit at night act as a beautiful beacon to attract visitors.

(Top left and right) Brisbane Exhibition Centre. The computer image shows roof structural system which was erected in a straightforward premeditated sequence with no adjustment or temporary works.

(Centre) The completed building is over 400m long, but the roof is broken up into more manageable units.

(Left) Two adjacent roofs emerge like enormous whales from the surrounding residential streets.

Perception of large spaces has changed since the days of cathedral building: comfort needs change with activity which is especially important in the design of large spaces like airports. To control comfort and conserve energy, passive environmental design sees enclosures as a microcosm of external climate and employs the explanations of climatology. Big spaces can now perform with no intervention or with minimally active systems, and their performance can be predicted using computer models.

Alistair Guthrie

Designing for climate control in large volumes

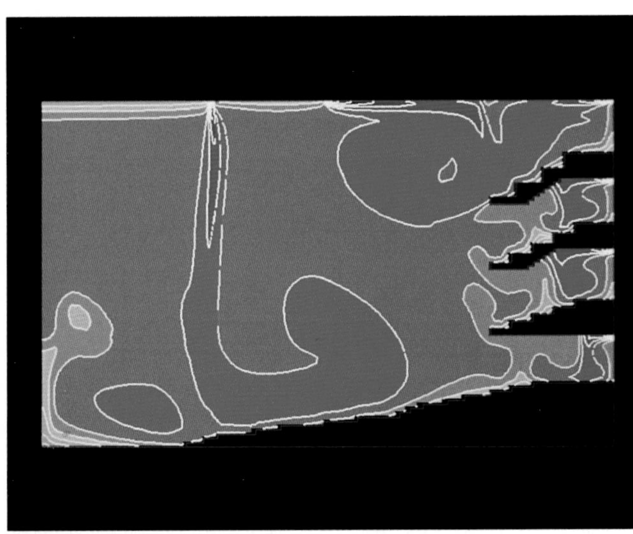

(Above)
Performing Arts Theatre, Escondido, California. A 2-D computer model showing air distribution patterns.

Civilisation has for a long time needed to construct large spaces and enclosures for the purpose of meeting, worship or leisure. These have ranged from the Pantheon in Rome, through the medieval cathedrals of Europe and 19th century train sheds, to present-day atriums, shopping malls, and arenas. In addition, construction techniques have changed from the traditional materials of wood and stone to the new materials of glass and steel. These have both become universal and have tended to usurp the traditional shapes and forms of large enclosures which have not always respected the climatic conditions of the location. In addition to this, the requirements of the spaces have become more demanding. Where people were previously content with an umbrella against the weather, they now perceive the need for a greater degree of protection from it. This now means enclosed spaces with a system controlling temperature, air movement, and possibly humidity. To this must be added the world-wide concern to reduce energy consumption.

The enclosure and the volume

We have come to realise that the most important parameter in the comfort design of these large spaces is air movement. This is not just because of its effect on the local air velocity but because air movement also controls the temperature distribution in the space. Air movement can be generated in two ways; firstly by using temperature differences and secondly by motive force from external wind pressures or by mechanical fans.

Temperature-generated airflows

Temperature-generated air flows are the driving force behind our weather systems and knowledge of these can be helpful in understanding air patterns within enclosed microclimates. Using this knowledge helps us create the best set of internal conditions for comfort with minimal energy consumption. The four most important climate characteristics are: the greenhouse effect; stratification; inversion; and finally moisture formation.

The greenhouse effect describes through an everyday metaphor how the atmospheric barrier of carbon dioxide and ozone reacts in a similar way to a

(Above) Nottingham Concert Hall. Air outlets in the ceiling at stalls and balcony levels of the auditorium were used to create a circular pattern of air movement.

glass enclosure. When the sun shines into a space through glass it heats up the enclosing surfaces. This process changes the short wave radiation from the sun into long wave radiation. Glass is opaque to that kind of radiation and it cannot escape; hence, under these conditions, a space will continue to heat up and will only lose heat by conduction or ventilation. This absorption of heat into the enclosed volume and its surrounding surfaces is beneficial during periods of the year when heating is required to maintain comfort, but during the summer this effect regularly gives problems with overheating.

Air which is warmed either directly by the sun, by contact with the earth, or by mechanical means becomes less dense and rises, causing the denser, colder air above it to replace it. In the external environment the density changes give rise to low and high pressure areas, which are the driving forces behind the movements of air and change in temperatures of our weather system. In large enclosed volumes, stratification creates a natural temperature gradient in the space with cooler air at low level and warm air at high level. If the bottom of the space is the occupied volume, the stratification will be beneficial during hot weather when the coolest air will be in the occupied zone. However, during the heating season it is difficult to maintain the required comfort conditions at low level as the heat escapes to the top of the space.

Inversion naturally occurs as a result of differences in buoyancy between cold and warm air masses. A cold mass of air accumulates over a warm mass of air. The two air masses then reach an unstable position and invert causing a rapid flow of air as the cold air replaces the warm at low level. In large enclosed volumes this behaviour leads to unstable air movements which from time to time will be perceived as draughts or more beneficially as cooling breezes.

As moist warm air cools, its ability to hold moisture decreases (the concept of relative humidity), until it reaches saturation. In the natural environment, this leads to precipitation, rain or snow. This situation can occur in an enclosed volume. The most striking example was the formation of vapour clouds at the top of the Saturn V rocket assembly shed. Normally this situation appears as condensation on the surface of the enclosed volume. This is because the surface is itself cold, the air adjacent to it is locally cooled, and when it becomes saturated it deposits moisture. This is a particular problem in large assembly spaces with lightweight roofs. The moisture concentration can be high due to the number of people, but the inner surfaces of materials such as glass or fabric can be cold. In these circumstances moisture can drip from the roof into the space almost like rain.

These climatic principles have been applied to a number of projects requiring the comfort control of large volumes, and in each case research was carried out to determine the appropriateness of the solutions and the physical parameters of the design. Some examples will illustrate.

Temperature-generated airflows in theatres and performance spaces

The auditorium and the stage house can be considered as two connecting volumes. The full auditorium has a large heat load from the audience at stalls and balcony levels and also a large heat load from lighting at high level. The stage house has a large lighting load above the

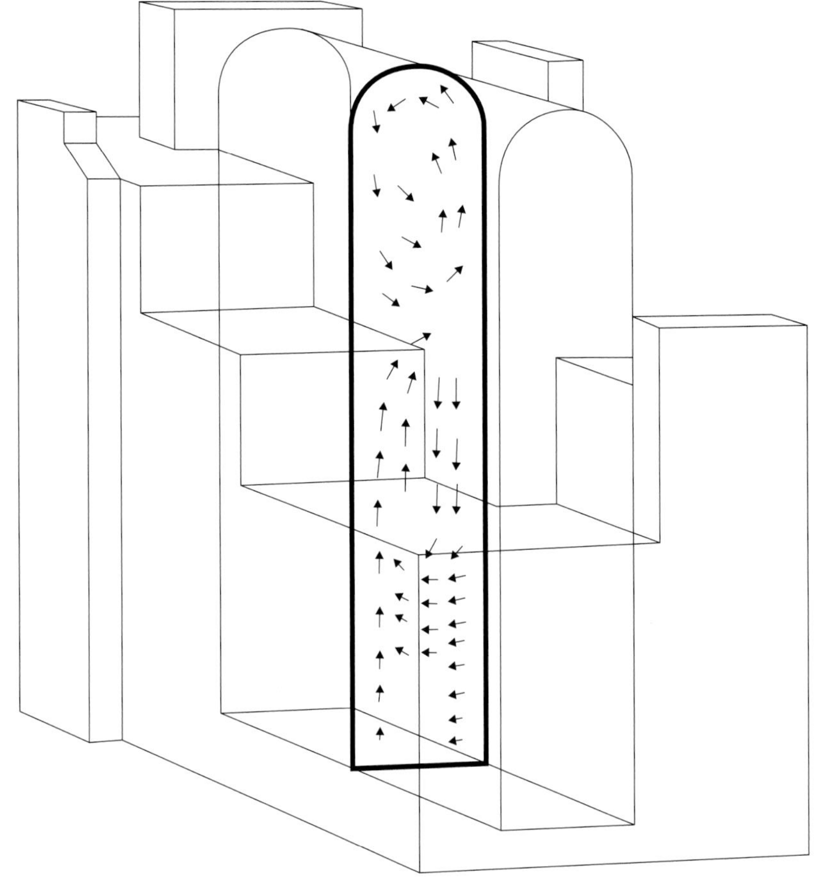

(Right) Lloyds of London. A section through the atrium showing the computer-simulated air movement on a cold day. Since the downdraft did not reach the occupied areas of the atrium, it was not considered to be a comfort problem and a system of compensation was not required.

stage. Studies of theatres show that under normal conditions with the curtain closed, the low-level air is heated and rises at the back of the auditorium, adjacent to the balconies, being the zone of greatest audience density. This in turn displaces some cooler high-level air, generating a circular movement with cooler air dropping from high level at the front of the stalls. If these air patterns are properly controlled, sufficient beneficial air circulation takes place; however, if not controlled, draughts will become a nuisance. In the design of Nottingham Concert Hall in 1978, these patterns were recognised and it was decided to introduce air for ventilation and temperature control into the space at high level by blowing down and forward mixing with the natural air circulation currents. Under most conditions this works very well but as the load in the space increases during some concerts, the body of warm air at low level fights the natural convection trapping the cool conditioned air at high level. After a period of time, the mass of the heavier cool air becomes unstable, overcoming the buoyancy forces and exchanging places with the warm air. This caused a large movement of air which was felt as a draught in the stalls. Thus the classic inversion pattern which causes winds in our weather system was repeated in the enclosed microsystem. This type of air pattern is unstable and its frequency and magnitude cannot be calculated using traditional steady state methods. This led to a search to find better ways to calculate these temperature-generated effects and also to investigate other solutions.

One tool increasingly becoming available was computational fluid dynamics (CFD) software. This had been used to model temperature and velocity distribution in industrial applications, but was seen as a potential tool to predict and model time-dependent distribution in enclosed spaces. One of the first uses was to model the tall central glazed atrium of the Lloyds Building. Investigation was required to determine whether cold air from the glass roof and walls would drop onto the open dealing floor at the base of the atrium, or whether the warm air rising from the open floors at the bottom would cause sufficient stratification to counteract it. A (CFD) model showed that although cold air did drop, it met the rising warm air above the occupied levels, setting up a circulation current which did not affect the temperature or air movement in the occupied spaces. Hence no countermeasures were designed and the space performs as predicted.

CFD techniques are valuable in determining temperature-generated air flows, but are best suited for reasonably simple 3-D shapes or 2-D sections, and are difficult to use in complex auditoria where no section is representative and the boundary conditions are difficult to define. CFD relies on reaching a stable converged condition after a large number of iterations. Relatively long-term instabilities like inversion, if properly modelled, will manifest themselves as non-convergent models. For this reason alternative techniques have been used alongside CFD to determine air flow in auditoria.

The replacement opera house at Glyndebourne was the first of a number of large auditoria spaces to use the principle of air displacement and to refine its application. Air was supplied at low level at low velocity close to the audience. As the air passes over the audience it heats up and its natural buoyancy takes it to the roof. Here it picks up additional heat from the lights and stage equipment before either being exhausted or recirculated. This principle gives two significant advantages. Firstly, the air is evenly distributed and driven by natural buoyancy forces avoiding the local areas of large air movements sometimes associated with other systems. Secondly, since the air is supplied close to the audience, it is only required to absorb the heat generated by them in order to maintain comfort. Thus although air is being introduced into the space, it is being done in such a way as to enhance, and not upset, natural stratification. This means that the air is just 2°C–3°C cooler than comfort

conditions and as such can be supplied directly from outside without cooling almost every night of the year, considerably reducing the energy bill. At Glyndebourne, the air is supplied from an underground plenum through a very low velocity diffuser which forms the pedestal of each seat.

Checks were made to establish whether the air would remain stratified under all conditions particularly when the curtain opened, to connect the high volume of the stage house with the auditorium. Tests were done at Cambridge University laboratories on a saline model. The differing densities of the air are modelled using correspondingly different salt densities which are dyed different colours. This method gives a good visual record of the flow of the air starting from zero conditions. However, it is difficult to model steady-state conditions and to predict possible instabilities, since after a while the fluids become mixed. After the building was completed, smoke tests and measurements were conducted in the building to measure temperatures in the occupied areas, and also to determine the effectiveness of the stratification of the air stream. Heat sources (light bulbs) were placed on the seats to simulate people. The tests demonstrated that stratification was achieved and that comfort could be maintained in the various occupied areas of the auditorium.

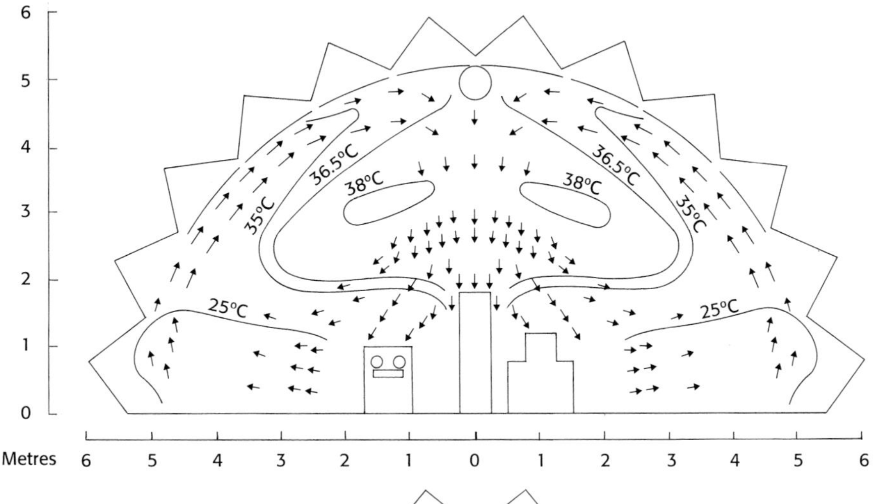

(Top left) Glydebourne Opera House. A saline model was used to predict air movement patterns.

(Top right) Air is introduced through a specially designed outlet in the base of each seat.

(Left and below) IBM Travelling Exhibition. Predictions of air stratification helped generate the building radius and environmental control system.

Stratification in glass enclosures

Natural stratification of the air within large spaces has, historically, been used to improve comfort. Tall spaces in hot

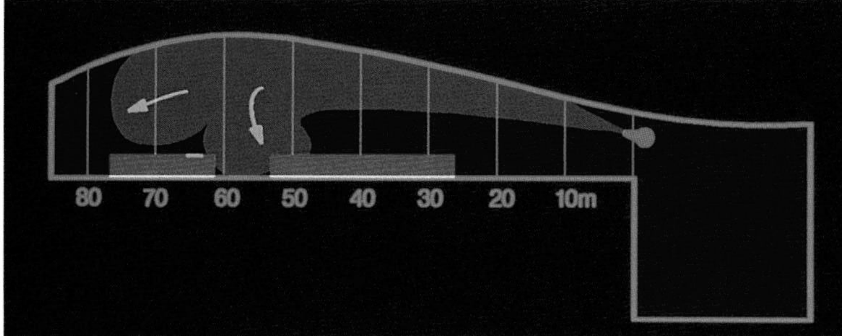

(This page and opposite) Kansai International Airport. Macro climatic control of the large spaces is achieved using large air nozzles. These blow the air across the space forming air circulation elements to provide clean air, temperature control and comfortable air movement.

climates perform well since the residual cold air of the building remains at low level, in the occupied zone, whilst heat accumulates at the top. This principal was used in the design of the IBM Travelling Exhibition. This half-cylinder of transparent polycarbonate housing the latest computer technology was to be located in different climatic conditions throughout Europe during the course of its travels. It was decided to introduce heated or cooled air at floor level and attempt to achieve a stratified environment so that the computers and people could be kept in comfort without too much concern for the temperature at the top of the space. A computer model was generated such that under the different climatic conditions the temperature at different heights within the semi-circular cross section could be predicted. Under the extreme summer conditions it was found that although stratification was achieved, the temperature at head height was too great. Further analysis showed that a relatively small increase in the radius of the cylinder was sufficient to raise the critical stratification temperature to a height of 2.4m which is above the occupied zone. In this case the building environmental analysis was used to feed back into the building shape and size.

A similar design was used for a VIP conference room on top of the Lingotto Building in Turin, Italy. The space consists of a glass hemisphere which acts as a greenhouse in summer. To counteract this, conditioned air is introduced into the base of the sphere.

In this project, although stratification could be achieved in the occupied zone, detailed computer analysis of comfort in various seasons showed that the radiant effect of the glass dome could not be offset by the air temperature. To solve this problem a partial partition of the space was achieved with a large umbrella separating the occupied zone from the top half of the space, enhancing the stratification and providing shade.

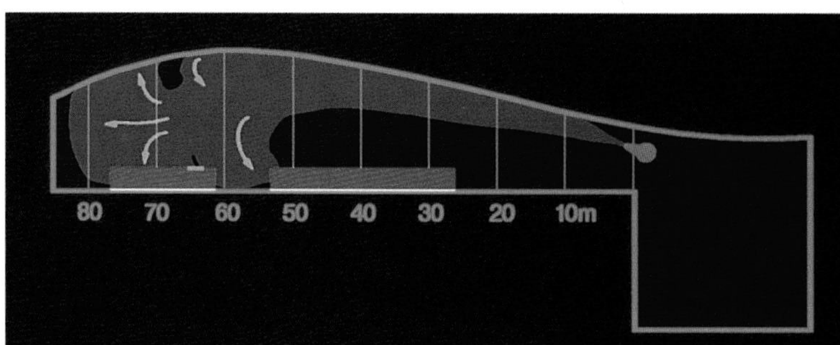

Airflows generated by motive forces

In some large spaces air movement has been used to control the temperature distribution. Controlled air movement can either be from forced mechanical fan systems or by the use of wind and wind pressure in the space. Heated or cooled air is blown into a space, mixing with or inducing room air to control patterns of air movement. These air patterns need to be controlled in such a way as to achieve the correct mixture of forced air and giving the right temperature for comfort which may run counter to natural air movements.

At Kansai International Airport the concourse is a large space 300m long, and 80m wide, with a covered roof which at its maximum is 19m high. An airport concourse needs to be sealed against dust, pollution and noise from the external environment, hence it was necessary to provide clean, heated or cooled outside air into the space. The concept was to provide a general level of conditioning to the whole space, the macro environment, and local additional cooling to areas such as check-in desks with a large number of people and hence larger cooling loads. The macro system consists of 19 large air nozzles

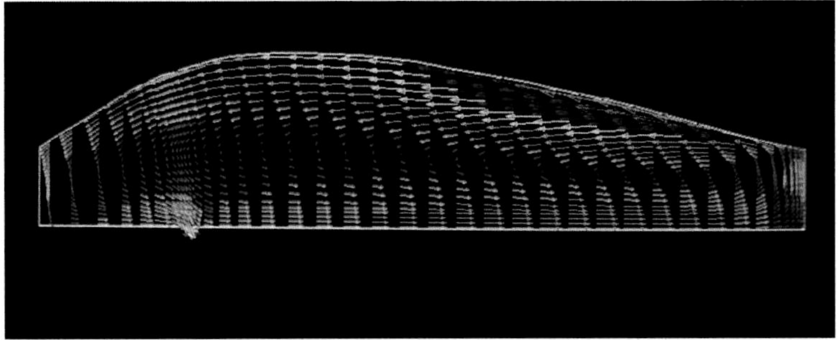

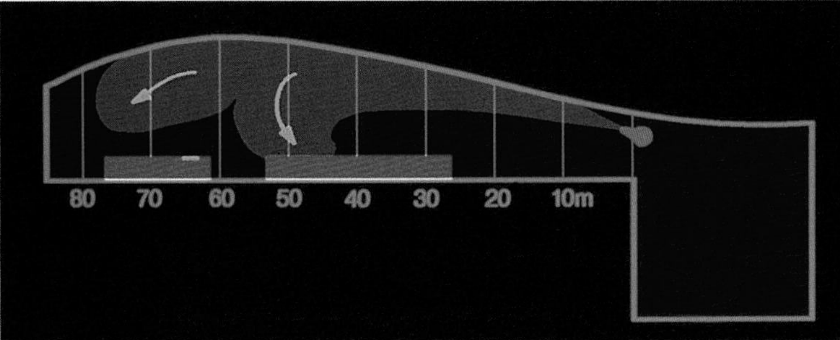

which blow across the 80m width of the space. The natural trajectory of the airstream generates a curve which gives shape to stretched fabric panels which hang below the roof. The air adheres to the fabric, entrapping room air, expanding and losing velocity as it crosses the space. This air stream contains all the clean and filtered new air to the space, setting up a circular air motion in the concourse and achieving good overall mixing of the air, as well as sufficient air movement to ensure that the space is lively without causing a draught. The aim was to give the feeling of a pleasant gentle breeze.

Initially, the analysis of the performance of the air nozzles used a three-dimensional CFD model to predict the air movement caused by the jet. At that time this was the world's largest application of CFD in buildings. A lot of adjustments were required to get the model to converge, but when this was achieved, the parameters for the performance of the jet, temperature exit velocity, and mass flow rate were established for different boundary conditions. The aim was to control the comfort parameters of temperature and air movement in the zone up to 2m above floor level using this large circulation pattern.

It was found that the most critical parameter in achieving this performance was the angle and exit velocity of the jet. A 1/10th scale model was built to refine the performance of the jet under various temperature, conditions and provide detailed measurements of velocity, temperature and air movement direction within the space. From these measurements comfort indices were calculated at regular intervals across the space. It was also possible to use smoke to visualise the performance of the jet. These tests and adjustments ensured that good air distribution and hence temperature was achieved in the macro zone without discomfort for the different conditions which occur throughout the year.

The goal of having one jet angle and velocity for all heating and cooling conditions was made possible by the use of a number of climatic principles. During the cold periods of the year the air is heated above space temperatures. The natural buoyancy of the airstream keeps the jet up until adequate mixing has taken place. This means that the air movement at concourse level is kept low which is necessary to maintain comfort under heating conditions. During the cooling season the supply air is cooler than the surrounding space. This shortens the throw of the jet and causes it to drop more quickly, mixing with the warm air rising from the concourse. This creates greater air movement at low level. Comfort is thus achieved with slightly higher concourse temperatures, coupled with increased air movement requiring less cooling energy.

A second example, Kanak Cultural Centre, Noumea, New Caledonia, uses wind as the source of motive force to achieve comfort by controlling temperature and air movement. Wind is an unreliable source, so a building needs to be designed to operate with very little wind, but also with winds as much as 100 times greater. The idea behind this scheme was to make use of the site on a peninsula, facing the sea, channelling the prevailing winds into the building to create comfort by air movement and temperature control. The design is a series of very large hut-type structures based on local traditions. After experimenting with a number of ideas using both numerical analysis and a simple physical model, a building shape was developed with four stages of wind control. In the first stage when the wind speed is greater than 8m/s the wind is too strong to allow it to blow directly into the building. The wind pressure on the chimney structures is used to induce air through the building and up the chimney. In the second stage (when the wind speed is lower) the wind is channelled into the building by its shape, and the amount of air entering is controlled through adjustable openings in the front of the structure. During the third stage, there is no wind; the chimney works by natural convection. The fourth condition is for the occasion of wind from the opposite direction when the chimney acts as a scoop channelling the wind down into the building.

Comfort conditions were set for the building such that they should not be exceeded for more than 5% of the time in any month. To determine this a scale model was built, and the various stages of wind speed and direction were simulated in a wind tunnel. The ratio of outside wind to internal air velocity was recorded for varying wind directions and at different points in the model. This data was then combined with a weather tape of data obtained from a nearby site. Hour-by-hour conditions were computed for varying wind speed, direction, temperature and humidity. These figures were corrected for the actual site and the ratios obtained from the wind tunnel tests to obtain the velocity at several points within the building. These were combined with the computed internal temperatures and humidities to arrive, using comfort equations, at hourly comfort conditions in the space. These were then statistically analysed to show that, for the weather year used, the overall comfort criteria were not exceeded. This extensive project-based research was necessary to take the design away from the traditional approach of providing an air-conditioned box.

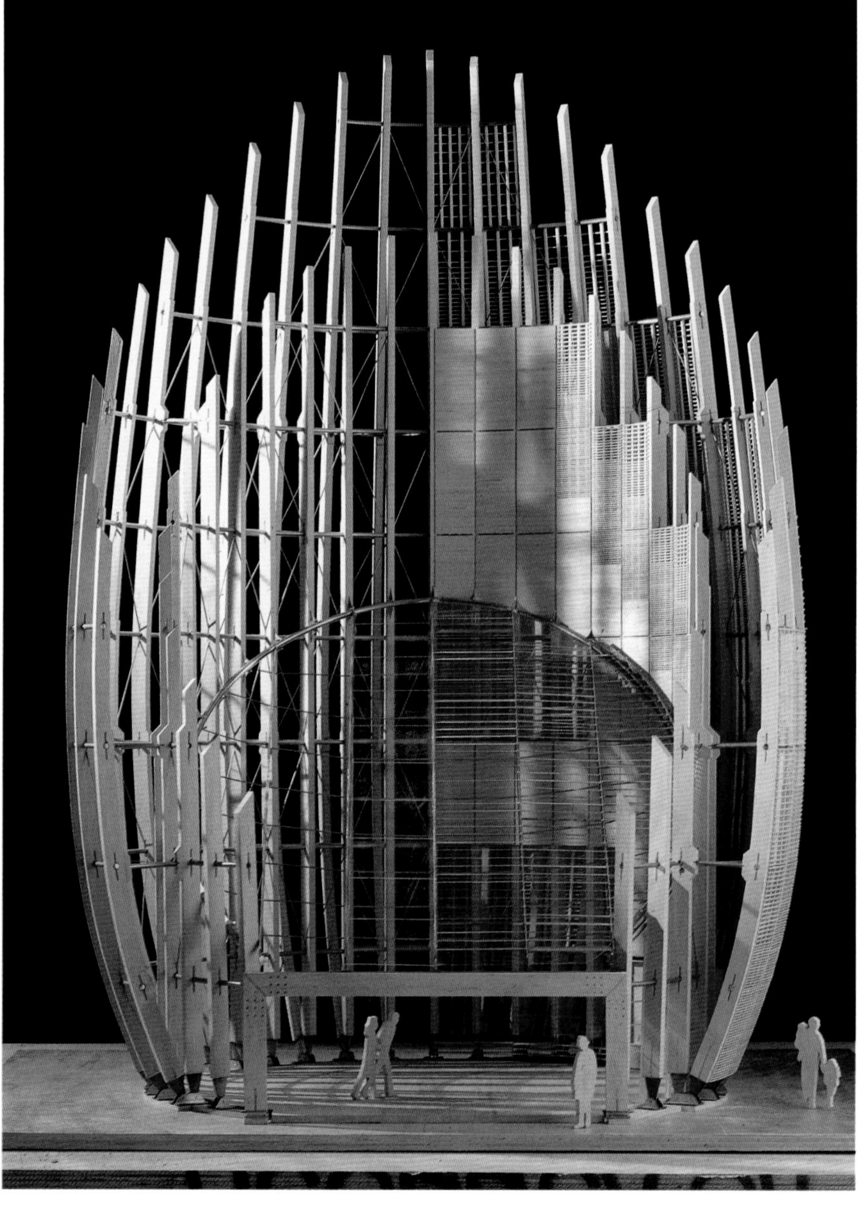

(Right) A model of one of the structures for the Kanak Cultural Centre in Noumea, New Caledonia. The structure acts to trap in sea breezes and channel them into the building giving comfortable conditions for most of the year. The structure acts in different ways according to the strength and duration of the wind.

(Opposite) Each structure houses a different function of the centre and has different environmental needs. The structures face into the prevailing wind.

In many projects it will be necessary to combine the effectiveness of temperature-driven airflow with some forced ventilation. One such project is the feasibility study for a large enclosure to house a living tropical rainforest exhibition in Osaka, Japan, which illustrates the concept of two climatic zones in one space. The upper climatic zone is naturally ventilated, relying principally on convection-driven airflow whilst the lower is controlled for both temperature and humidity. The exhibit structure is an 80m diameter glass dome. Inside the dome, conditions need to be maintained to simulate a rainforest, since living plants and animals will be exhibited. The enclosure must be glass to allow enough light for the trees and plants to grow, but this tends to produce excessive temperatures in summer and requires a large heat input in winter. An additional problem is that due to the high internal humidity, condensation could easily form on the inside of the glass during cold periods of the year. This condensation reduces light transmission, increases maintenance, and prevents visual connection with outside. The proposed solution is for two skins of glass separated by the structural frame, the

outer skin to comprise sealed double units, whilst the inner skin is a single sheet of glass forming a 2m wide cavity between. The structure is thus also protected from the birds and animals inside the enclosure. This cavity can be closed and heated in winter reducing the risk of condensation, and opened top and bottom in summer to reduce unwanted heat build-up. The volume itself is divided into two spaces, the space below the tree canopy and the space above it. In a natural rainforest these two spaces have a very different climate condition, the lower space being hot, humid, and relatively free from air movement, whilst the top space is drier and has air movement generated by the rising air from the forest, causing a breeze in the air above the canopy.

A preliminary analysis has been carried out to determine whether it is possible to reproduce these conditions in the enclosed volume. During the hot humid summer in Osaka, the outside unconditioned air is allowed to enter the top half of the space from the cavity. It is heated up by the sun shining through the glass (greenhouse effect) and passes out through the top. This means that the top half of the space can be kept at reasonable temperatures by natural convection alone. The bottom half of the space is heated or cooled to maintain the required temperatures whilst humidity is maintained by water from misting nozzles located throughout the volume. These conditions have been modelled using CFD to investigate the distribution of air, temperature and humidity and how much migration occurs between the two spaces. The results show that minimum migration will occur although it has to be acknowledged that assumptions have had to be made on how to model a green canopy. This result is extremely important as part of a feasibility study since it reduces the energy needed to maintain the space to less than half required for a similar enclosure treated as one space. It will also reduce the capital cost since the size of fans and heating and cooling equipment will be reduced.

Whether they are atria in office buildings, malls in shopping developments or sports stadia, there exists an increasing demand to enclose very large indoor spaces. Most of us are demanding more comfort from our work and leisure spaces, and simultaneously we expect structures to provide this comfort for increasingly lower running costs. We are beginning to understand how these spaces can be made to work by harnessing natural forces within large volumes, and we are increasingly aware of the dynamics of the external climate and the microclimate buildings create inside. By designing within these developing parameters, we can hope to reduce the massive energy wastage that has sometimes characterised this type of building enabling us to create the exciting new spaces in the future. I am convinced we will see more inside/outside spaces including the covering of whole streets and perhaps, one day, Buckminster Fuller's dream of a city under glass!

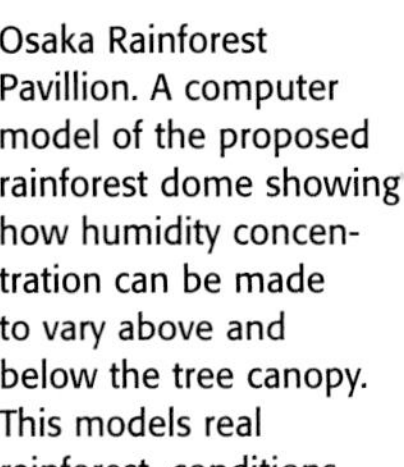

Osaka Rainforest Pavillion. A computer model of the proposed rainforest dome showing how humidity concentration can be made to vary above and below the tree canopy. This models real rainforest conditions.

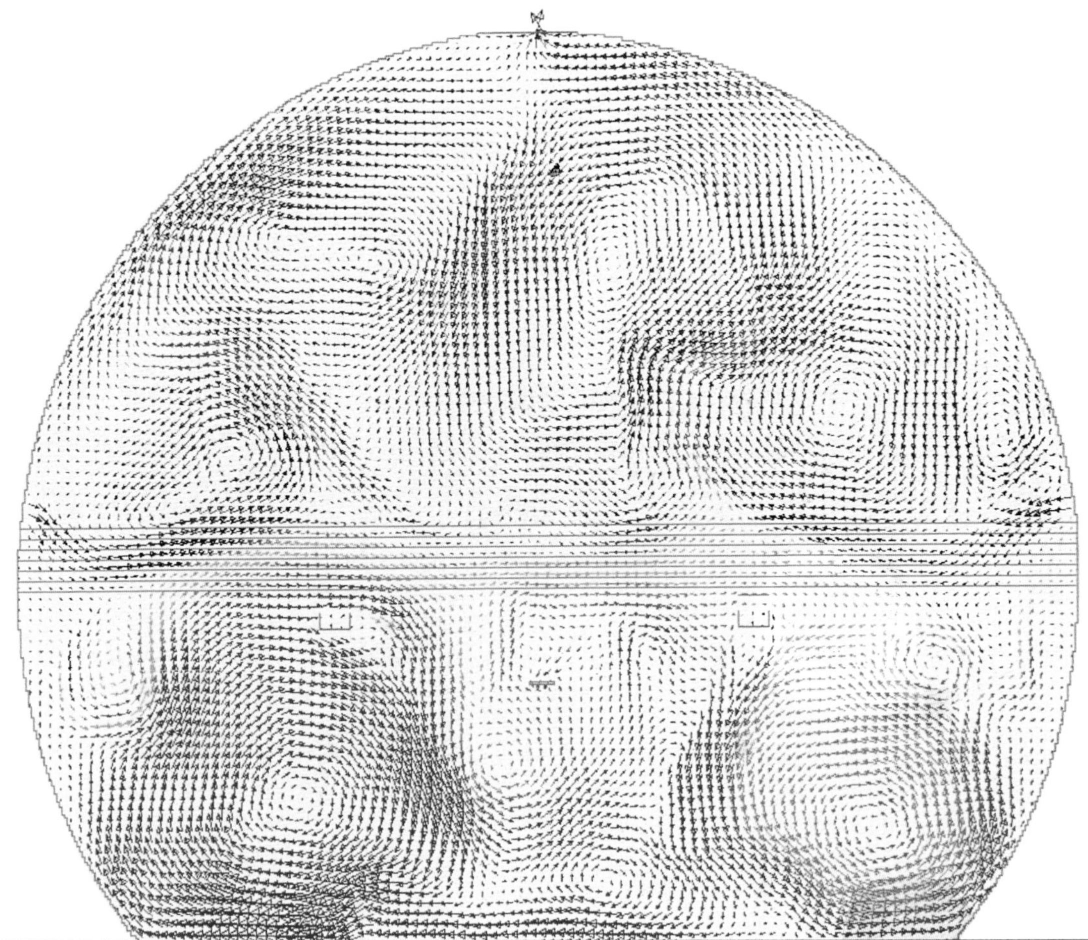

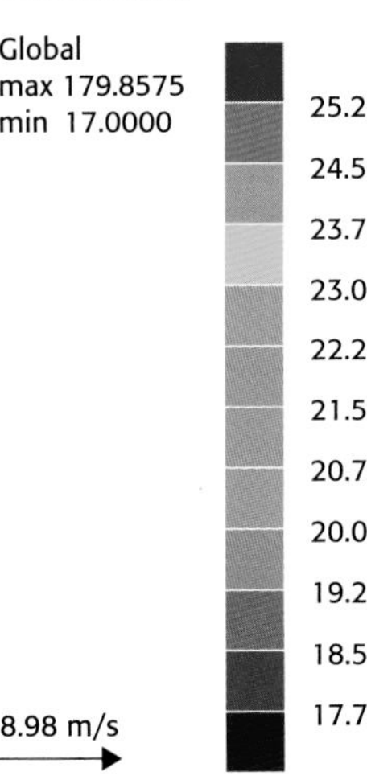

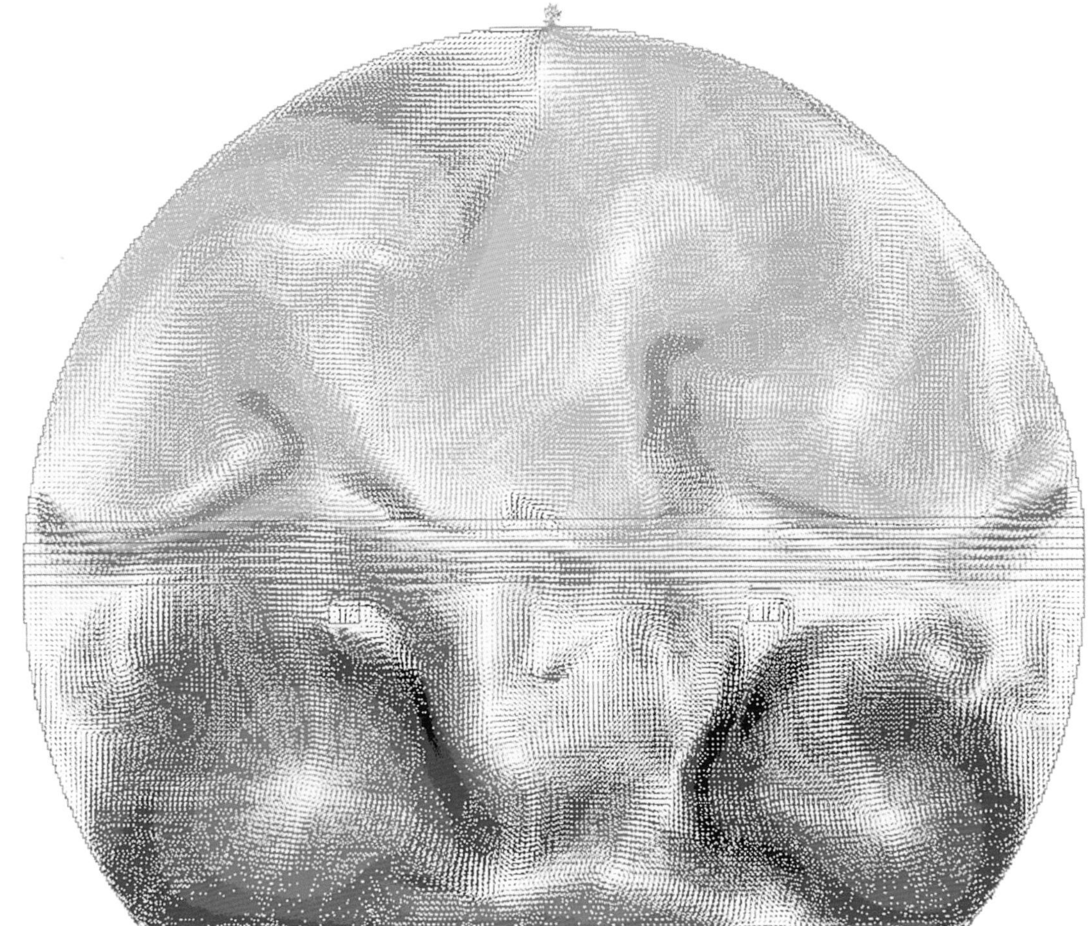

A computer model illustrating the temperature differences above and below the tree canopy in a simulated rainforest. Temperature control of the space above the canopy is achieved by natural ventilation.

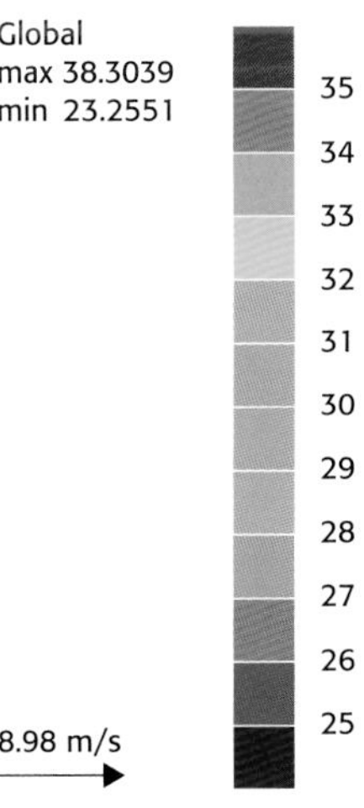

Services are expected to be neither seen nor heard, an objective deriving perhaps from that most highly serviced space, the Victorian house full of servants. Now building services present particular problems as expectation has increased of the invisible power sources which they supply. At one extreme office services must keep pace with miniaturisation; at another concert halls can be constructed over railway lines with no transfer of vibration. In both cases cultural expectations have to be weighed against the knowledge that the future will require change. Increasingly the building services engineer has to look at design as if she or he will be in charge of the building through all the vicissitudes of its life.

Tony Marriott

Neither seen nor heard

(Right) UCLA Biochemistry Laboratory. The familiar presence of well-ordered services is highly visible.

Some of the most highly-serviced spaces were designed and built a century or more ago. The servicing in those days was not discernible from drawings, which show only the fixed elements of the building, but from novels and diaries of the time. These show that there was a continual stream of people taking coal to fireplaces and laying the fires, riddling the fireplaces, removing ash, lighting, snuffing, and replacing candles, boiling and carrying pathetic quantities of hot water for washing, shaving, and so on. The energy consumption of the servants in any self-respecting household was about half a kilowatt (several times that of a central heating system pump) and it was provided by fuel (food) of high cost per Joule. The servicing of the installations was pretty disruptive too, having to be done from within the occupied spaces – and the servants occupied a rather greater proportion of the building than does a modern air-conditioning plant. The results that such systems could produce were limited and so were the expectations of the owners.

Now, the servicing requirement of almost every kind of space has increased enormously, as has the capability and reliability of the installations. The disruption expected during maintenance has shrunk to the point where most of the occupants of buildings do not even realise that maintenance is needed or done. Sadly, that applies to some owners as well, but that is a different story. It is almost a truism that building services, more disciplined than Victorian children, are expected to be neither seen nor heard, though they have to work as hard as any chimney sweep ever did.

One major task of the designer of highly-serviced spaces, therefore, is to preserve the serenity of the visible space, even though intense activity may be taking place just out of sight; very much like a swan swimming upstream. When the services are concealed in ducts, behind ceilings, within floors, or in plantrooms, the parallel would be the swan swimming in murky water, and the activity is covert. It is possible in these circumstances to get away with a certain amount of visual disorder, while still ensuring functional order and thermal elegance. However, when the water is clear the swan, if it is not to lose its reputation for serenity, must ensure rigorous order below the water line as well as above. Similarly, when the engineering becomes exposed to view, or overt, in addition to preserving the functional quality of the installation, it is important to control the appearance.

This is often seen initially by the engineer as an additional problem, but the inexorable logic of proper visual order usually turns out to have considerable advantages for the user and maintainer of the installations. It often helps the installer as well. Great adaptability is usually expected of building services during the life of the building, and the greater the clarity with which the logic of the servicing is expressed, the more natural it is to follow that logic when extensions or alterations are made. That is where there is an advantage to the user, in that the systems after a number of years should preserve the economy of layout and the carefully thought-out logic of the initial installation. A further advantage is that the house engineers, being human, will put more care into an installation of which they can be proud than into one which looks like an afterthought. These engineers put great effort into preserving the safety, health, and comfort of the occupants

LAB COLD WATER
LAB HOT WATER

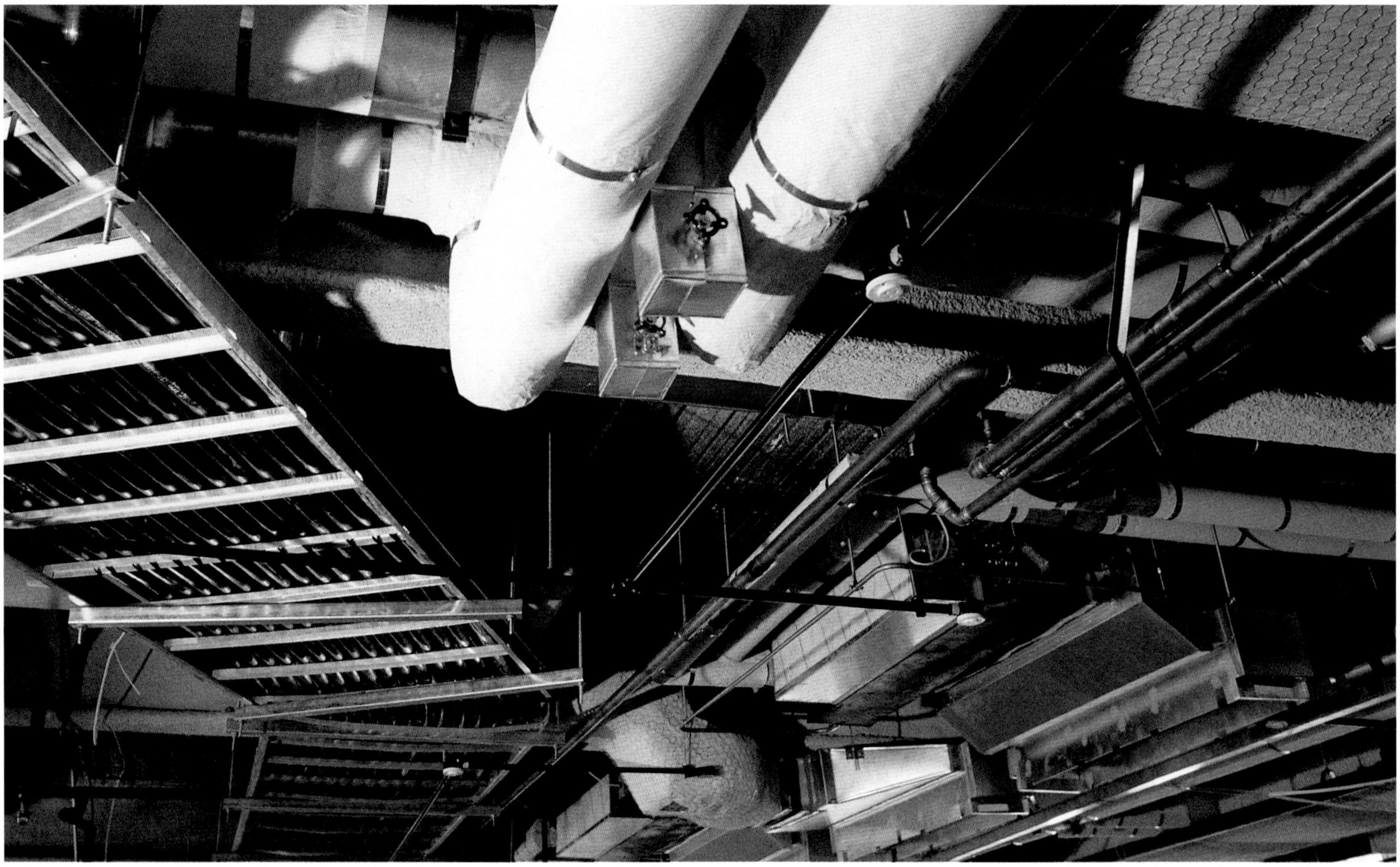

(Opposite page, top and bottom) No.1 Finsbury Avenue. Above and below the ceiling. Whether or not it is visually ordered, we are still looking for the new technologies which will eliminate this cabling (right).

and in providing the systems they need to do their work satisfactorily. It is a great pity that so few people who work in commercial or industrial buildings ever go to see the installations or the people who ensure their continued efficient working environment, but in many instances the systems are so well-integrated into the building that nobody realises they are there. Rather like the music for a film, they are only noticed when things go wrong.

To take another analogy, that of a transport system (and most services installations are transporting energy or fluids into, around, or out of a building), the first step towards keeping the task within manageable bounds is to minimise the load to be transported. The second is to organise things so that the load is transported through the shortest distance, and the third is to ensure that it is transported in the most appropriate containers. The fourth is to organise as steady a demand as possible to maximise the load factor of the systems.

The designer of the services for any space should pay heed to those principles, and as the intensity of the service requirements increases it becomes ever more important to do so. Minimising the peak load and spreading the demand through the day by maximising thermal mass, maximising the use of 'free' sources of warmth or coolth like the diurnal temperature cycles, can accomplish much. So can the careful use of daylight, both for task visibility and also for modelling interesting spaces, but there will still be a need for mechanical, electrical, and communication systems to almost any occupied space, whether the occupants realise it or not.

Most people know something of building services for offices, even though it may not be conscious knowledge. Few will have had the opportunity to look above false ceilings or under raised floors. The illustrations of a dealers' room in 1 Finsbury Avenue, and those showing what lies above the false ceiling of the dealers' rooms, make the point that there is an intensity of life support systems which are not (and need not be) discernible by the inhabitants while they are at work. Dealers' rooms need, to be fair, rather extreme examples of office services, and when that building was designed, dealers' desks were just a twinkle in someone's eye. It is some achievement that the building operates very satisfactorily for a purpose for which it was not designed: an extreme case of adaptability.

However, the communications requirement of many more usual types of working area is surprisingly intense. There has been the promise for many years of new technology which will make present cabling look dated, but it is not here yet. Even when the universal slim cable or fibre becomes common, it may be short-sighted to remove the apparently redundant servicing space, though that view will, no doubt, be regarded as reactionary at that time. It is worth remembering that as 'the electronic office' swept through our cities in the 1980s, the change in office technology was more easy to accommodate in older buildings than in newer ones because there was just that little extra space. When the buildings were designed, the space was there for other reasons, generally relating to the limitations of the services installations at that time, but in many cases that little extra space

Isolation undercroft

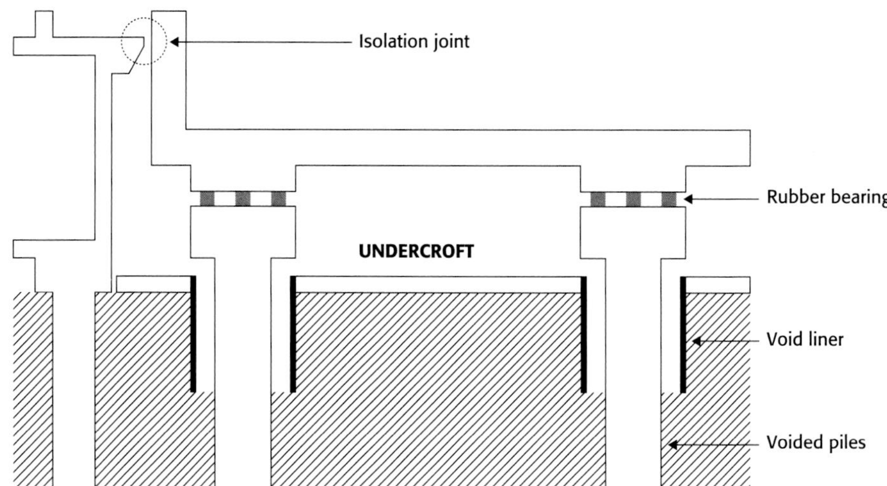

(Right and below) Birmingham Symphony Hall. Illustrating the location of the Hall near railway vibration and the means of isolation.

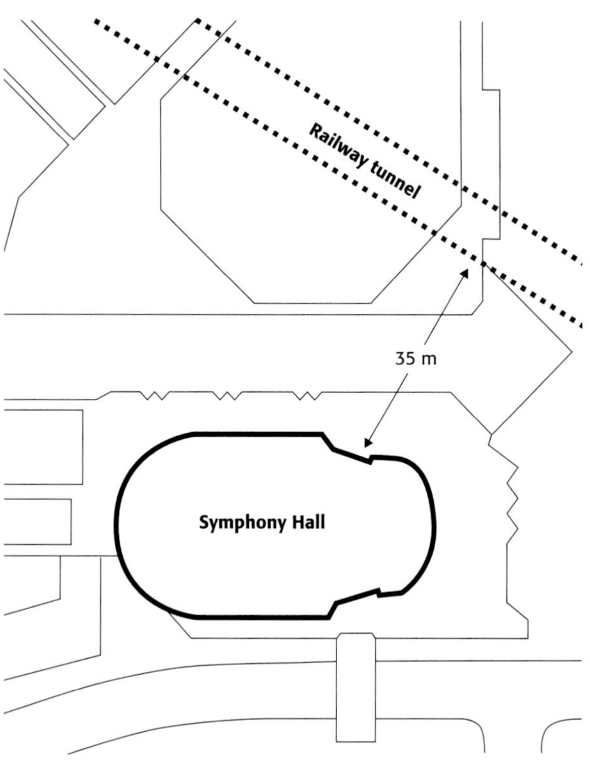

allowed buildings to be modernised rather than replaced. To allow no space for the future suggests a confidence that the miniaturisation of services will proceed faster than the increase in demand for increased capability. History indicates that this hope is unlikely to be fulfilled; indeed, anyone who made that assumption at any time in the past would have been wrong! Even if the cables and ducts do somehow become smaller in the future, the maintenance personnel will not, and their performance will be impaired if insufficient access space is allowed.

The real test of a good building is that of time. It is almost certain that within 10 years of moving into a building, the firm will be doing something different from what is was doing when it moved in, or will at least be doing the same thing differently. While nobody expects to be able to move floor slabs or columns about to accommodate a new business method, adaptation is an everyday story of building services folk, and the facility to change should be part of the initial design thinking. Many clients will be responsible for specifying one building in their working lives, and will therefore have little or no experience of what to ask for, particularly when it comes to describing a 'reasonable' allowance for the future. In a building services designer's armoury, analysis and calculation tools are vital for success, but the ability to design for change and adaptation, stemming from an appreciation born of hard-won experience of how people will use, misuse, abuse, and alter their buildings, is crucial.

More expert and experienced clients present a challenge of a different kind. There are examples in all the building services disciplines, but the one which is probably least often thought about is noise or the relative absence of it. There are few better examples than concert halls: the acoustic environment is not as extreme as radio broadcast studios, but is also achievable only with great care. The Birmingham Symphony Hall is a good example. It may not feel like a heavily serviced space to the audience or even the musicians, but that is just because it is a very good example of covert engineering. Built almost over a railway line, it presented a challenge in noise and vibration isolation the Hall where the delivery to the listener of the quality of 'the silence between the notes' was considered as important as that of the sounds between the silences.

Although the railway was a primary problem, solved by placing the concert hall on rubber 'springs', surrounding it with a gap and taking many other, more subtle precautions, the problem remained of services passing into or out of the hall. They all had to cross the gap and they all had the capability of transmitting noise and/or vibration into the hall. It was essential that services entered the hall to provide ventilation, heating, cooling, lighting, power, communications, security, fire alarms, and so on, and it was equally essential that no noise or vibration crossed the gap with them. A combination of springs, inertia plates, flexible pipe sections, and changes of direction were needed to provide the required separation and attenuation.

The question of the appropriateness of building services depends not only on the technical demands of the spaces to be served, but also on culture and climate. Although these considerations are the correct starting point for many aspects of design, and help to make similarities and differences logical, they are not always the prime considerations. For applications like laboratories, despite considerable differences in climate the paramount technical, regulatory, and safety requirements of the business dictate the nature of the solution. The American examples are interesting because they show how two applications, similar technically, can look so different, depending on the degree to which the structure and services are exposed. Matters of aesthetics still apply, and all the functional requirements must be satisfied in either case. What goes on behind the ceiling is just not seen; another case of swans swimming in clear or opaque water.

The UK examples are very much constrained by regulations, as they are for higher hazard use. Industrial researchers have to deal every day with hazards to which students are not allowed access. Again, what is needed for the researchers is in principle the same, but the degree to which they need to be given protection by building services systems from the hazards of their work is the determining factor. Although these are pharmaceutical laboratories, the organisation of the building and the services would be almost identical for many industrial processes where similarly hazardous

(Right) Birmingham Symphony Hall.

Research facility. (Above)Services which will be concealed by a smooth ceiling.

(Right and opposite) Research facility. Two areas from the same project illustrating two distinct modes – well-ordered visible services and well-ordered invisible services.

materials, gases and so on are used. From our experience in designing many highly serviced spaces and helping the owners for years afterwards, some principles emerge. Obviously, no subject of this complexity can be contained in a few rules of thumb, but the following principles apply.

Firstly, one should try to turn a potentially highly-serviced space into a less highly-serviced one by using all the natural assistance from the structure, weather, and anything else that can contribute passively to the result. Remember the old saying 'what's not there can't go wrong; what's there will go wrong one day'. It is obvious, though sometimes forgotten, that natural forces or passive energy measures are natural and therefore uncontrollable, in the true sense of the word. Like actors working with children and animals, the outcome is never completely predictable, but the reward is worth the effort and occasional mishap.

Secondly, keep some manoeuvring space – it may appear wasteful initially, but one day your client or whoever comes after him or her will thank you. Remember, a good building is one which is capable of sustaining activities which had not been invented when the building was designed; the others form the next decade's workload for the designer by rapidly becoming difficult to use, being pulled down, and replaced.

Thirdly, the designer should try to think of himself as a new operator and maintainer of the building, on the receiving end of all the 'customer complaints'. Thinking out a design from that standpoint will prompt a whole new set of questions to improve the brief. There is no guarantee of a perfect solution, but it is more likely if it is in response to the perfect statement of the problem.

Working within a controlled and planned economy in a transition towards a more industrialised form requires engineering thinking which uses the materials to hand. Thus in Zimbabwe, structures minimise steel and maximise local materials, and it is necessary to think about buildings which of necessity use passive environmental controls rather than air-conditioning.

Lotte Reimer and Ignatius Dube

Relevant technology in Africa

Or, to be precise: relevant technology in Zimbabwe, although it would probably be relevant for other parts of Africa, too. 'Relevant technology', as we have experienced it in Zimbabwe, covers a wide range, from mud brick constructions to the latest state-of-the-art, computer-modelled, environmentally-friendly, office complex. What do mud bricks have in common with a smart office complex, you might ask? They are both examples of practical technology!

To be practical it has to be readily available, easily produced, constructed with the resources available, affordable, locally maintainable, and it must 'do the job'.

Zimbabwe has been a closed economy for a long time, first through sanctions during the UDI period starting in 1965 and then, after independence in 1980, through being a controlled planned economy. This control has been eased in the last couple of years but the fact remains that in Zimbabwe 'we use what we have'. There is a saying here that 'if it cannot be repaired with reggan (old inner tube) it cannot be fixed!' Reggan is extensively used – for making sandals, as washers, for fixing leaking pipes and taps, for catapults, furniture straps and much more. So next time you have a puncture and decide to change the inner tube, keep the old one – it might come in handy!

In the building sector 'we use what we have'. If we cannot get large steel sections, we can make them up from smaller sections. If we cannot get fancy building materials, we use concrete. If we cannot get smart cladding, we use bricks. If there are no factory bricks available, we use farm bricks. If we have no theodolite, we use a piece of garden hose and some water. There is always a way round the problem. Like the driver who had a leaking petrol tank and used instead his windscreen fluid bottle to contain petrol – it worked fine, he said, 'as long as it is filled up every 10km'.

Clay bricks, known as farm bricks, are produced all over Zimbabwe. The best kind is made from old termite soil. The bricks are usually burnt in brick kilns and the better quality bricks compare well with factory-produced common bricks.

Arups in Zimbabwe have been involved in a number of projects using brick vault structures made from a combination of farm bricks and normal bricks. The first of these projects was, as it happened, not in Zimbabwe but in Botswana. It was Chobe Safari Lodge built in 1970 in Chobe Game Park about 90km from Victoria Falls.

Due to the location of the game park it would have been difficult and costly to build a conventional concrete structure. There were no local skilled artisans, no crushed stone in the area, and the local Kalahari sand was unsuitable for concrete. However, Kalahari sand mixed with river sand can be used for sand cement bricks with a crushing strength of 7MPa. The project was therefore designed and constructed using local skills and locally made bricks to minimise the need for transported materials.

Prefabricated hollow clay blocks form arches which are supported by brick cross walls. The end result is a structure which appears completely in harmony with the surroundings because it literally grew out of the ground.

The largest brick vault construction we have undertaken to date is Monis Mukuyu Winery, situated about 120km from Harare.

The reasons for choosing this kind of structure were:

- The building had to look and feel like a winery. It was to be used to advertise the product and attract the visitors.
- A structure with a high thermal capacity, that would not respond immediately to the mid-day heat, would be ideal, as grapes are picked at the height of the hot season.

(Above) Chobe Safari Lodge. The lodge situated on the banks of the Chobe River in Botswana was built using local skills – 'literally grew out of the ground'.

(Opposite) Prefabricated clay block roof units overcame the need for formwork to the roof.

(Above and opposite) Monis Mukuyu Winery. A 5m clear span in unreinforced brickwork vaults using a mixture of farm and conventionally fired bricks. The construction provided the high thermal mass and aesthetics required for the winery.

• Farm bricks were locally produced and extensively used in the area.

The structure adopted was a series of unreinforced brick vaults supported by brick cross-walls, corbelled out at the top to receive the vaults. The cross-walls are at 6m centres and the clear span between the corbelling is 5m. The building, which was completed in 1978, is both aesthetically and climatically pleasing, and the client is extremely happy with the way it functions.

Bricks, however, can also be used for underground structures.

In 1990 we were appointed civil consultants for the Zimbabwe Posts and Telecommunications Mashonaland Digitalisation Project. This consisted of an underground network of telecommunication cables with underground manholes for jointing and pulling of the cables.

At this time Zimbabwe was experiencing a shortage of cement, causing serious delays to most constructions. It was therefore decided to design the 2m wide x 3m long x 2m deep manholes as brick structures for speed of construction. The brick walls (230mm) were designed to withstand lateral earth pressure for depths up to 2m by reducing the length of the brick panel with the help of precast, prestressed, vertical concrete lintels. Zimbabwe as a developing country is short of technically skilled manpower. This poses a problem, not so much in designing services, but more so in maintaining them.

This was a particular concern during the design of Shabani asbestos mill, the largest concrete building in Zimbabwe, completed in the year of independence, 1980.

The control system selection was based on the level of high tech maintenance required. The design team therefore opted for the simplest system, using electro-mechanical relay logic, over the more sophisticated computer-based equipment and solid state systems. The type of system selected was less sensitive to disturbances, required no special knowledge for application and would be reasonably simple for maintenance staff to understand.

Despite Harare's latitude, its altitude of 1500m above sea level tempers climatic extremes, and in terms of humidity and temperature the climate

(Above and left) Eastgate. Floor void construction where the supply air absorbs or rejects heat; also note the ceiling barrel vaults designed to serve as light reflectors and to channel warm air towards the central core to exhaust through vertical ducts.

is very benign. It has, however, become the norm to design any reasonably prestigious commercial building as an air-conditioned building. Much foreign currency has been spent on importing air-conditioning plant and equipment and much energy has been wasted in running these air-conditioning systems.

The new Eastgate development in Harare, a shopping mall cum office block, ignores this 'norm' to break new ground in Zimbabwe. After many discussions with the client, Old Mutual, close design co-operation with architect Mick Pearce, and assistance from a building engineering group at Arups in London, with computer modelling, the first prestigious, non air-conditioned office development in Zimbabwe is now under construction.

The building has been designed to maintain acceptable comfort conditions without the use of air-conditioning. It is based on the ancient technique of thermal storage in the structure, as known from old Arabic buildings in the desert. With computer technology we are now able to quantify the benefits of thermal storage, and use it for optimum design.

Natural ventilation is normally associated with openable windows and problems of noise, dust, control of winter heat loss, and security. For Eastgate these problems have been solved by using a filtered mechanical supply of air, in conjunction with the building's thermal mass and careful control of internal and external heat sources.

Capital and running costs for this system, not to mention foreign currency content, are a fraction of those for a conventional air-conditioning system. In addition, the system is simpler to operate and easy to maintain.

Another development, the new National University of Science and Technology (NUST) in Bulawayo, has adopted a similar system and we hope that these projects will encourage other developers and designers to follow suit. Zimbabwe has in drought years experienced extensive power cuts and there is a growing awareness of the need for energy saving.

(Above) Hunyani pulp and paper mill. Water flume and jack ladder for log transportation from storage area to the chipper.

The population of Zimbabwe is approximately 10 million and there is a great need for development of low-cost housing.

There are many different systems, from rammed earth construction to fast track, galvanised steel-stud wall structures. All systems require infrastructure, which is a prohibitive cost factor in these projects. One way of cutting this cost is to develop 'community-based infrastructure'. The community, or co-operative, provides the labour for excavation of water and sewerage and the plumber, the only skilled worker necessary, lays the pipes under supervision. We are exploring this with Zvakatanga Sekuseka Co-operative in Hatcliffe and, on a grander scale, with the Ministry of Public Housing and Construction.

We are heavily involved in industrial development and it is always a challenge to explore ways of saving cost and foreign currency – as well as easing maintenance problems – by finding a local solution.

A good example of this is the Hunyani pulp and paper development woodyard solution. The woodyard incorporates all aspects of timber handling from receiving the logs to delivery of small chips to the chip silo.

The logs are predominantly delivered by rail to the marshalling yard, but allowance for road delivery has been made. They are then stored on the log storage platform to accommodate delivery and demand fluctuations, and thereafter fed via a water flume and jack ladder conveyor system into the chipper. Finally, the chips are screened and conveyed to the chip silo.

Traditionally the logs are hand-loaded onto the chain conveyor feeding the chipper. With Hunyani's new development, a safer, mechanised system was required.

Looking at sophisticated western machinery, which would have to be fully imported, would require imported spares, and would need highly skilled operators, it became clear that this was not the right solution.

Instead a system using a three-wheeled tractor ('telle logger') for log handling, combined with a water flume for log feeding the chipper, was adopted. The 'telle logger' was imported at one 10th of the cost of a hydraulic grapple crane, while the flume was designed using locally available materials (concrete, asbestos cement piping, and a pump) and construction techniques. It all added up to another practical solution 'using what we have'.

Although the economy is opening up to imports of all kinds, there will be a need for 'using what we have' for a long time to come. It is a valuable skill which will, hopefully, not be forgotten with new developments, but be part of them.

Where codes do not cover the design, then judgements have to be made based on testing and modelling. When traditional materials are used in ways not predicted by codes covering their projected use, the engineer has to rethink the basis of calculation and design from first principles. The argument is illustrated with recent projects using brick and stone.

John Thornton

New uses of traditional materials

A structural engineer at the end of the 20th century is accustomed to a situation in which many aspects of his work are closely constrained. Although he has complete freedom in proposing the principles of a structural solution to a problem, codes, generally, will specify minimum requirements for the analysis. They will also, for the principal materials, give detailed guidance on element design. There is continuing debate about the extent to which structural engineering has become a matter of satisfying codes. However, codes do represent a consensus view of the profession's experience in research, design, construction, and performance. Regularly updated and properly balanced, they are a valuable means of transmitting knowledge and provide a common basis for agreement.

The idea that codes inhibit creativity is misguided. On the contrary, an agreed basis on which design can proceed frees the engineer to concentrate on the more creative parts of his work. However, an inherent problem with codes is that they have been framed with some degree of preconception regarding their scope of application. The engineer should therefore understand the technical background to the code to know when he is moving outside its limits. Understanding the background to the code will also enable him to comply with its intention rather than the letter of the code by demonstrating how specified objectives will be achieved rather than just following a set of detailed rules.

There is no doubt that reliance on codes is now so completely ingrained in our thinking that it is an unusual experience for most engineers to carry out a significant piece of design which is not code-based. Even in these circumstances comparison will almost certainly be made with similar, codified, situations.

I have been referring to codes, but our reliance on codes is, to an extent, a manifestation of our reliance on textbooks, research papers, and established practice. Published information has respectability. Following established practice not only reduces the risk of failure, it also gives the engineer the defence that he performed his duties as well as could reasonably be expected. When there is no information, or it does not answer his questions, he is obliged to exercise his judgement to a greater degree than normal and consider the possibility of failure. It is easy to suppose that this situation is most likely to arise with new materials, but the same may be true when a well-established material is used in a new way.

When an engineer passes the limits of previous experience, he must face the issue of how to decide what is reasonable. Sometimes it is possible to think of the problem in such a way that it can be compared with previous experience; at other times it is necessary to judge the proposition entirely on its own merits. This situation is perhaps the same as that in which some of our predecessors worked but may be rather disconcerting for those trained by a technical education system which teaches facts and standard solutions.

While these two issues – the possibility and consequences of failure, and how to judge what is reasonable – should be of fundamental concern to the engineer, he is largely protected from them by compliance with existing practice. One of the most interesting aspects of our work exploring new uses of traditional materials is how it has brought these concerns into the foreground.

Early in the design of Glyndebourne Opera House the architects, Michael and Patty Hopkins, concluded that the

(Above) Glyndebourne Opera House.
(Above left) A flat arch.

building should be of brick to fit into its context. The use of brickwork in modern non-domestic buildings is largely confined to the cladding of a steel or concrete frame, sometimes simply as a decorative facing to a precast concrete panel. This is revealed by the bonding patterns, frequent soft joints, and the use of bricks in improbable situations such as soffits to slabs. This runs contrary to the whole basis of the Hopkins' architecture, which gains strength from the discipline imposed by insisting that the components of a building should do what they appear to do, and by accepting the limitations of each material. It was thus inevitable that Glyndebourne should be a 'traditional' load-bearing brick building, not a brick-clad frame, and that the brickwork should be solid and not cavity wall.

At this point our expectation was that we would be using a fairly high strength cement – lime:sand mortar, today's conventional solution. However, such a mortar is brittle and requires frequent movement joints to prevent cracking due to expansion of the brickwork and contraction of the mortar.

The design for the external wall began to develop as a series of flat arches spanning onto brick piers, and it soon became obvious that it would be very difficult to introduce movement joints. Furthermore, joints would undermine the visual integrity of the wall as a load-bearing structure rather than a decorative skin. Another objection to joints was that they would be a potential weakness in the acoustic isolation which the massive inner 'fortress wall' provides to the auditorium and stage areas.

To avoid the joints, we chose an 'unconventional' solution, which was to return to an older tradition and use lime-putty mortar. In the last 50 years lime:sand mortars have virtually disappeared. Cement:sand mortars set more quickly and cement is cheap. However in order to be workable they must have the proportions of about 1:3 cement:sand which leads to unnecessarily high strengths and a large drying shrinkage. These factors lead to cracking. The situation is improved if a proportion of the cement is replaced by lime, giving the traditional cement:lime:sand mix.

Lime mortars have a slow strength gain which, unlike cements, is due to carbonation by carbon dioxide from the atmosphere rather than reaction with water. This process can take centuries and means that the mortar retains a degree of flexibility. Lime putty is even better than hydrated lime in this respect, since it retains more water which improves the bonding between mortar and brick and gives greater flexibility. However, it is not commonly available and is more difficult to use.
The design team was not experienced in the use of lime-putty mortar and so it was necessary to talk to those who were and to carry out a literature search. We also thought seriously about our design and the relevance of historical precedents.

We were encouraged by the performance of long Victorian terraces with frequent openings and returns, but Glyndebourne is a large building and its fairly highly-stressed piers and flat brick arches place demands on the brickwork which are different from traditional buildings. In addition many traditional buildings can be seen to have cracked and moved in ways which would not be accepted now, and which would not be tolerated by the structural system we proposed.

Our solution was not covered by modern codes or conventional good practice, but we did have the benefit of advancements in the knowledge of materials, structural

The brickwork at the back of the Glyndebourne Opera auditorium during construction.

Inland Revenue Centre, Nottingham. (From left to right) Brick piers under construction and nearing completion in the works; the 'terracotta army'.

(Opposite page) lifting a pier into position; the completed assembly; the finished building.

behaviour, and analysis. Analysis of the interaction between the brickwork and the concrete slabs and the configuration of the walls led us to conclude that joints were not necessary given the low expansion characteristics of the bricks selected and the flexibility of the lime-putty mortar. A most important factor was realising that solid masonry is constrained by the structure behind, unlike the outer skin in conventional cavity construction, and so has different movement characteristics. In the end, after all the analysis and historic comparison, our decision was based on engineering judgement.

At Glyndebourne we used a neglected traditional material in a large, highly-engineered structure. At the New Inland Revenue Centre in Nottingham the brickwork used conventional cement:lime:sand mortar, but in a new way.

The structure of the office buildings consists of a thin precast concrete, folded-plate floor which spans 13m onto load-bearing brick piers. Speed of construction was important and building the piers in situ was obviously a critical activity. We suggested that the piers should be prefabricated off-site. This not only took pressure off the programme, but also had the advantage of making it easier to achieve good quality and the accuracy necessary for the assembly of the precast floor.

Making the piers required additional skills to those for in situ brickwork: the production of shop drawings, scheduling of production, and the ability to move and store completed units. To gain maximum benefit from off-site manufacture, the work should take place under cover. All of these could be supplied by a precast concrete contractor; he only needed to engage some bricklayers. There were also advantages in placing responsibility for the supply and erection of the whole frame with one contractor.

Working conditions for the bricklayers were ideal: dry, heated and well-lit. A system of templates and lines helped achieve great accuracy and no time was lost through bad weather. The system was equally successful on site. The piers were placed on levelled packs and plumbed. Landing the floor units was simple, and the structure was immediately stable. It only remained to grout joints and stitch slabs together.

In this instance the innovation was not in the revival and re-application of a traditional material, but in developing a new way of using a common material. Although this was a response to concerns over programme, it also arose out of recognising changes in circumstances. We no longer have available armies of lowly-paid bricklayers, and our abilities to lift and transport are far greater than even 50 years ago.

An important factor in this solution was the development in architectural design since the construction of the nearby Victorian brick buildings. In those buildings the external wall was a load-bearing skin perforated by doors and windows whereas in our Inland Revenue buildings the wall consisted of three components: the piers, the exposed slabs, and infill storey height glazing. This separation of elements was fundamental to the development of the concept of prefabrication, because it was never necessary to join brick elements to each other, which simplified construction and made the system feasible.

It is interesting to compare Glyndebourne and the Inland Revenue. Both used a traditional material: solid brickwork. At Glyndebourne, where it was in walls and expansion was the problem, we used a flexible lime-putty mortar. At the Inland Revenue it was in discrete piers. Expansion was not of concern but strength and robustness during handling were, so a conventional mortar was more appropriate

These two projects showed us how standard practice preconditions our thinking. We should always remember that particular technical solutions exist in a broader context. We become familiar with ways of using materials and take them for granted without questioning their validity when circumstances have changed. Brickwork in buildings at present is almost invariably laid in situ as the outer skin of a cavity wall using relatively high strength mortars, so that is how we think of using brickwork. In these two very different projects we realised

that a key factor had changed which made it possible to reappraise the nature of brickwork and how it should be used.

The New Parliamentary Building at Westminster is also to have precast concrete folded-plate slabs supported on piers; the difference is that the piers are of stone. Most are sandstone but eight piers which support a roof-top transfer structure are granite. Each pier will consist of a stack of large blocks such that the full load of the pier is carried through each block. Horizontal stability of the building will be provided by frame action in the external elevations, whereas at Nottingham it is provided by concrete cores. The piers are thus subject to considerable bending stresses.

We had previously used structural stone piers at Bracken House as a plinth to support the load-bearing cast gun-metal façade. The programme was short and there was relatively little stone. We recognized that *BS5628, The structural use of masonry*, had been written primarily for brick and concrete block walling and was thus not really appropriate for our circumstances. Nevertheless, there was no option but to design using what we considered to be a sensible interpretation of the code.

In the New Parliamentary Building all the vertical structure is to be stone and it is important that the size of the piers should be no larger than necessary. In this case, however, there was time for research during the construction of the new Westminster Underground Station which lies below the building.

We based our initial design on the assumption that large stone blocks with relatively thin joints would show the benefit of increased strength that *BS5628* suggested but did not quantify. A series of tests was carried out to verify this assumption. Because cube testing would be used for stone selection and quality control, we tested wet and dry, standard-sized cubes and third-size blocks to establish the relationship between cube strength and unit strength. We then tested pierettes both concentrically and eccentrically to establish the relationship between cube strength and masonry strength. We discovered that when the pierettes were loaded eccentrically, they split, so that what remained was concentrically loaded, which meant that eccentric testing on the finally selected stone was unnecessary. Apart from that, the results were disappointing because they showed that the pierettes were considerably weaker than we had anticipated. It was clear that a second series of tests was needed to understand more about the effects of a number of variables: block proportion, joint thickness, the benefits of joint reinforcement, moisture content, and joint material strength. We also looked at the significance of bedding or rift direction, because this affected the size of block which could be used, and the effects of surface-finishing techniques on granite.

The results of the second series of tests were also disappointing since the evidence on some of the variables was inconclusive and tended, if anything, to show that the increased strength due to thin joints and large blocks predicted by the code did not exist. The one positive conclusion was that joint reinforcement was a benefit.

This was a relatively short-term test programme for a particular project. We could draw only general conclusions on relatively limited data, and thus it was impossible to study the full interaction of the variables. We could test only one sandstone and one granite so we did not know whether our conclusions were general or particular to a type of stone.

With the amount of testing we were able to carry out it was difficult, in some cases, to be certain whether we were seeing the results of varying parameters or simply the variability of a natural material. There was certainly not enough data to postulate general rules which would allow extrapolation into other situations. We could only establish a particular relationship between cube strength and design strength. Subsequent tests on the stones selected for construction, which had similar cube strengths, confirmed the validity of this relationship.

Apart from the quantitative results of the testing, there was another substantial benefit. We became more aware of the failure mode of the stone and the possible consequences

(Above) The stone piers at Bracken House. (Above right) A load test on a sandstone pierette for the New Parliamentary Building. (Opposite) Tenth-size model of the New Parliamentary Building.

of irregularities in the form of the blocks which could create planes of weakness and initiate failure. This was particularly important since, unlike brickwork which is made up of a number of small units, our piers were formed of a stack of individual large blocks. Failure of one block meant failure of the pier. We were also very conscious that our testing could not replicate the actual design conditions. As a result we became more cautious about introducing holes and other discontinuities.

Ideally we would have load-tested full-scale prototypes of the components, but the necessary crushing loads would have required the construction of special rigs at great expense. We had to proceed on the basis of our limited data. There are parts of codes and other guidance which are also based on limited data but the engineer can reasonably expect to be judged to have behaved competently if he has followed the code appropriately. We, however, had moved from a 'codified' design to an 'engineered' design.

Earlier engineers or builders were also faced with a lack of codes and guidance. However, their designs were frequently based on long practical experience and, when they were not, society's expectations of success were less. We, on the other hand, work in a society which expects technology to succeed and have been conditioned to believe in theory and analysis rather than relying on practical experience. The use of stone as an engineering material places us in an unusual position. Although stone is familiar to us, we do not have the data and experience of using it as an engineering material. For this reason it is, in engineering terms, in a sense a new material. However, whereas we would expect a new material to exhibit consistency and predictability, stone is a variable natural material.

In our use of stone on the New Parliamentary Building, the structural system is different from that of traditional stone buildings and the stresses are higher, but perhaps the most interesting difference is the change in perspective arising from our confidence in analysis and a belief in precision. We and society have a different expectation of technology.

Each project I have described demonstrates a new use of an old material. Doing something new inevitably makes one more aware of what might go wrong. At Glyndebourne, if our judgements were wrong, the risk was that the building might crack, whereas at the new Inland Revenue Centre the consequence would have been that prefabrication would not have shown the expected benefits. However, at the New Parliamentary Building our concern was the basic strength of the material and the possibility of structural failure.

New uses of traditional materials are interesting from the technical view point. Also, in the evolution of our understanding of things we thought we knew well, they can focus attention on two essential concerns of the engineer: the consequences of failure, and how to justify decisions without inhibiting progress. In this sense the situation is the same as with new materials. The difference is that, with inherently less predictable materials, it is more difficult to have the same confidence in analysis and calculation, and this places a greater emphasis on engineering judgement.

When exploring new ground we carry out research, testing, and analysis, to satisfy ourselves that the design is sound. There is clearly greater risk of an error of judgement in this than when we are following a code, but if we work correctly the risk is low. The real risk is taken at the moment we decide to follow an idea because time and effort will be wasted if it turns out to be flawed.

How much 'risk' an engineer will take depends on his technical experience, his personality, and his previous exposure to risk. The 'threshold of risk' moves forward as experience is gained. A novel design requires an approach which is based on a clear understanding of the problem itself rather than the rules. It also requires the engineer to face the possibility of going up a blind alley. Whether the materials are new or traditional, the same holds true.

The advent of technical supercomputers in the mid–1970s opened the door for a new generation of numerical simulation techniques which then evolved in the field of structural mechanics and stress analysis. Since that time, those techniques have been developed and refined so that it is now possible to account for, and predict, almost every type of linear and non-linear structural behaviour. Computer simulations will soon be so comprehensive that all need for judgement in building the mathematical models will be removed, and the input data will be generated automatically.

Will this curtail the opportunities for intellectual challenge… or will it lead us to a more enlighted, confident, and productive design environment?

John Miles

Towards total simulation

In 1979 David Dowrick and I were sent on a speculative visit to the (then) CEGB in Gloucester, despatched by Mike Shears with a brief pep talk. 'Jack knows someone at Barnwood who's asked if we know anything about impact. You've done some of that sort of thing before, haven't you? Go down and see if there's anything we can do. Something to do with rail wagons.'

We chatted in the car as we drove down and decided there was little we could do to prepare ourselves. It wasn't at all clear how a firm of structural engineers could help the rail transport people at the Electricity Generator: we decided to listen politely and play it by ear.

At that time, the transport of spent fuel from the UK's nuclear power stations to the reprocessing plant at Sellafield in Cumbria was becoming a public issue. The CEGB had operated a round-robin rail service to each of the 13 nuclear power station sites for many years. The service delivered fresh fuel to each station once a week in a large steel flask (cuboidal in shape, measuring approximately 2m along each edge and weighing around 50 tonnes). The fresh fuel was unloaded and the spent fuel was loaded and returned to Sellafield two days after fresh fuel delivery. It was, and still is, the largest programme of nuclear fuel movements operated anywhere in the world. The safety record was first class – no significant accident had ever occurred over a period of more than 20 years. Nevertheless, the potential consequences of a major accident were sufficient to cause alarm to some sections of the public, and those known as 'the protestors' were vociferous and dedicated to moving the issue steadily up the list of public concerns. The CEGB decided that the time had come to prepare itself for a public debate and the process was initiated by inviting reputable outsiders to come in and assist the Board in developing its safety-related arguments. I had previously written a paper which described the application of non-linear computer analysis techniques to the design of impact-resistant transport vehicles, and our initial brief was to investigate the protective capacity of the rail wagons used for flask transport. Shortly after that, in early 1981, we were invited to bid to carry out a more wide-ranging, two-year, programme of research and development into the general subject of flask performance during accidental overload. The brief included both impact and fire events and the intent was to become authoritative in the field and put the Board in a strong position for the public debate which was anticipated.

Out of these circumstances was born Job No 10798, 'Additional Flask Work' – highly sensitive from a public relations viewpoint, titled with a careful anonymity to protect the innocent. During its life it was to become the biggest job with a time-basis fee which the firm had ever conducted. We explored the frontiers of impact and fire safety, spending over a quarter of a million pounds on computing and several millions more on reduced-scale and full-scale impact testing. The principal aim was to develop an ability to predict damage to the flasks resulting from severe impacts and fires, validating our chosen methods at every step by comparison with carefully-controlled tests conducted to our own specifications.

The main problem, we soon learned, centred on the detailed design of the flask. A $2m^3$ box, forged from steel over 300mm thick, is intrinsically very robust.

Indeed, the only weak point in the structure lay in the bolted interface between the body and its lid. As a consequence, it became essential for the analytical approach to take account of the strength and ductility of each individual bolt, the clearances in the bolt holes, the clearance between the lid spigot and the main body, and the friction at the lid/body interface. These factors introduced significant non-linearity into the structural behaviour of the flask and the search began for a means of simulating three-dimensional, non-linear, dynamic transient events. For the solution of the general class of problem, this dictated the use of computer-based techniques.

The Board's internal experts uncovered a little known piece of public domain software called DYNA 3D. This software had been developed by Dr John Hallquist at the Lawrence Livermore Laboratories in the USA as part of the US Weapons Development Programme. The Board recommended that we evaluate its potential for solving the flask impact problem and so, during the course of the 'Additional Flask Work' project, we learned how to use this software and became proficient in its application. It did, indeed, help us to address the acutely non-linear problems we were up against but, more importantly, it paved the way for some much wider developments which were to follow.

Hallquist's software was special because it led the first generation of computer software that could genuinely handle three-dimensional, non-linear transient dynamic problems in the fields of structural and stress analysis. It was able to do this because it adopted a very simple, but computationally very intensive, approach to solving the governing equations of motion. The approach, known as the 'explicit' technique, solves the problem by breaking the structure up into many small elements and then introducing a load at the boundary where first contact occurs in the real physical problem. The calculation then progresses in very small steps of time, tracking the progress of the stress wave through the structure, along with the gradual build-up of stress behind it, in the manner illustrated by

(Above) A 2m³ box, forged from steel over 300mm thick and weighing around 50 tonnes is, intrinsically very robust. The CEGB's Magnox flask can withstand huge impact forces, as witnessed by this drop-test conducted in a lid-corner attitude onto an unyielding target at 30mph. Predicting damage at the lid-body interface demands very detailed calculations.

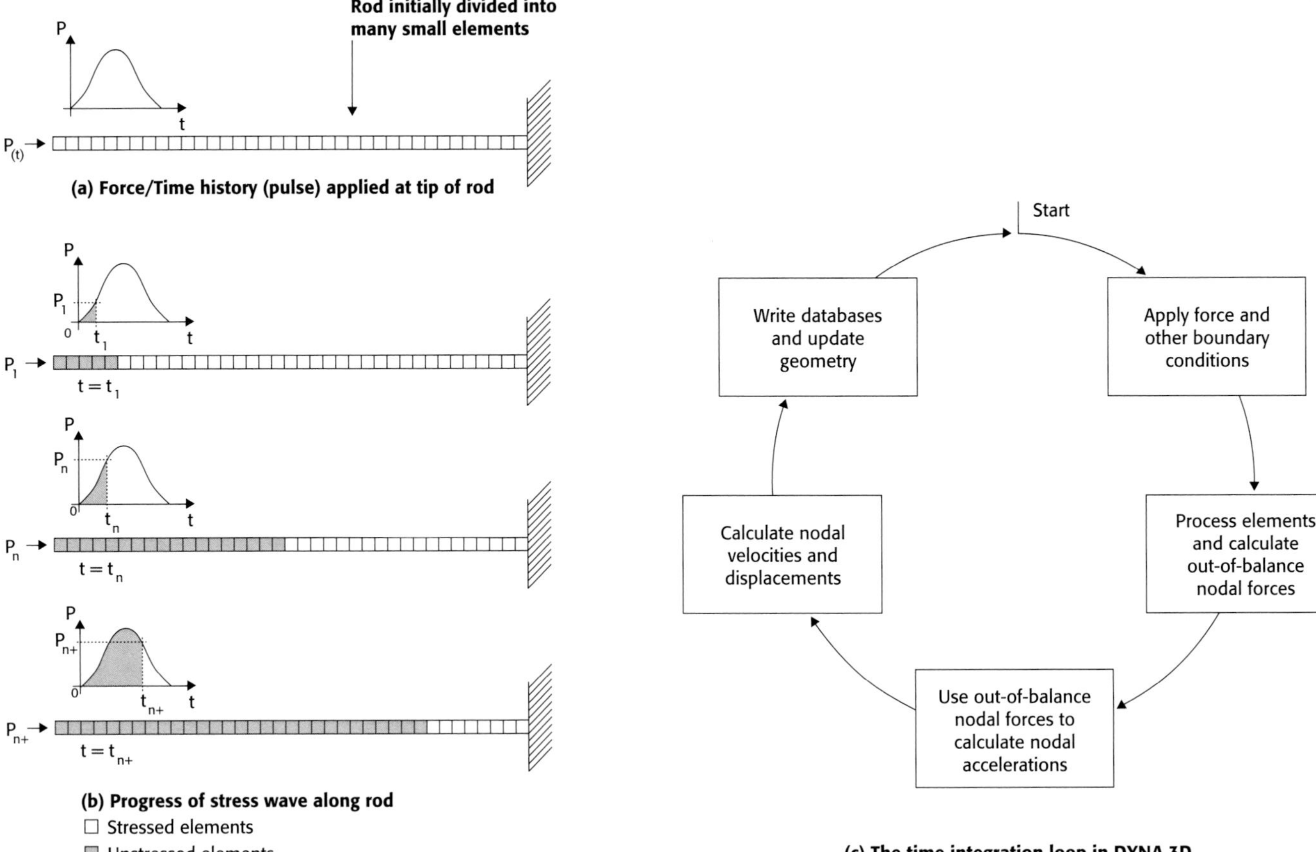

Figure 1.
Cantilever rod subject to an axial pulse load. Diagram illustrates the cyclical nature of the sequence of calculations which is fundamental to the explicit method. Note that the stress front proceeds at a rate of one element per time step from the point of initial excitation. Elements further distant than this are unstressed at that time.

the uni-axial bar problem in Figure 1. The key to the technique is to perform a very large number of simple calculations in preference to a relatively small number of complex ones. The integration algorithm was a central difference scheme and the only elements available in the original versions of the program were 8-node bricks with a single gauss point. As shown in the diagram, the shock front is propagated through the mesh in a manner that is directly analogous to physical reality. Here lies both the beauty of technique and its main drawback: because it directly reflects the manner in which all structures work, it is possible (in principle) to simulate any type of physical situation; linear or non-linear, static or dynamic. (Remember, of course, there is no such thing as a statically applied load: there are just dynamic loads applied very slowly). However, for the same reasons, the calculations become exceedingly long-winded for anything of longer duration than a typical shock event. This is because an event that settles down to near-equilibrium will require thousands of passes of the shock wave to achieve that state. Normal approaches to structures and stress analysis (the so-called implicit techniques) by-pass these problems, recognising that it is not necessary to represent the passing of the shock waves if you are only interested in the final equilibrium position.

Two camps in the field of numerical structural mechanics thus began to emerge in the late 1970s; the 'explicit' camp and the 'implicit' camp. Which should one choose for any given problem?

The explicit technique is clearly at its best when the problem at hand is associated with a short-lived transient. The flask was an ideal example – a very rigid structure being dropped onto an equally rigid steel target. Impact durations were typically in the order of ten/twenty thousandths of a second. However, events don't have to be this short for explicit techniques to be attractive. If the problem shows a marked degree of non-linearity, explicit solvers will still be competitive, even if the event duration far exceeds the transmission time for many passes of the shock wave. Thus, events like motor vehicle crashes (which typically will last anything up to two tenths of a second for cars and lorries) will be simulated very effectively using explicit techniques, because the degree of non-linearity engendered by sheet metal buckling, folding, tearing, and multiple surface contacts, will be very high.

As a consequence, we have subsequently been able to simulate in great detail a variety of automotive and other mechanical engineering impact events, using exactly the same analytical approach as was used for the flask problem. Indeed, we have not needed to limit ourselves to the structure itself; the whole point of designing vehicles to withstand impact is to protect the occupants. It makes sense, therefore, to include the occupant in the mathematical model (or, at least, a representation of the dummy, or manikin, which is specified as a substitute for human beings in the regulatory tests). By using a combined vehicle/occupant model, we may simulate the response of the driver and passengers, and predict directly pass/fail conditions based on parameters such as head acceleration, chest deflection, femur load, etc.
Modelling the human body, or a

dummy designed to represent it, poses some new and interesting problems. Although the body can be regarded as a structure, the materials involved have properties which are unfamiliar to most practising structural engineers. Flesh and other tissues, for example, are characterised by properties which are non-linear, hysteretic, and highly rate-sensitive. It is difficult to obtain reliable materials data from the laboratory to provide input to mathematical models of this type but, in terms of simulation at least, it is now possible to attempt a fairly serious representation of a variety of bio-mechanics applications. The examples of a shock wave passing through a brain, and the opening/closing of a heart valve, represent two recent examples of where we have conducted this type of investigation usind the DYNA 3D computer program.

But, to return to the question of choice of technique (implicit or explicit), it is becoming apparent that unequivocal rules cannot be laid down. Any preference must be conditional on the timescale of the event and the degree of non-linearity inherent in the problem. Traditional (implicit) methods of structural analysis have always struggled to cope with severely non-linear behaviour. Even for essentially static situations, the computer power required to solve markedly non-linear problems escalates dramatically as implicit solution algorithms begin to struggle with convergence and stability issues. Thus, the best method of solution may not be immediately apparent (even for static problems), but the explicit technique is always attractive because of its close resemblance to physical reality; in principle, there is no source of non-linearity, or quirk of dynamic behaviour, which it cannot handle. The only penalty is that the computer power required to solve the problem rises in direct proportion to the duration of the event. This is a key issue.

DYNA3D was one of the first computer codes to exploit the explicit technique in a fashion that was applicable to real engineering problems. An essential background (or enabling) technology, however, was the advent of vector supercomputers in the mid/late 1970s. Machines such as the CRAY-1, and its peers, enabled the huge numbers of simple calculations demanded by the explicit codes to be carried out in a realistic fashion. Since their appearance, computer price/performance ratios have consistently fallen by an order of magnitude every four years or so. With every movement of this sort, the objections to carrying out bigger and bigger computations become smaller and smaller. The range of problems which can be usefully addressed using explicit techniques becomes correspondingly wider and wider.

How far can we extend applications of the technique using current computer hardware? Can it, for example, be applied to non-linear problems in earthquake analysis (which are very long period)? What about the extremely non-linear behaviour of soils? Experience is beginning to show that events lasting several tens of seconds (problems that are very nearly quasi-static) can be addressed competitively when compared with traditional techniques if, once again, the degree of non-linearity is marked.

Despite the fact that a typical vehicle impact event will last an order of magnitude longer than a flask impact, the same simulation methodology may be successfully applied.

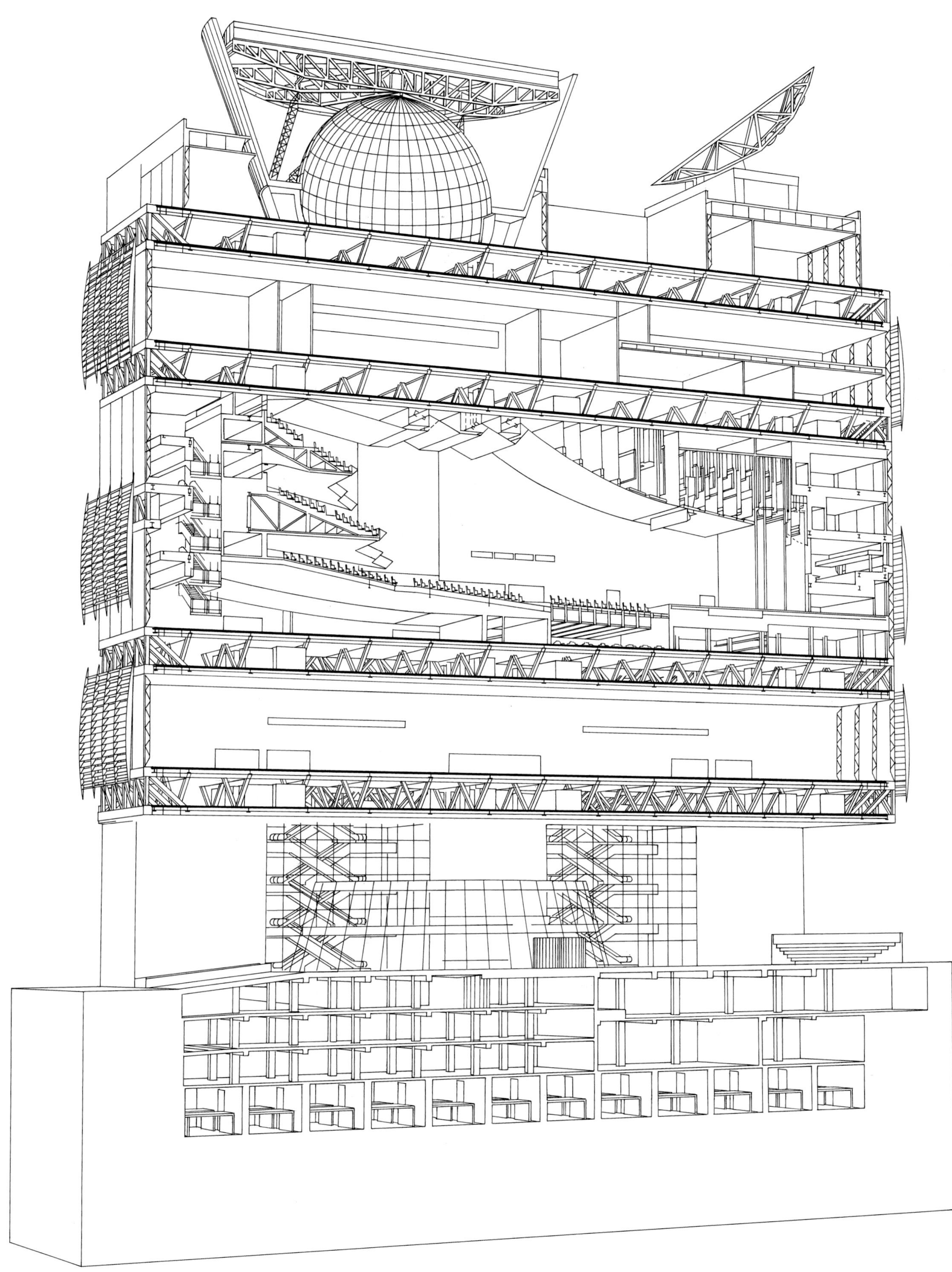

Osaka Conference Centre – a tall structure which Japanese regulations require to be analysed beyond the point of initial yield. DYNA 3D was used to carry out both static 'pushover' analyses and a comprehensive 3-D non-linear time history analysis.

As a consequence, problems of earthquake analysis in which the response of the building superstructure must be evaluated beyond its elastic limit are now amenable to this approach. The most recent example of this type of application is the Osaka Conference Centre analysis. Both 'static pushover' and full, three-dimensional non-linear transient time history analyses were completed for this structure using DYNA 3D.

A more extreme example of a 'slow' problem is the static collapse analysis which was carried out to guide improvements to the Class 465 Networker Express train structures. In this case, the goal was to predict the static load/deflection characteristic of the standard carriages under axial loading and then re-design the structures so that improved crash performance could be achieved. The carriages were of welded aluminium construction and, although the parent material was relatively ductile, the welded connections were not. The key to being able to predict the correct load/deflection characteristics for these structures lay in the ability to represent the fractures as and when they occurred in the connections.

However, before a calculation of this type begins, it is difficult to say which connections will fracture, and it is particularly difficult to say in which sequence they will fracture. Unfortunately, the fractures will have a marked influence on the collapse behaviour of the structure and, as a consequence, it was necessary to adopt an analytical approach which did not require the analyst to make *a priori* judgements of this type. In the end, this was achieved by modelling the material in the major connection welds directly using a fine mesh of elements with low ductility. In this way, the calculation was left to work out for itself where failures would occur; it was not incumbent on the engineer to specify predetermined points of failure in the analytical model and an opportunity for 'judgemental error' on the part of the analyst was removed. Very good agreement with laboratory tests was achieved, thus demonstrating that explicit techniques can be used very effectively to solve essentially static problems.

And this leads to some concluding observations which rise above questions of detail and technique. It has not been the purpose here to extol the virtues of the explicit technique *per se*; no doubt, in future, new and better approaches will rule the roost. Rather it has been the intention to highlight our increasing ability to simulate problems in applied mechanics with a similitude that approaches physical reality. Through a combination of improving software and dramatically increasing computer power, we are now fast approaching the point where two quite important watersheds will be reached.

First, we can see the time coming when it will not be necessary to exercise engineering judgement in the construction of our mathematical models. We will simply define models which replicate every detail of our proposed designs and then allow the calculations to work out, for themselves, where the boundaries of structural performance lie. This could open the door to an interesting consequence; in the past, analytical models have only ever been capable of predicting classes of response which are anticipated at the outset by the analyst. (A slender column will only be shown to buckle under compressive axial load if the governing equations formulated by the analyst in the first instance recognise that possibility.) On occasions where serious structural failures have occurred, and the fault has been traced to a design flaw, it has almost always reflected a situation where the engineer did not recognise the possibility of a particular mode of failure and had therefore (accidentally) excluded it from the design considerations. If, rather than making behavioural judgements for the purpose of simplifying the calculations, the analyst is in future able to allow for all possible failure modes by virtue of the verisimilitude between the mathematical model and physical reality, a major source of error will have been eliminated from our work. The mathematical model will, at last, have become a real prototype for the final structure.

Second, as the quest for price/performance drives computers towards yet another order of magnitude improvement, the latest generation of technical machines are the (so-called) massively parallel computers. These machines have not one Central Processor Unit (CPU), but maybe 1024 or more working in parallel.

The operating system and applications software on these machines must divide up a problem and farm it out to the various CPUs for processing, recalling the data from each CPU at regular intervals for co-ordination as the calculation progresses before farming out the next stage (ad infinitum) until the simulation is complete. If the same problem, with the same initial conditions, is run 10 times on such a machine, the statistical chances of each element of the calculation being carried out by the same individual processor in the same sequence each time is close to nil. The random nature of arithmetical round-off error which is present in all numerical calculations, plus the random route which the calculation will follow through the CPUs, will ensure that the analyst receives a different, but similar (and computationally correct) answer to the problem each time. The variance between the answers will represent the sensitivity of the calculation to minor perturbations in the round-off errors.

At first sight, this will provoke horror within the hearts of the analytical fraternity. But perhaps it should bring a warm glow to the hearts of the design fraternity. For, in reality, the 10 different but correct answers could be seen to reflect the 10 different but correct answers that would be obtained if the structure was tested repeatedly in the laboratory. Better, it could give a clue as to the sensitivity of the design to random imperfections. A slender structure will be prone to buckling under compressive load; if tested repeatedly in the laboratory, the spread of test results for that buckling load will be relatively wide (much wider than, say, the results obtained for the fully plastic moment in a rectangular solid steel section under pure bending). By the same token, the equations describing a buckling-sensitive structure will be more sensitive to round-off and other errors in the numerical calculation process than would the equations for the plastic hinge. The spread of calculated results when repeated 10 times will correspondingly be wider for the slender column than for the hinge. Never before have we been in a position where two answers coming from the same calculation on the same computer could be accepted as being correct… and yet different.

And so, the future is full of exciting possibilities: a place where the geometric detail of our models will reflect the real structures we seek to build; where meshes exceeding one million elements are automatically generated from 3-D CAD data within a matter of hours, so that data preparation is no longer subject to judgement, and the time required to prepare it is no longer an issue. Computer models will account for dynamics and non-linear behaviour as a matter of course; they will have the ability to detect classes of behaviour that were not foreseen by the analyst at the time the model was created. They will warn the engineer of problems he never thought about.

A frightening place? One in which the engineer is de-skilled by the computer and left with no room to exercise intellect and judgement? Not at all… indeed, precisely the reverse. It will be a place in which the engineer is freed from the need to make approximations (and introduce scope for errors) simply in order to make his design calculations tractable. Instead he will be able to spend his energy and creativity exploring the effects which changes have on the design, and then moving towards the preferred solution with a confidence mirroring that which should have been gained if a physical prototype had been constructed and proof-tested.

In short, a world in which analysis has once again become the servant of design, rather than its master.

That sounds nice, doesn't it?

M6033
MIRA
MIRA
MIRA
MIRA
MIRA
MIRA
M6033

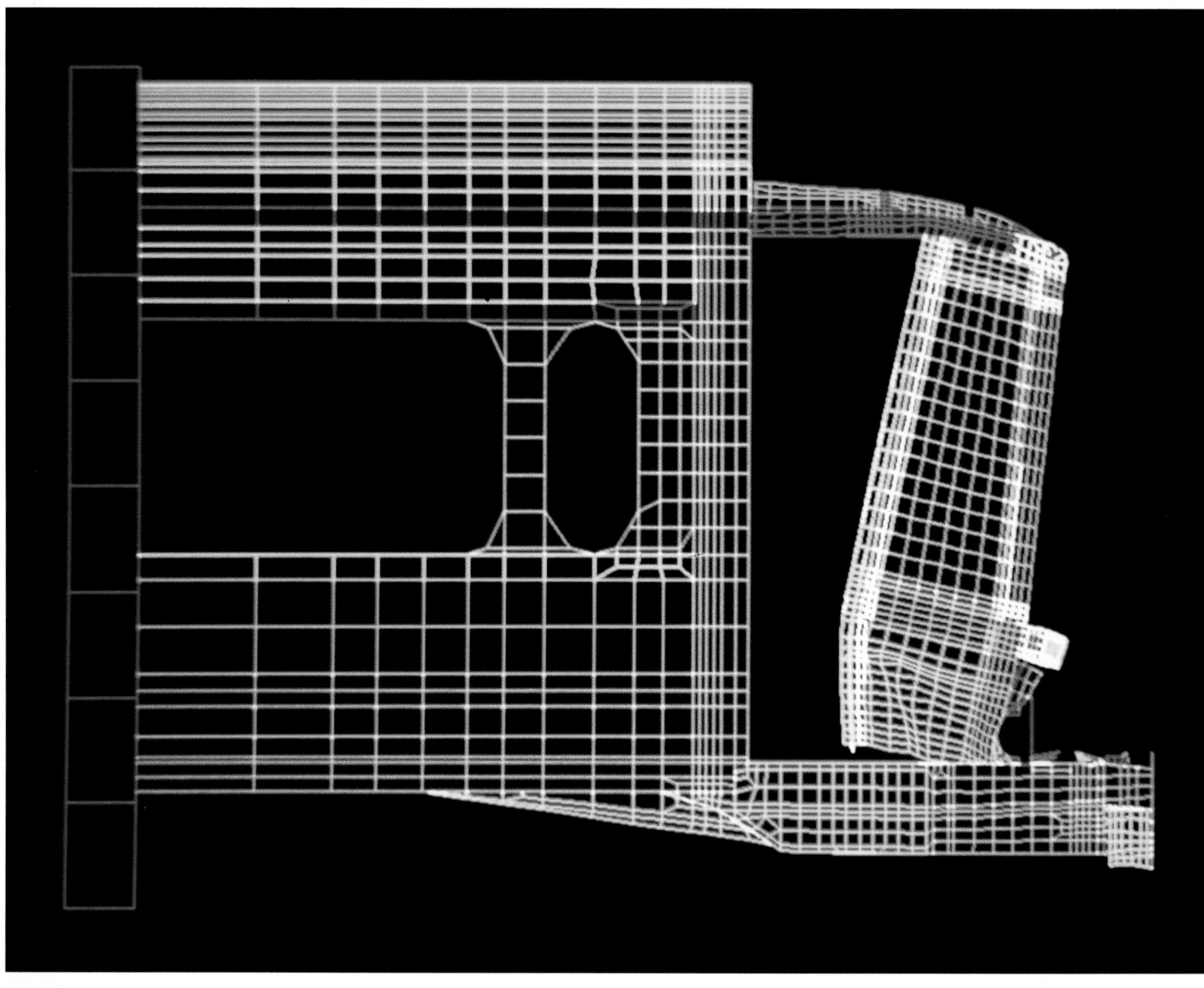

The new regulations governing the structural design of railway rolling stock demand that passenger carriages meet certain minimum criteria under static load tests, which are carried out to destruction. The key to successful analysis in the case of these aluminium structures lies in being able to predict the weld failures which lead to the onset of catastrophic collapse. Note the similarity of collapse mechanism which has been achieved between simulation and reality for this static loadcase.

The story so far

Previous essays in this book discuss some of the burgeoning activities of the firm. In this independent account, the distinguished historian Robert Thorne analyses the growth of Ove Arup & Partners, from its founding in 1946. He concentrates his account on three important factors: the political and economic context of the post World War II construction industry, the role of architecture within the firm's growth and, finally, the intellectual and moral input of Ove Arup himself. Thorne's commentary is provocative, critical and salutary, and serves to frame the text which founds the firm's activities – Ove Arup's 'Key Speech'.

Robert Thorne

Continuity and invention

An account of Ove Arup, the man, and his influence on the firm.

Ove Arup in the early years of the firm.

Those who establish a new institution or firm inevitably do so in the light of their past experience. However progressive they may claim to be, what they set out to do is fixed in what has gone before. The power of that retrospection varies. Sometimes it takes a stranglehold on a new organisation, which maintains its vigour for only so long as its past inheritance is relevant. But in other cases retrospection is only the starting-point for a process of evolution which soon leaves that inheritance behind. To set up an organisation which is rooted in the past, but which has a propensity for growth and change, is a delicate task. Few of those who have the courage to venture out on their own get the balance right. This account explores the development of Ove Arup's ideas and, in particular, their effect on the firm in which he worked actively for over 25 years. Subsequent historians and biographers will want to draw upon the rich archive of oral history and other material which the firm is starting to compile as a comprehensive record of its achievements.

Ove Arup was 51 when he founded the practice which became Ove Arup & Partners. He had more experience behind him than most people who decide to start a new organisation, and he drew heavily on that experience. In later lectures and articles he chronicled his early career in order to illustrate the drawbacks of the traditional building industry[1]. According to his view of what happened, the frustrations and obstacles that he encountered between the wars led inevitably to what followed.

If excessive retrospection is a hindrance to new organisation, Ove Arup's constant contemplation of the past should have had a stultifying effect on the newly-founded practice. In fact quite

1. For instance, 'The World of the Structural Engineer' (1968 Maitland Lecture) and 'A Society for the Built Environment' (lecture at the Institution of Civil Engineers, 1972), both reprinted in *The Arup Journal*, Vol. 20, No.1 (Spring 1985), pp. 2-10, 37-44.

the reverse has happened. From beginnings in a modest Soho office Arups has become a world institution, with offices in more than 30 countries and projects yet further afield. Starting with building structures, which were Arup's main interest and for which the firm is still best known, its name is now associated with almost every aspect of the man-made and material environment; from the intensely tangible world of civil engineering works to the almost ethereal world of simulated spaces. For many people, the word Arups is the only one that comes to mind when engineering is mentioned. As with a grand, well-staffed hotel it is a name which seems to conjure up a world in which there is nothing that cannot be provided.

To understand the position that the practice has reached after 50 years means going back to Arup himself and the spirit in which he saw the world after the Second World War. Judging by what has happened since, it seems that the inheritance he gave the practice has been wholly beneficial; that he set it up on the right lines and nothing has gone wrong since. But that conclusion, whether it is right or not, leaves a huge amount unsaid. Most of all it omits from the account the detail of how things happened, how others helped make them happen, and how at any point along the way the momentum might have faltered or stopped. To assume the inevitability of success is never a good approach to history. It is particularly unhelpful in dealing with the turbulent world in which consulting engineering has to exist.

One way that the founder of a practice can spread his message to those around him is through simple maxims. Ove Arup became a master of that method. People remember him for talking in one great flow of unfinished sentences and abstractions, but his written lectures are clear and to the point, and are memorable for the pithy way he put over his ideas. 'Good design should embody a practical way of building': architecture is 'a way of building that delights the heart': 'engineering design is creative accountancy'. These remarks, or versions of them, recur in almost everything he said. They remain a consistent litany in his pronouncements from first to last.

Above all Arup used the two phrases 'total design' and 'total architecture' to sum up what he meant. Both were for him interchangeable ways of referring to the need for synthesis in the world of design and construction. For Arup they could be applied at every level and to every aspect of the way the built environment is made. Starting with the need for the integration of structure and architecture, Arup moved naturally to the need for collaboration between architects and engineers, and to the requirement that design and construction should be thought of as parts of a single process. And for him the quest for synthesis stretched yet further, to the relationship of building and technology to society as a whole. His search for a sense of totality knew no ends.

Arup's obsession with the process of synthesis in design and construction, including the very words he used to describe it, seem to link him to one of the principal sources of Modern Architecture, Gropius's Bauhaus Foundation Manifesto of 1919. Indeed Arup admitted his admiration for the ideas promoted by the Bauhaus[2]. On taking over the school at the end of the First World War, Gropius had infused its programme with his vision of the integrated building team, based on an idealised

2. Ove Arup, 'Art and Architecture. The Architect:Engineer Relationship', *RIBA Journal*, 3rd ser. Vol. LXXIII (August 1966), p. 355.

understanding of the medieval building world. 'The ultimate aim of all visual arts is the complete building!', ran the Manifesto: 'Architects, painters and sculptors must recognise anew and learn to grasp the composite character of a building both as an entity and in its separate parts'[3]. A few years later, in *Idee und Aufbau,* Gropius again argued for an all-encompassing, total architecture, based specifically on a synthesis of art and technology.

3. Walter Gropius, 'Programme of the Staatliche Bauhaus in Weimar' (April 1919), reproduced in Tim and Charlotte Benton, *Form and Function* (1975), p. 78.

However, just as important as the similarities of approach that link Arup to Gropius are the differences in what each of them understood by 'total design'. The primacy that Gropius gave to the intuitive perceptions of the artist meant less to Arup, for whom equal collaboration of all those involved in the design process seemed most important. And whereas Gropius looked upon the finished building as the symbol of totality, for Arup it was the process by which the building was created that mattered most of all. This difference of emphasis points to the fact that Arup's understanding of design started from an engineering viewpoint. Its philosophical basis was directly related to what his own experience had taught him about how buildings and structures are conceived and built.

Ove Arup's blend of philosophy and practicality derived from the way his early career developed. He was born in 1895 at Newcastle-upon-Tyne where his father had been posted as veterinary commissioner to the Danish Government. Within a year the family moved to Hamburg, and his childhood was spent there until at 12 he went to Sorø Academy in Denmark. Later in life he recalled having an adolescent's inordinate and undisciplined curiosity about the world, which was to lead to his reading philosophy at the University of Copenhagen: 'Once I had discovered what could and what could not be discovered then I would be ready to decide what to do next.'[4] In fact his main discovery was that academic philosophy, though fascinating, was too dry and specialised to be satisfying. Architecture was his next choice, but there he had doubts about whether he was enough of an artist to be able to excel. So, following the example of a cousin, he enrolled in 1916 at the Polytechnisk Laereanstalt, the Danish Technical College: 'I could at last become an engineer. I was good at mathematics, physics, and all that. Perhaps I could become an architect later on, once I knew all the technical stuff.'[5]

4. Ove Arup, 'Philosophy and the Art of Building', in Patrick Morreau ed., *Ove Arup 1895-1988* (1995), p. 11.

5. *Ibid*, p. 14.

In a sense the practice that Ove Arup founded owes its origins to his brisk and seemingly rather arbitrary decision to opt for engineering. What was equally important in the long term was that he studied at a college which specialised in reinforced concrete, the material in which the practice made its early reputation. Having graduated in 1922 he was taken on by the leading Danish contracting firm of Christiani and Nielsen, who sent him first to Hamburg and then to London: in 1925 he became the Chief Designer in their London office. Christiani and Nielsen were specialists in the reinforced concrete construction of marine and industrial structures, for instance jetties, piers, bridges, coal bunkers, and cooling towers. Ove Arup published a series of articles on reinforced concrete jetties and piers, especially their resistance to impact from ships; and in 1931 he took out a patent for cooling tower design. But the chief lesson he learned from his early working years concerned the practical realities that design has to obey. He became expert in the cost of materials, estimating,

Jetty at Fawley, Hampshire (1928), typical of Arup's early interest in reinforced concrete structures.

and what happens to a design during its execution on site. The aphorisms about the links between design and construction with which he peppered his lectures came from the heart of that experience.

In the 1920s and early 1930s the structural engineering of buildings was mainly handled as a sideline by civil engineers, who did what architects asked them to do and no more. For many architects that was sufficient, but those of a Modernist persuasion could see the promise in a much closer collaboration with engineers of the right outlook. The functional aesthetic of early Modernism implied the expressive use of materials linked to a natural system of structural design.On all counts this pointed towards a greater reliance on engineers; and anyway had not Le Corbusier lionised the engineer as a kind of freeborn genius unencumbered by traditional conventions? The joint exploration of new materials and structures offered the chance of creating more satisfying and socially responsible forms.

Most important of all, few architects had a detailed understanding of new materials to match their enthusiasm for using them. So it was that Berthold Lubetkin, the Russian-born architect who arrived in London from Paris in 1930, turned to Christiani and Nielsen to help build his cylindrical Gorilla House at the London Zoo (1932-3). Arup, who was then still with that firm, brought to the project his knowledge of thin-walled silo construction and a method for building the structure using a climbing shutter. Lubetkin had found an engineer of the right expertise and sympathy, and Arup had found an architect 'who taught me how good architecture was produced, and what a serious business it was.'[6] Together they created a new kind of collaboration, leaving far behind the convention of the obliging technician at the service of the gentleman artist. After the Gorilla House they worked together on the Penguin Pool (1933-4), with its swirling ramps in imitation of an ice floe, and on other buildings at London, Whipsnade and Dudley Zoos. At Highpoint One (1933-5) Arup provided the means to realise one of the most potent images of the emerging Modern Movement – the tower in the park, raised above the ground so that space could flow beneath. The walls were of load-bearing concrete, with special allowance for transferring the loads to the ground floor columns, and the system of climbing shutters was repeated from the Gorilla House project. As Highpoint One neared completion Arup, again working with Lubetkin and his Tecton partners, showed how the same principles could be applied to mass housing: their scheme won first prize in the Cement Marketing Company's 'Working Men's Flats' competition.

6. Ove Arup, 'Art and Architecture', p. 353.

The Gorilla House, London Zoo.

The complex reinforcement for the ramps of the Penguin Pool, London Zoo.

Highpoint One had another kind of significance, in addition to the fact that it associated Arup with a building which attracted wide attention. His old firm of Christiani and Nielsen were reluctant to tender for the project, which seemed to be too far from their world of civil engineering, so Arup moved to the contractors J.L. Kier & Co. on the understanding that they would undertake such work: Highpoint One, the Penguin Pool, and other Lubetkin projects were done with them. However even with a sympathetic firm – Olaf Kier was a fellow Dane and an old friend of Arup's – there remained the fundamental difficulty that engineering design, particularly for buildings, was marginal to the business of contracting. They might suggest an engineering solution to the architects but that would not

Highpoint One, Highgate, North London, under construction, with climbing shutters.

guarantee them the work unless their tender was the cheapest. Most frustrating of all, their design input could be used in sending the work out to tender and thus be appropriated by another contractor. If the structural design of buildings was to be properly recognised and rewarded, it had to be established as a separate activity in its own right.

In 1938 Ove Arup's cousin Arne, an English-born businessman, offered him a partial freedom from dependence on the world of contracting. Together they set up Arup and Arup, a design and construction firm, but alongside it was Arup Designs Ltd offering separate design services, and Arup himself also started practising as a consulting engineer in his own name. Arup later recalled how this curious trinity was meant to run: 'If people chose to come to me as a consultant, neither Arup & Arup nor Arup Designs could bid for the job. That established clear lines. Of course the two personae were quite clearly one and the same firm, but the advantage for our clients was that they could employ, as consultants, engineers who had a practical experience of buildings, or on the other hand employ a building firm that was run by engineers.'[7]

Finsbury Health Centre, London.

7. Ove Arup, 'Philosophy and the Art of Building', p. 23.

With the approach of war, and even more once war had broken out, Ove Arup's ingenuity as an engineer came into its own. Wartime necessities helped eliminate the barriers of suspicion and caution which previously had held back the application of new technologies and ideas. That is not to say that all of Arup's proposals were automatically welcomed. Despite persistent lobbyings his ideas for large-scale deep air raid shelters, designed so that they could be re-used in peacetime as car parks, never met with official approval. Other schemes for saving scarce materials were also disregarded. However the design of wartime facilities kept the small tripartite office busy, and at least one of its projects is counted as having played a significant part in the way the war was won. This was the system of fenders that was used to protect the pierheads of the famous Mulberry Harbours from the impact of berthing ships. The fenders, which were supplied by Arup & Arup, were a model of all that was best in wartime engineering: a system of crank-shaped blocks that looked simple but were in fact the result of a carefully-calculated design for a highly unusual set of conditions.

The war forced contractors to think imaginatively about the design of structures and the use of materials, which suited Arup well, but equally it brought restrictions which made the world of contracting more aggravating than ever before. Price controls, shortages of materials and centralised supervision took the thrill out of competing for a job, and left only marginal opportunities for profit. Contracting became, thought Arup, 'a thoroughly uninspiring, even degrading business.'[8] And in such circumstances the conventional prejudices against contractors as tradesmen looking for short cuts in everything came even more to the fore.

8. Ove Arup, 'The World of the Structural Engineer', *The Arup Journal*, Vol. 20 No.1 (Spring 1985), p. 5.

Increasingly discontent with the state of the building industry was the main factor which persuaded Arup in 1946 to terminate the contracting side of his activities, and to practice solely as a professional consulting engineer. In the long term there seems no doubt that this was the right thing to have done, yet in a sense it represented a defeat for his ideal of total collaboration in the building process. The variations on design and build that he had experimented with disappointed his expectations, yet that didn't stop him from continuing to insist that design should never be detached from the practicalities and economic constraints of building. Part of him continued to long for a more all-embracing organisation than the one he finally founded.

On a more positive note, the years of experiment allowed Arup to discover that he wanted above all to be involved in the structural aspects of architecture. The youthful sense of artistic inadequacy which prevented his training to be an architect still coloured his view of what an engineer could contribute to the making of a good building. By temperament and experience he could never be just a subservient technician, waiting to be summoned to provide an engineering interpretation of an architect's most fanciful ideas. Yet he was prepared to defer to an architect's feeling for form and space which, if it excited him, he would go out of his way to help achieve. Working with Lubetkin and the Tecton group gave him his first taste of that excitement. After countless projects with many other architects he found it yet more intensively in Jørn Utzon who, at the outset of the Sydney Opera House saga, seemed to be just the kind of architect he had always been waiting for.

Architects for their part delighted in his willingness to engage with what they were trying to achieve. The leading spirits in the Modern Movement had either recently arrived in England like Lubetkin, or were followers of the new architectural and social ideals that were coming out of continental Europe and Russia. Arup got to know them through the network of new architectural thinking centred on the Architectural Association, the MARS Group, and the Architects' and Technicians' Organisation. He was welcomed partly because he could claim an authentic European pedigree including a European upbringing and a command of the latest foreign engineering ideas; plus he had a strong Danish accent. The architectural friendships he made between the wars were with many of those who were the most influential in peacetime reconstruction.

A number of other factors made it auspicious for him to set up independently. One was that he had met a number of engineers whom he persuaded to join him and who subsequently became partners in the practice. Ronald Jenkins had been with him at Kiers and then had been Chief Engineer at Arup & Arup. A brilliant analyst, he had assisted F.J. Samuely with the daunting calculations for the ramps of the Penguin Pool, and had shared in the design of the Mulberry Harbour fenders. Enigmatic and apparently aloof, he made many colleagues including Arup feel slightly uncomfortable. But people were prepared to put up with a lot from him, knowing that his expertise, particularly in the calculation of statically indeterminate structures, was crucial to the early reputation of the practice. Geoffrey Wood had worked at Christiani & Nielsen and Kiers, and came to Arups in 1947 after formidable wartime experience with the Royal Engineers in India. Peter Dunican joined the tripartite Arup organisation in the middle of the war and went on to become the first drawing office chief in the practice and eventually a kind of master-at-arms figure watching over every detail of its organisation. In 1948, when Arup announced the formation of a partnership, the first partners alongside himself were Geoffrey Wood, Ronald Jenkins and Andrew Young (who later left to join the Coal

Rosebery Avenue flats, London.

Store Street bus station, Dublin.

Board). Peter Dunican became a partner in 1956. Arup's authority in the practice stemmed partly from his experience and contacts but equally from the fact that he was at least 10 years older than those he gathered round him. This age difference cast him in a fatherly role, but he was certainly not a stereotypical paterfamilias, stern and unyielding. Having picked the right people he was generous to a fault in giving them their heads. 'If you could write you could sign your own letters', as a later recruit put it.[9] What he couldn't tolerate was people who failed to share his stubborn curiosity about anything and everything. As well as being one of the main characteristics for which Arup is remembered, this reflected the tremendous sense of expectation of the post-war years. So much seemed possible if new modes of understanding were allowed to flourish.

9. Interview with Michael Lewis, 15th March 1995.

Chassis production factory, Inchicore.

A number of things gave the new practice a strong head start, of which the most important was its expertise in reinforced concrete. Since the turn of the century steel and concrete had been jockeying for primacy in civil engineering and large building projects. In the 1930s steel largely had the upper hand: its industry was better organised and consulting engineers found it easier to deal with. By contrast concrete still carried the aura of being slightly foreign, a hangover from its original promotion through licensed systems imported from France and America, and its specialist contractors were thought unreliable. Ove Arup was one of those who stood out from the crowd in his enthusiasm for applying the lessons of reinforced concrete from civil engineering to building structures, which at the time of Highpoint One and the zoo buildings was still quite unusual. But by the end of the war priorities had been reversed. Steel was in extremely short supply and architects, drawing on the lessons of Lubetkin and his continental contemporaries, relished what could be done with concrete. The fact that in using concrete they were dependent on steel reinforcing bars did little to reverse this swing of architectural opinion. The use of steel in buildings remained restricted by government controls until 1953.

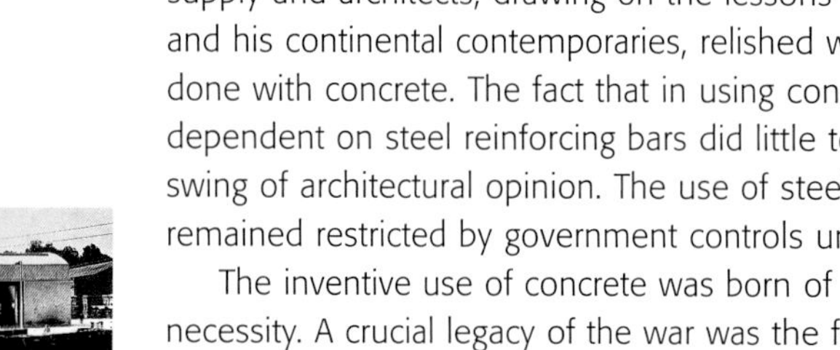

Donnybrook garage, Dublin.

The inventive use of concrete was born of more than just necessity. A crucial legacy of the war was the faith in new technologies that it engendered. The boffin mentality, which had proved itself in conflict, could be applied to the architectural problems of post-war reconstruction. And as happened so often during the war, mistrust was swept aside by enthusiasm for the benefits of innovation. This ethos, combined with the fact that contractors had not yet systematised the new methods of construction, created ideal circumstances for consulting engineers of the right outlook.

Nothing showed this change of mood more clearly than the popularity of both concrete shell construction and prestressed concrete. Cylindrical shells and domes were not new. They had been used before the war in Germany and Denmark and had been widely published, yet in Britain their potential had hardly been recognised. After 1945 architects and engineers had a chance to see for themselves what had been built in Germany and the reports of their visits, made just at the time when steel shortages began to bite, transformed the appeal of shell construction: by 1952 some 500 shell concrete roofs had been completed.[10] Arups played a crucial role in this

Sish Lane flats, Stevenage.

10. Marion Bowley, *The British Building Industry. Four Studies in Response and Resistance to Change* (Cambridge,1966),p.129.

turnabout, both through the design of some of the most well-known examples and through the publication of Ronald Jenkins's dazzlingly erudite exposition of the calculations that made them possible, Theory of Cylindrical Shell Structures (1947).

The building that occasioned that publication was the Brynmawr Rubber Factory (1946-51), a desperately optimistic project to revive a Welsh industrial town by introducing a new kind of democratic, visually appealing workplace. The circumstances of the scheme whereby the directors of the firm chose the design team but the government paid the bills allowed its innovatory aspects to be explored to the full. Working with Architects' Co-Partnership, Arups arrived at a design involving nine two-way curved domes covering the main workspace, springing from bow-strung trusses supported on giant V-shaped columns. Subsidiary rooms had shell barrel vaults, and the boiler house had its own parabolic shell concrete roof to keep the rain off trucks delivering the coal. Finished in the year of the Festival of Britain, Brynmawr summarised in one building what astonishing engineering could come from conditions of austerity.

Brynmawr Rubber Factory. The nine shell roofs cover the main production hall. In the foreground are the entrance and administration block.

Busaco Street flats, London.

After Brynmawr, shell roofs were a recognised part of the Arup vocabulary, leading ultimately to the shells (which looked like shells but weren't) of the Sydney Opera House. Their peak in popularity was in the early 1950s, just before the wider availability of steel made the economies that were claimed for them seem less evident. The Bank of England Printing Works at Debden in Essex, completed in 1954, was designed as a huge, unobstructed printing hall with an asymmetrically curved roof made up of prestressed arched ribs linked by north light concrete shells. At Kidbrook Comprehensive in South London the need for an assembly hall to seat the entire school of 1500 resulted in a single 75mm thick concrete dome with prestressed edge beams. Perhaps the most astonishing Arup shell roof of that era was Ronald Jenkins's design for Peter and Alison Smithson's entry in the1951 Coventry Cathedral competition, a roof square in plan tilted towards the altar on massive inclined struts. Arups were also the engineers for the successful entry by Sir Basil Spence, but that had nothing to match the single-minded geometry that the Smithsons championed.

Kidbrooke Comprehensive School, Greenwich, South London (above and below).

The architectural virtues of shell construction continued to attract attention until other forms of roofing, especially in steel, became more freely available. By contrast prestressed concrete, having been introduced for many of the same reasons as shell concrete, endured much longer, especially in bridge design. As with shells it had a foreign pedigree, traceable most clearly to the work of Eugéne Freyssinet who had presented a paper on prestressing in London before the war. Since it not only made up for the weakness of concrete in tension but required less weight of steel than reinforced concrete, it is hardly surprising that prestressing 'swept the country' after 1945, as Arup put it.[11] Prestressed components were combined with shells at the Bank of England Printing Works, Kidbrooke Comprehensive, and elsewhere, and prestressing was crucial to Arups' unbuilt design for roofing the Centre Court at Wimbledon. In prestressed bridge design the Arup contribution started with a footbridge at the Festival of Britain connecting Waterloo Bridge with the Royal Festival Hall terrace, running straight across the site where the firm later helped to build the Queen Elizabeth Hall.

11. Ove Arup, 'The World of the Structural Engineer', p. 5.

The main hall of the Bank of England Printing Works, Debden.

However the greatest opportunities for concrete construction were in public sector housing, where new technologies were harnessed to achieving the welfare aims of post-war reconstruction. From the time of its genesis in Victorian anti-slum campaigns, local authority housebuilding had become a permanent part of overall housing provision, but was more innovatory in its financial than its technical aspects. Except on rare occasions, neither cottage estates nor deck-access blocks of flats strayed far from conventional construction. The destruction of houses by bombing, followed by scarcities of materials and labour, helped throw caution aside, especially in the case of cities that wanted to maintain pre-war densities within their boundaries; and architectural formulas for a new kind of light and healthy living were well advanced. Flats seemed to be the only answer, at first five-storey traditional blocks but eventually rising to four or five times that height as a way of exploiting inner urban sites. Across the country as a whole high-rise blocks never played a major role in local authority housebuilding, even at the peak of their popularity in the mid-1960s, but as symbols of the technical transformation of housebuilding they were crucial. They were synonymous with the modernisation of the building process – standardised, mass-produced, and reliant on a greater degree of central planning than ever before.

Hallfield Primary School, London.

Hunstanton School, Norfolk.

Arups took naturally to public housing projects, sharing the ideals they stood for, and that commitment helped provide the mainstay of the practice during its first decades. An early sign of the importance of housing was the Housing Group under Peter Dunican, established at an office in Finsbury close to the Tecton projects there. The long slab blocks of the Spa Green and Priory Green Estates were derived from the 1935 Working Class Flats project, but structurally simplified by the use of load-bearing cross walls above first floor level.

Where most pre-war deck access flats had relied on cellular construction, with small rooms to match, cross-wall construction allowed a more open internal layout, naturally lit from both sides. The walls could be cast a storey at a time using interlocking sheet steel formwork. The same essential principles were used at the Hallfield Estate in Paddington (1947-55), executed with Drake and Lasdun after the Tecton group broke up. By the time that was completed, cross-wall construction was commonplace, and instead what gained attention was the obtrusive pattern-making of the façades.[12]

12. Reyner Banham, 'Façade', *Architectural Review*, Vol.116 No. 694 (November 1954), pp. 303-7.

The replacement of traditional cellular construction and framed construction by designs based on load-bearing concrete walls was a significant step in the mechanisation of house-building, but it still represented a largely site-based approach. At the end of the war something much more radical had been anticipated, and indeed the factory-based production of houses had been achieved in order to meet the housing emergency. Prefabricated houses, delivered to site from 'kit-marshalling centres', may have looked architecturally unbeguiling but they seemed to be a logical peacetime application of the production methods that had helped win the war. Arups had been involved in the development of the Arcon Mark V, one of the most widely-used prefab types. Once tower cranes became available in the early 1950s, the dream of using similar methods on a yet larger scale became possible: a crane that could lift shuttering and materials across a site could just as readily handle prefabricated components manufactured elsewhere. So poured concrete methods lost their prominence in favour of large panel concrete construction.

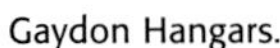

Gaydon Hangars.

TUC Memorial building, London.

At Arups no-one was more enthusiastic about industrialized housebuilding than Peter Dunican, for whom it was a social and political as much as a constructional issue. In engineering terms it had the potential to fulfil the Arup ideal of integrating design, manufacture and erection, but the scale of operation it required meant that it could only work properly with political backing. Dunican later became on good terms with Labour politicians and with the architects of lively Labour-run authorities, especially the London County Council. If demand could be planned well ahead, mechanisation and repetition could be exploited down to every detail of house design. Dunican recognised that the proliferation of rival contractors' systems made a nonsense of standardisation, and he played a key part in the National Building Agency, set up in 1964 to co-ordinate the demand for different systems. The Ronan Point collapse four years later involved a block which Arups had nothing to do with, and Dunican was incensed by the official response which implied that the same kind of failure in the floor-to-wall joints could happen with blocks built on other systems.[13] The inevitability of industrialized housing still seemed obvious to him, even though well before Ronan Point its apparent economic advantages were fading and with it the political sponsorship on which it depended.

13. Peter Dunican, 'Ronan Point: Swallow or Summer', *Ove Arup & Partners Newsletter* 41 (March 1970), pp. 565-8.

Alton East, Roehampton, London.

The beginnings of the Arup involvement with large panel concrete construction were to be seen in a prototype estate at Picton Street in Camberwell (1953-5), which Peter Dunican worked on with Cleeve Barr, of the LCC Architects' Department. There the cross walls were still cast in situ, but everything else was prefabricated to be slotted into place with the aid of a tower crane as the walls went up. By the early 1960s, prefabrication had progressed much further, involving the proliferation of rival systems with curious acronyms and titles. Amongst the many home-bred systems, Arups was particularly associated with one developed by a Yorkshire consortium of local authorities, known as the Yorkshire Development Group Mark 1. At the same time the practice was instrumental in introducing the Jespersen System from Sweden, where Victor Kemp, a stalwart on the housing front, had seen it on a visit. 12M Jespersen, as it became known, found favour with the Government as the system best suited for general use across the country.

The Picton Street Estate, Camberwell, South London.
A key project in the evolution of large panel system building.

From Arups' point of view the most important of the systems was one developed with Wates the contractors, distinguished by the fact that its components were produced in site-based mobile factories rather than larger regional production centres. This was widely used, especially by the LCC and London boroughs. Ultimately the relationship with Wates took on an international dimension with a joint entry to a US Government housing competition. 'Operation Breakthrough', as it was called in America, led to the completion of a single block in St Louis but never produced a sequence of projects of the kind that London had seen.[14]

A Wates on-site factory at Feltham, Middlesex.

14. *The Arup Journal* Vol. 8 No. 1 (March 1973), p. 23.

Part of the problem of industrialized housebuilding was that many of the sites where it was used were too small to allow its benefits to make themselves felt. At the opposite extreme, some of the largest post-war comprehensive redevelopments involved designs which, although repetitive, did not lend themselves to ready-prepared systems. The architectural idiom of these large projects was usually so strong that they had few imitators, or else their influence was dulled by the

Park Hill Housing Scheme, Sheffield.

The Barbican, City of London, looking across the Arts Centre and Frobisher Crescent to the residential towers.

Princess Margaret Hospital, Swindon.

University of East Anglia, Norwich.

Gymnasium, Mauritius.

time they took to complete. Two cases on which Arups worked were Park Hill in Sheffield and the Barbican in London. The Park Hill flats, snaking along the hillside just east of the city centre, were too complex in their configuration and section to be candidates for full system-building. But their concrete frame and brick infill is the key to their overall texture.

At the Barbican, Arups were appointed in 1960 and experienced every twist and turn of that daunting, not to say tiresome, project. From the outset the plan by Chamberlin, Powell & Bon involved a complex integration of functions, circulation and levels, with the added complication of an Underground line through the middle of the site. Putting some of the residential element in towers of over 40 storeys (higher than any of their kind in Europe) allowed breathing-space elsewhere, but created onerous problems in foundation design. As if that were not enough, the decision five years later to turn the concert hall and theatre as originally planned into a fully-fledged Arts Centre prolonged the ordeal yet further. Nothing about the Barbican was straightforward: even the relative simplicity of elements such as the cross-wall construction of the long terraces of flats was lost in the tough, mannered architectural finish that was adopted. As at Park Hill, the straightforwardness of structural repetition was overwhelmed by other demands put upon the scheme.

In tandem with housing went work on the other publicly-funded building types of the post-war period, especially hospitals, schools, and the university buildings of the 1960s' boom in higher education. The ideal of modular, prefabricated construction applied to many of these projects as well, for instance in the precast units of the ziggurat residences at the University of East Anglia (1965-8). But they were seldom as reliant on the widespread use of contractor-sponsored systems. Because each project was more localised, together they have seldom received the same attention (favourable or otherwise) as their housing counterparts.

The story of Arups in its first 20 years or so is a microcosm of what happened to architecture and construction after the war, with a vigorous public sector allied to progressive ideals in building. Arups was not the only practice which flourished in that exhilarating atmosphere and, allowing for some significant differences, a number of other practices could be used to illustrate the same general tendencies. But from the outset Arups was set on a path of a yet more adventurous kind. One major characteristic of the way the practice developed was the continued hankering after a closer integration of all those involved in the building process – the pursuit of total architecture – which culminated in the setting up of Arup Associates. Another was the creation of overseas branches, of varying strength and independence. To some people these developments may have seemed incidental, but in the long term they have assumed a crucial importance.

Though the establishment of Arup Associates may have seemed inevitable, particular circumstances brought it into being. By the late 1950s, Arups had what was known as the Building Group, a multi-disciplinary team of architects, engineers and quantity surveyors offering a comprehensive design service. Its forte was factory design, traditionally a sphere where clients want quick, uncomplicated decisions. Thus CIBA Ltd, a chemical firm specialising in resins for bonding timber, came to Arups for its research laboratories and processing plant. But if a glue factory might seem an appropriate project for an engineering practice, an Oxbridge Ladies College was not. When Philip Dowson, the leading architect in the Building Group, was approached about an extension to Somerville College, Oxford, the time seemed right to set the team on a different footing.[15] Arup Associates was announced to the world in autumn 1963, with Dowson, Ove Arup, Ronald Hobbs and Derek Sugden as its first principals. Peter Foggo was one of the founding associates.

15. Philip Dowson, *RIBA Journal*, 3rd ser. Vol. 88 No. 8 (August 1981), p. 60.

Ahmadu Bello University, Nigeria.

However much it was claimed that there was nothing new in what was being said and done, some architects were furious at the emergence of Arup Associates from the ranks of an engineering practice. At least one architectural partnership declared that it would never work with Arups again. To overcome this professional antagonism meant demonstrating that this development was more than just a ruse to get more work for the firm as a whole.

The early buildings of Arup Associates were emblematic of the ideals the team stood for, above all the disciplined integration of structure and services. The way this was handled related to the 1960s fascination with industrialised building and the engineering approach which that demanded, and was given an architectural expression which left no doubts about how the parts had come together. Clarity in the frame and grid, sometimes to an exaggerated degree, were the hallmarks of the team's designs. The full effect of this approach was revealed in the Mining and Metallurgy Building at Birmingham University (1963-6), where the heavy ductwork necessary for servicing the laboratories dictated the building's anatomy. Each of the four blocks of the building is laid out on a 20ft grid, with the vertical services carried up within the clusters of columns supporting the laboratory floors. Horizontal services run within the precast floor slabs. The form of the façades depends entirely on this disciplined system, punctuated solely by the ventilation units rising like watch towers above the perimeter columns.

26, St. James's Place, London.

The muscular logic of this Birmingham project was soon repeated in a less rigid way elsewhere, especially in university residences put up at the height of the mid-1960s higher education boom. At Leckhampton House for Corpus Christi, Cambridge, and at the Sir Thomas White Building for St John's College, Oxford, the discipline of the concrete frame, defining the configuration of student rooms, was broken by the staircase towers: a system still exposing the bones but more obviously flexing the joints. Simultaneously a string of projects of a quite different kind, focused on conservation and musical performance, was initiated by Benjamin Britten's commission in 1965 for the conversion of the Snape Maltings to a concert hall for the Aldeburgh Festival. No conversion is straightforward especially when, as at Snape, a disastrous fire requires the work to be done twice over. But at a time where conservation was still regarded as a marginal and eccentric activity, Derek Sugden and his colleagues showed how much pleasure and use could be got from adapting a building that was already to hand. From Snape followed a succession of projects for upgrading and converting concert halls and theatres to meet the more exacting demands of audiences and performers.
In contrast to the creation of Arup Associates, which was

Coventry Cathedral.

Shabanie Mill, Zimbabwe.

Standard Bank, Johannesburg.

St Catherine's College, Oxford (above and below).

Ninewells Hospital, Dundee (below).

the logical outcome of the message that Ove Arup had long been preaching, the pattern of overseas expansion seems less premeditated. In terms of Ove's own attitudes and ambitions this overseas blossoming reflected the liberal and occasionally opportunist way he saw things, plus the fact that he was more international in outlook than most of his contemporaries. Also in his formative years with Christiani and Nielson he happened to have worked with one of the major European engineering firms that operated internationally. After the war British construction efforts naturally focused on home demand, but the by the mid-1950s large British construction firms were beginning to look abroad, particularly to the African colonies; the great era of decolonisation was yet to come. Like such firms, the way that Arups expanded overseas tended at first to follow the flag. The most obvious connections were in that direction, helped by ties of language and common codes of building practice.

One way in which Arups' overseas branches were seeded was through foreign-born engineers who had worked in the London office returning to their home country. Michael Lewis and Jack Zunz both trained in South Africa and came to London to further their experience: in Zunz's case, to be thrown straight into working on the Smithsons' Hunstanton School. When they decided to return to Johannesburg, Ove Arup characteristically suggested that instead of setting up in their own name they should establish an outpost of Arups. The office at Lusaka in Zambia was founded by Fraser Anderson and others in much the same way.[16] Arups in Johannesburg grew to be over 200-strong, successfully wresting work from the dominance of design-build contractors. Projects such as the Standard Bank (1964-70) and the Carlton Centre (1966-75) easily matched in complexity anything being done in London: indeed the suspended floors of the Standard Bank Tower were as bold as any office design of their time.

16. Ove Arup & Partners, *London Newsletter* No. 15 (September 1963), p.1.

The more usual way for an Arup office abroad to be started was through the request of a local architect for Arup's assistance. That was how the Dublin office was set up, straight after a legendary meeting between Michael Scott and Ove Arup in 1946 concerning his Dublin Bus Station and Offices (1946-53). In Nigeria the architect Maxwell Fry had been appointed Town Planning Advisor to the Resident Minster in 1943, and a year later had produced 'Village Housing in the Tropics', a bible for colonial administrators. From then on he and Jane Drew designed numerous concrete-framed schools, plus colleges and government buildings, along the whole West African coast. The engineering design for most of these was done by Arups in London, but work on Ibadan University required a resident engineer: the Nigerian office dates its origins to that posting.

The generation following Maxwell Fry found work beyond the Commonwealth connection, especially in North Africa and the Middle East. Arups staked out the same territory – Algeria, Tunisia, Libya, and eventually Saudi Arabia. In 1966 Trevor Dannatt won a competition for a hotel and conference centre in Riyadh and asked Arups to join him. Almost immediately came a similar request from Frei Otto and Rolf Gutbrod for a similar project at Mecca, with a single-curvature, cable-hung roof over the conference centre part. These Saudi Arabian projects, both started in 1968, symbolised the future internationalisation of construction: designed in London and Stuttgart, built by firms based in Paris and Rome, and reliant on subcontractors in France, Germany, and Lebanon. By the late 1970s the outreaches of Arups were almost as far-flung: all were founded in connection with specific projects.

St Cross Library, Oxford (above and below).

In the catalogue of overseas offices founded on the back of a particular architectural project, the link between the Australian offices and the Sydney Opera House is pre-eminent, though in the Australian story the opening episodes have tended to overshadow all that has followed. Arups' Sydney office opened on the Opera House site in 1963, the year originally assigned for the building's completion. Not only was that event still a long way off: instead of being a celebration of the completion, the year was marked by the first serious breakdown in relations between the engineer and architect. From then on the legends of what happened are a distraction from understanding what was ultimately achieved. People talk about the affair, not the building.

The decision to build a music centre on a promontory in Sydney Harbour was made in 1954, and the competition for its design was won three years later by Jørn Utzon. His drawings, though 'simple to the point of being diagrammatic', were sufficient to convince the assessors that they had found a proposal 'capable of becoming one of the great buildings of the world'.[17] Utzon took no structural advice in devising his entry; nor did the assessors include an engineer. After he had won Utzon visited Ove Arup in London, who recognised from the published sketches what an intriguing project this was. Utzon was, felt Arup, the ideal collaborator, with all the qualities he looked for in an architect including a beguiling sense of passion; and of course he was a fellow Dane. The promise of a fruitful partnership was symbolised by the shell roofs of Utzon's design, in which structure and architecture were indistinguishable. The fact that there was no immediately obvious way of building them was no reason for turning away from the scheme. As Arup later recalled: 'the "shells" would be solved somehow – this scheme I wanted to go in for, with all I had'.[18] Arups were appointed in 1957, and acted as the principal agents for the first two stages of the construction of the Opera House.

17. Michael Baume, *The Sydney Opera House Affair* (1967), p. 120.

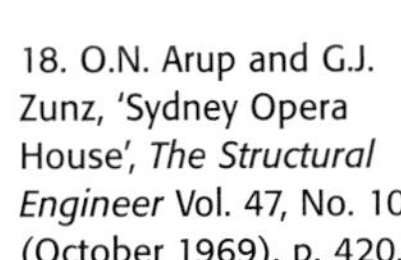

18. O.N. Arup and G.J. Zunz, 'Sydney Opera House', *The Structural Engineer* Vol. 47, No. 10 (October 1969), p. 420.

Sydney Opera House. The glass screen enclosing the end of a vault being erected (above); Detail of the glazing system showing the ball-jointed horizontal tie (below).

The fact that the building acquired a legendary status well before it was finished partly reflects the time it took to complete, plus the very public nature of the altercations that occurred along the way. But that is not enough to explain why the Sydney Opera House is the one building in Australia that everyone has heard of. What the competition assessors realised was that the Utzon scheme, with its shell roofs leaping from a deep podium, provided a powerful and attractive symbol for the Sydney waterfront. The roofs were the immediately recognisable aspect of Utzon's submission and early on attracted a loyalty which helps explain why, long before they were finished, people were anxious to see them completed as he had intended. They were also by far the most visible part of the project, and as their profile took shape their symbolic importance was confirmed – well before the spaces beneath them were fitted out.

The roofs were also the crucial part of the project for Arups, a challenge which took almost six years to resolve and stretched the practice to the limit. Essentially the design went through three stages. The roofs were first thought of in the

University of Sussex, Brighton.

way they have always been described, that is as shell structures gaining strength from their curvature. Ronald Jenkins, the master of shell construction since Brynmawr days, led the design at that stage. However the shape of the roofs was too geometrically undefined to enable shells of such a size to work, and anyway they would have been formidably difficult to construct. The first major alternative was to have two shells, strengthened by web members, but this too produced problems of geometry and construction. The eventual solution, which allowed the roofs to be built more easily, involved shaping the shells from segments of the same sphere. This produced a much simpler geometry, and allowed the precast concrete ribs which made up each shell to be of similar overall form regardless of their length. It is not entirely clear who thought of this solution first. Utzon claimed that it was his, and thus it became cited as an example of his helping the engineers out of their difficulties; yet there is an even stronger case for saying that it derived from a radical rethink at Arups, which Utzon quickly recognized as a breakthrough. In the long run the matter of attribution is probably not as important as the fact that the spherical geometry at last allowed the project to proceed.

Maidstone Technical College.

Were it not for the absorbing fascination provided by the roofs, there are a good number of other aspects of the Opera House project which would be the centre of attention. One is the approach steps and concourse by which the pedestrian approach to the building is separated from the route for vehicles. The steps and platform are carried on immense prestressed beams of spans varying between 41m and 56m. Another is the glass walls, which as well as sealing the ends of the shell vaults fan outwards to resolve the difference in area between the roofs and the podium below. The glazing is a mullion system, which as well as filling the full height of the openings has to take account of the complex curvature of the shells and podium. This is handled through the use of ball-jointed horizontal ties which connect the mullions across the full extent of the curve, controlling the way the roof spreads outwards over the podium. Finally there is the accommodation beneath the roofs, and its servicing, which had to be tailored to requirements that changed many times during the course of the project. The fact that the Opera House is not what its title claims (though opera can be produced in the smaller of the main auditoria) hints at how drastically the design had to be re-shaped before the end was reached.

South Bank Arts Centre.

The determination that made Arups stick with the project through to its completion in 1973 had two main causes. The first was that, having started, there was no way of turning back, despite the agonies of pressing onward and despite the realisation that much of what was happening was beyond their control. From the premature decision, on political grounds, to start work on the substructure before the rest of the design was resolved, the project had an illogical momentum. But secondly there was the conviction that despite many torments along the way the result would be a masterpiece; indeed the difficulties could be seen as proof that something out of the ordinary was being achieved. As Arup said, 'this was no ordinary building, ordinary rules did not apply, and it was almost an advantage to be unimpeded by precedent'.[19] However it gradually had to be admitted, more reluctantly by Arup himself than by other members of the team, that the work would have

19. Ibid, p. 421.

to be completed without its original creator. Whatever the immediate causes, ultimately Utzon's resignation was an acknowledgement that he was overwhelmed by the complexities of the project. Masterpieces are usually attributed to a single individual, and the inspiring contribution of Utzon to the Opera House has never been denied; but more than usually so, this was a building which owed it successful completion to the perseverance of others.

OCBC Centre, Singapore, under construction 1975. Three tiers of floors are supported off transfer girders spanning between the service cores.

As long as the Opera House saga continued, the separate Sydney office provided the chance of a respite for engineers wearied by its enormity. What was on offer was an increasing range of projects, with a specialisation in high-use buildings but extending into highway engineering, transport studies, and other urban issues. Many were in places well beyond Sydney, and so led to the establishment of separate offices in Canberra, Perth, and Melbourne. By the time the Opera House was finished there was an Arup network across Australia, which had good links with previously-established offices in Malaysia and Singapore.

Ife University, Nigeria.

In 1970, with the Silver Jubilee of the practice approaching, Ove Arup was persuaded to give the talk known as the 'Key Speech' to summarise the principles which he believed should guide its development. As the title it acquired suggests, this was soon regarded as the most significant testimony of what the practice stood for, not just in its engineering aims but also in its social and moral outlook. It is indeed a remarkable pronouncement, not least because engineering practices are not given to making philosophical statements about their aims; and if they did so they would be unlikely to produce such a broadminded and jargon-free message. The speech can also be read as a valedictory statement at the point which initiative in the firm passed from its founder to the next generation.

The narrowly engineering aspect of what Arup said was brief; a reaffirmation of the idea of total architecture, and of the need for collaboration to achieve it. More than previously he acknowledged that collaboration was best obtained by the inclusion of skills within the practice. So the diversification into planning, geotechnics, and other aspects of construction, even into architectural design, that had occurred before 1970, was inevitable and not to be resisted, provided that the different strands could still be properly interwoven. Thus far Arup's message was a familiar one, but then a tone of anxiety creeps in. If the penalty of success, and of diversification, was constant growth, how were the values he held dear to be maintained? In part it was a matter of getting the right people and keeping them, by making sure they had rewarding and creative work. But also it meant maintaining a collective sense of the quality and social usefulness of engineering. In Arup's words: 'Unless we feel we have a special contribution to make which our very size and diversity and our whole outlook can help achieve, I for one am not interested.'[20] The area where the sense of unified endeavour was most put to the test was in the overseas practices, about which Arup was particularly anxious. Setting up branches 'in various exotic places' was always a temptation, but with each extension of the Arup empire it became potentially more difficult to maintain and police a common outlook.[21]

One clear sign of what Arups stood for was the way the

20. *Arup Journal* Vol. 5 No. 4 (December 1970), p. 3.

21. Ibid, p. 4.

partnership was organised, which is only lightly and apologetically touched on in the 'Key Speech'. Since 1948 the practice had been a conventional partnership: Ove never let people forget the £10,000 he had invested in it. Up to a point it worked very well, but the concentration of ownership in the partners was oddly out of tune with the ethos that governed other aspects of the practice. More specifically it had the financial disadvantage that the partners were taxed as if the profits of the practice were part of their personal income, plus the organisational drawback that it focused too much power at the top. The 'Key Speech' alluded to the fact that a search was in progress to find a different form of partnership, one which would be more democratic and would protect the practice from the threat of takeover by outsiders. With no obvious model to imitate, this reorganisation was as complex and innovative as the boldest piece of engineering. The eventual result, announced in 1977 after almost a decade of discussions, involved the termination of the old partnership and the transfer of ownership to a new Ove Arup Partnership controlled by two unlimited companies, one essentially a charity and the other responsible for the physical assets of the practice. Separate from this were to be the operating companies, principally Ove Arup & Partners and Arup Associates, each with its own Board of Directors. Profits were to be shared amongst staff after a proportion had been paid into the Ove Arup Partnership companies.[22]

School of Oriental and African Studies (SOAS), London University.

Emley Moor Tower.

The long-term significance of this change is that it has secured Arups' financial independence as a practice. At the same time, under the reorganisation more individuals than before gained a voice in its running, allowing management by meritocracy to emerge. Roger Rigby, who masterminded the transition, remembers 'the tremendous freeing of energy that resulted'.[23] The value of diversification had been foreseen, perhaps more by others than by Arup: now the administrative means had arrived which allowed it to happen.

In the 'Key Speech' Arup made no reference to the fundamental changes that were affecting the building world, particularly in Britain, at that time. Yet the fact that he felt the need to reaffirm the principles that should guide the practice reflected the threat that these changes presented. Arup's belief in an integrated building industry dedicated to moral and social ends reflected exactly the aims of the post-war welfare state. He was to engineering what Beveridge and Tawney were in the wider political sphere. By the end of the 1960s there were clear signs that the economic system which had underpinned the development of the welfare state since 1945 was coming under severe strain, and that this was having major consequences for new construction. The certainties which many people had taken for granted since the war were coming to an end.

St John's College, Oxford.

This transformation in the British economy was most acutely felt in a decline in public sector work, which started in 1969 and since then has never been reversed. In 1969 the public sector commanded 51 per cent of new construction: by 1985 that proportion had fallen to 32 per cent.[24] Within that decline new public housing dipped lower than other publicly-funded projects. At the time of Arup's 'Key Speech' it might still have been possible to believe that the recession was short-term, but the oil crisis which started in 1973 put an end to that hope.

These changes were of profound importance in determining how Arups developed in the following years, more so than

22. Ove Arup & Partners, *Newsletter* No. 105 (1977).

23. Interview with Roger Rigby, 21st August 1995.

24. Michael Ball, *Rebuilding Construction* (1988) pp. 101-7.

anyone at the time could have foreseen. In more than one sense they helped justify the virtues of diversification. As public sector work, with which the practice had been so closely associated, went into decline, there was the capacity to switch to other kinds of work. And as the oil crisis took hold in Britain, sending architectural and engineering firms scurrying for work amongst the oil-rich countries of the Middle East, Arups had the benefit of having been there already.

Western Bypass, Gateshead.

At a deeper level Arups were able to adapt to a changing perception of the engineering contribution to building. In the predominantly public sector projects with which the firm grew up, the ideal was to achieve production targets and standards through continuity from one project to the next, particularly through the use of industrialised systems. Though their role was crucial, architects and engineers characterised themselves as agents in the service of a general cause. Anonymous corporate effort was the keynote, in line with the scientific ethos of Modernism. However from the early 1960s there was also evident an alternative way of looking at the relationship between architecture and industrialised technology, in which structure was treated expressively for its own sake and in the design of which the engineer was promoted to a far more heroic status.[25] When the old mode of industrialised building began to fall from favour, because the degree of state intervention that made it possible could no longer be sustained, construction projects became more fragmented and episodic. In such an environment the ideal of an all-embracing Modernism faded, to be replaced by one of a far more varied, if not arbitrary kind.

Mosque, Mecca.

25. Peter Buchanan, 'Nostalgic Utopia', *Architects' Journal*, Vol.122, No. 36, 4th September 1985, pp. 60-9.

Royal Exchange Theatre, Manchester.

Within a matter of just a few years this transition took the structural engineering side of Arups from an association with the most sober of post-war public sector projects to a reputation as the engineering brains behind the most exciting high tech architecture. The reasons for this leap are not hard to find: propelled from within by the search for new challenges, and pulled from without by architects with new ambitions. And in some respects it is a leap which can be exaggerated. For every spectacularly expressive project there was another of a much more humdrum kind, in which the lessons of industrialised building were applied in the cheap and efficient use of everyday components. This sub-theme of gradually increasing use of prefabrication in construction is the unglamorous aspect of what happened to the post-war architectural dream.

Nothing signalled the turnabout of the early 1970s more clearly than the revival in the use of steel, and the building which best symbolised that revival was the Pompidou Centre. Even after the lifting of restrictions in the 1950s, the use of steel was handicapped by inadequate promotion and by anxieties about the fire protection that steel frames required. Improved methods of steel production, including the ability to fabricate complex shapes, helped bring steel back into favour, but in the revival architectural preferences counted for even more. The language of structural expressiveness – the loose-fit frame as a container for adaptable spaces – implied machine technology and thus the use of steel. Architectural points of reference confirmed that choice, whether the railway trainsheds and industrial monuments of the past or the science fiction cities of the future. When it came to building that vision on a parking lot in central Paris, steel was the material that came immediately to mind.

Like the Sydney Opera House, the Centre du Plateau Beaubourg (to call it by its original name) was conceived of as a multi-functional cultural centre, and its design was put out to competition. The late Ted Happold, then working in Povl Ahm's group at Arups, was sufficiently fired by the competition brief to persuade Renzo Piano and Richard Rogers that though the odds were crazy this was one worth going for. The key ingredient that most excited them was the idea of a palace of culture that would be unashamedly open, popular, and constantly evolving. Here was a chance for the collaborative exploration of what a flexible, communicative architecture really meant; even if their design wasn't chosen they would have learnt a lot along the way. The eventual announcement in 1971 that Piano and Rogers were the winners stunned the team. Even more it astonished the jury, who had never heard of that young partnership. But then the citation continued, 'in consultation with Ove Arup & Partners, Consultant Engineers'. 'Thank God we are all right', someone is reputed to have said.[26]

26. Nathan Silver, *The Making of Beaubourg* (1994), p. 42.

From the outset the building structure was conceived of as a framework providing highly adaptable spaces, with the services and means of circulation pushed to the outer edges and fully expressed – escalators and walkways on the front, mechanical services at the back. In the competition entry the pairs of columns supporting the floors created a forcefully rectilinear frame, saved from monotony by the signage and circulation systems. Having won, the team began to rethink that aspect of the design in the search for a form of structure of a more tactile, sympathetic kind. At Peter Rice's insistence the idea of using cast steel was brought into play. On the evidence of what he had seen of that material, especially as used for the joints of a spaceframe structure for the 1970 Osaka World Fair, it seemed that it offered a means of combining the machine aesthetic with the sculptural qualities of more traditional materials. At the Beaubourg the place where cast steel could be most effectively employed was at the junction of the floor beams and the external columns, in the immediate sight of people threading their way up and down the building.

Pompidou Centre, Paris: the outer columns and gerberettes.

Hence arose the gerberettes, named in honour of the 19th century German engineer Heinrich Gerber who had developed and patented a cantilever girder for bridge design. These are cast steel beams which pivot on bearings on the main columns, connecting the 44.8m truss girders of the suspended floors to the vertical tension ties on the outer edge of the building. They simplify and dramatise the structure, turning it into a positive and easy-to-appreciate statement about how the building works. Every element of the gerberettes and their bearings, connections, and ties is on full view, and they succeeded as everyone hoped they would in making the building feel special. Such huge biomorphic shapes could only have been achieved in cast steel, finished by fettling with a power-grinder to give each one an individual effect.

From many points of view the Pompidou project sounds like a re-run of the Sydney Opera House – the young architectural team chosen by competition, the Arup involvement, the constant glare of publicity, and above all the political vicissitudes that it had to survive: seven different Ministers of Culture came and went during the course of the work. But in terms of speed there is no comparison. The 11 internal frames plus the two end frames, and the concrete floors, took just eight months to complete, and the building as a whole was finished within six years.[27] Like the Crystal Palace, an obvious analogy in inventiveness and speed of construction if not also in appearance, its reputation long preceded its completion, and amongst the cognoscenti its influence was felt from the day the competition drawings were first published.

27. *Ibid*, p. 132.

Fleetguard factory Quimper. The suspension structure supports the roof and buildings services (above).

The movement towards the expressive use of steel which the Pompidou Centre helped legitimise coincided with the 1980s' demand for out-of-town factories and science facilities, for which the visibility of high tech structures was a bonus regardless of whatever else was claimed for them. The well-serviced, brightly-finished marquee, seen from far around, superseded industrial structures of a more understated kind. For Arups some of the classic examples were the Fleetguard Factory at Quimper in Brittany (1979-81), the Renault Warehouse and Distribution Centre at Swindon (1981-2) and the PA Technology Building at Princeton, New Jersey (1982-84); all of a type in terms of having suspended roofs, plus frames designed to suggest their infinite extendability. PA Technology, an Arup-Richard Rogers collaboration, comes closest to the Pompidou Centre in having its services showing externally between the bipod masts which support the roofs, and in exploiting the image of that conjunction to make a point about the innovative activities the building contains. Beneath the services runs the spine corridor, with offices and laboratories arranged in flexible formation down either side. Less explicit, but no less important, the platform carrying the services gives lateral stability to the masts.

PA Technology, Princeton, New Jersey, USA. Externalised structure used as symbol of innovation and flexibility (above and below).

The use of the gerberettes at the Pompidou Centre as a way of softening the repetitive angularity of the structure had few equivalents in the projects that followed. Though often highly crafted, their components are used in a way that suggests machine assembly, an effect compounded by the modular grid of the building layout. Yet the idea of biomorphic form, implying something more complex, luxurious and friendly, refuses to go away. There are the trees and branches of the Stansted Airport Terminal (1986-91), still on a grid but suggestive of natural growth, and more recently still there are the great asymmetrical trusses of Kansai International Airport (1988-95). In a sense Kansai represented a return to where the story started, since it involved Renzo Piano and Noriaki Okabe from Pompidou days, working with an Arup team led by Peter Rice (one of his last projects before his death in 1992). Knowing of that connection it is easy to detect how issues first explored on the Pompidou project re-emerged at Kansai. The Airport stretches out along its artificial island in Osaka Bay, with the main building merging directly into the long arms of the boarding wings. Above the different sections the roofs curve and combine, sharing a common geometry and a disciplined system of glazing and stainless steel finishes, yet providing a distinct sequence of spaces for people to pass through. Where Pompidou has its gerberettes, Kansai has tubular steel ribs, props, and connections which play a direct part in how that sequence is experienced. From entering the Airport beneath the sawn-off cantilever of the departure hall, to passing through the props of the main trusses and along the inward curving wall of the boarding wings, the structure

Theatre Royal, Plymouth.

shapes a sympathetic environment for this enormous complex.

Using steel in new ways was part of the general surge of interest in old and new materials in the late 1960s, engendered by the changing demands that were being put on the building industry as well as by the liberating influence of computer analysis. An even clearer indication of this new tone of engineering inquiry was the focus on lightweight structures. This rather catch-all term has been applied to the use of various forms of prestressed membrane in the double role of structure and covering, normally in combination with cable systems or some type of air support. Such structures have seemed ideal for covering very large spans like sports stadia and exhibition spaces, where the functions of the roof are less onerous than for buildings in continuous use. They are in effect circus tents on a grand scale and that association, suggestive of improvisation and festivity, is part of their appeal. Of course tents and awnings are amongst the earliest recorded forms of enclosure, but contemporary applications of the same idea had to wait until materials and techniques were available to enable the basic principles to be translated onto a new scale. In particular, until the arrival of computers it was difficult to achieve an adequate understanding of structures of such a non-regular kind, though models were of some help. And in the search for covering materials of the right stiffness, durability, and translucence, the arrival of PVC-coated polyester textiles and Teflon-coated glass fibre, in some cases as the by-product of space exploration, transformed what could be done with traditional forms.

Sulphuric acid plant, Kafue.

In the overall genre of lightweight structures the cable net designs of Frei Otto are especially renowned. Inspired by the saddle-shaped cable net roof of the Raleigh Arena in North Carolina he went on to explore much freer forms, with what at the time seemed astonishing results in the German Pavilion that he and Rolf Gutbrod designed for the Montreal Expo of 1967.[28] As well as having an irregular plan, this displayed an astonishing profile formed by cable nets hung from masts of different heights and angles. While that structure was being completed, Gutbrod and Otto made their first contact with Arups in connection with the Mecca Conference Centre, with its single curvature cable roof. From that collaboration came Arups' Lightweight Structures Laboratory, set up in 1973 as a think tank and workshop by Ted Happold and Frei Otto. The dividend of that development was that through allowing the exploration of visionary schemes (to take two extremes, 'The Arctic City' and 'Shadow in the Desert') it helped more modest applications of the same ideas to take root. The Schlumberger Research Laboratory at Cambridge (1983-5) is one example of the outcome, using a cable-hung membrane roof to provide a high, naturally-lit covering to the central space between two single-storey research wings. And as the technical problems have been answered, the symbolic potential of lightweight construction has come into its own, nowhere more so than in the Arups-Renzo Piano project Il Grande Bigo at Genoa (1992). There the tent coverings of a reclaimed wharf, hung from tubular steel masts like ships' derricks, are entirely appropriate to the nautical setting.

Staatsgalerie, Stuttgart.

28. Brian Forster, 'Cable and Membrane Roofs – A Historical Survey', *Structural Engineering Review* Vol. 6 (1994), pp. 161-67.

It would be wrong to give the impression that concrete was the loser in this diversifying use of materials. It is true to say that confidence in the durability and attractiveness

Il Grande Bigo, Genoa.

of unfinished concrete had begun to wane by the 1970s, contributing to a revision of the post-war faith in its capacities. Turlogh O'Brien, doyen of Arups' research and development group, warned about the intractability of its defects in 1970, well before the concrete repairs business came of age.[29] But that did not stem the exploration of new ways of using concrete, as two examples from the opposite ends of the engineering spectrum can show.

29. Turlogh O'Brien, 'The Changing Face of Concrete', *The Arup Journal* Vol. 5 No. 3 (September 1970), pp. 20-23.

Kylesku Bridge.

At the Menil Collection Museum in Houston (1981-6) the demand for a form of indirect natural lighting led Peter Rice and Tom Barker from Arups, working with Renzo Piano, to devise a system of roof-top louvres. The sun striking the back of one would be reflected onto the underside of the next and thus into the gallery space below. The ideal shape, like half a ship's hull, suggested the appropriate material to use, the ferro-cement of boat-building. Each louvre was made by spraying cement onto a thin mesh in a mould, and was hand-finished. Because the louvres were too long to be self-supporting, and required a glazed roof above them, they were conceived of as the lower membrane of a truss, the top half of which was designed to be made from ductile cast iron. The effect of this combination, as felt by the gallery-goer, is of light filtering through the leaves of a broad and generous tree.

Swimming pool, Peebles.

Exchange Square, Hong Kong.

In total contrast, as the Menil Museum neared completion discussions began which led to the installation of the Ravenspurn North gas processing plant in the North Sea 80km off Flamborough Head. From the earliest days of North Sea oil and gas exploration, Arups argued that concrete was a valid material for the extreme environmental conditions that the various installations would have to withstand. American engineers, drawing on their experience from the Gulf of Mexico oilfields, favoured tubular steel structures on steel piles, and introduced that thinking to the North Sea projects. The Arup alternative started from the assumption that the major problem to be addressed was the choice of a suitable form of foundation, and that a gravity structure would be easier to anchor and ultimately to remove. This seemed to point logically to the use of concrete, which also promised benefits in ease of construction and maintenance. Throughout the oil boom years of the 1970s the American approach predominated in British offshore fields, despite a number of bids based on the Arup alternative: the Norwegians meanwhile showed a much greater interest in concrete gravity structures. So Ravenspurn North represented a breakthrough in British North Sea installations. Designed as a caisson with three shafts to support the operating platform, the whole structure except the platform was constructed in dry dock before being floated and then towed into position. The cellular form of the caisson made it easy to submerge through the gradual flooding of the cells: conversely, when no longer needed in its first location it can be re-floated and towed elsewhere.

Lloyds building, London.

A last word in this roundup of the developing use of materials should go to the revival of traditional methods of construction. In conservation work they had hardly gone away, but as long as an ideological division existed between working on old buildings and designing new ones, the chances of absorbing lessons from the past into the mainstream of practice were slender. It is a further significant aspect of the transformation of the building world that has occurred since

Old Vic restoration, London.

Seletar Interchange, Singapore.

Yulara Village, Ayers Rock, Australia.

Burswood Island Resort.

Sydney football stadium.

Newcastle Theatre Royal.

1970 that the value of existing buildings, whether designated as historically important or not, has gradually been recognised: thus the need to understand how they function is no longer regarded as an esoteric byway. Arup Associates' creation of the Snape Concert Hall symbolised the beginnings of that change of heart within the practice. As a lesson in dealing with existing buildings it was all the better for not being thought of as just a scholarly exercise in conservation. With succeeding projects of a similar kind it has become increasingly possible to think unashamedly in terms of traditional forms and methods, not just for dealing with existing buildings but also on totally new projects. For instance at the new Glyndebourne Opera House (1992-4) the relationship to the adjoining family house suggested the use of brick, but whereas a few years ago that would have meant a brick skin on a framed structure, the design chosen was predominantly for loadbearing brick in lime mortar. The adoption of this method signals a major reconsideration of the relationship between progressive innovation and traditional methods.

In an account of what has happened at Arups in the quarter century since 1970 it is easy to overemphasise the significance of structural engineering. That bias follows naturally from the way the practice first made its name, as well as from the simple fact that buildings attract attention and debate far more than most other forms of engineering. Within recent memory the major development at Arups, on top of keeping abreast of dynamic changes in building requirements and design, has been the expansion of activity into a whole range of related spheres, for many of which the practice is now as well-known as for its original expertise. Other essays in this book show how many technical disciplines it now encompasses. Through this process, total design has come to mean the provision of a comprehensive range of engineering services, plus other appropriate skills, so that like a well-stocked department store almost everything can be provided under one roof. As with a store, the problem becomes one of keeping the totality together, ensuring that all the parts support each other and that none take on a life of their own that detracts from what the practice as a whole is doing.

In a sense, civil engineering was there from the start, by virtue of Ove Arup's personal background and the world the practice grew up in. It also offered an obvious opportunity to play the role of prime agent in total design. However the way civil engineering came to prominence is also interesting as an example of how work overseas changed the complexion of the practice: the beginnings of a fundamental change of balance between the offshoots and the centre. Road design was significant to the West African practices almost from the start, and by 1971 over 2000 miles of Arup-designed highway had been completed in that region. That experience fed into work back home and elsewhere overseas, from the Gateshead Western Bypass (1971-4) – the first Arup project of its kind in Britain – to roads conceived on an heroic scale such as the Trans Saudi Arabia Expressway and the Bukit Timah Expressway in Singapore. The related transportation studies have always been an added strength of the overseas offices, especially in Australia, often operating in a political environment where strategic thinking comes more naturally than it does at home.

In the light of that experience abroad it is easy to understand why the emergence of a strategic project such as the Channel Tunnel Rail Link has seen Arups prepared to enter the fray over its planning and construction.

Tracing back the patterns of engineering diversification, in retrospect the linkages seem obvious but some might not have developed as fast or as strongly as they have done were it not for the enthusiasm of individual engineers. Thus bridge design is an essential part of transport engineering, but was promoted in its own right by Povl Ahm and Bill Smyth. Similarly, mechanical and electrical engineering had their beginnings in Arup Associates, plus work on telecommunication towers in South Africa and elsewhere, but it took some persuasion from John Martin and Jack Zunz for it to be recognised that the practice of total design was incomplete without building services. If there was scepticism about this particular development, it was because services had traditionally been regarded as a technical extra rather than an integral part of building design, and engineers who specialised in services looked upon themselves as playing a secondary role. Almost as soon as it was known that Arups had adopted the cause of building services, large projects in that sphere came their way such as the Carlsberg Brewery in Northampton and the first part of the National Exhibition Centre, Birmingham. In the long run the move proved fully justified, though it took some time to establish the principle of designing every aspect of a building with their environment and climate in mind.

The multiplication of specialisms gathered pace in the early 1970s and has hardly let up since. Arup Fire, Arup Geotechnics, and Arup Acoustics are just three examples of groups which have emerged from the work of the practice to acquire quasi-independent reputations. With fire engineering it had become obvious by 1970 that changes in the scale of buildings, and the spaces within them, along with new methods of construction particularly of façades, demanded a reconsideration of conventional fire safety measures. It made no sense to treat the Pompidou Centre as if it was a traditional building. Developing a new approach, meant not just adjusting the old rules to new methods and materials but thinking afresh about how fire and smoke behave, and how humans react when a fire occurs. Without the acceptance of an alternative approach a building such as Stansted Terminal, based on the ideal of an unimpeded concourse beneath a high ceiling, would not have been possible, and fire safety in other contexts such as offshore installations, would have been understood quite differently. The urge to take a broad and sometimes radical view has also characterised Arup Geotechnics, associated in its formative years with David Henkel whom Peter Dunican wooed from academic life in 1970. Henkel's interest in the relationship between the shape and form of a site and the geological processes at work has been fundamental to the success of deep excavation projects and schemes in geologically complex locations. In a similar way, Arup Acoustics, established in 1980 by Derek Sugden, Richard Cowell and Peter Parkin, has been drawn into issues of engineering new types and sizes of building. To most people acoustics means how an orchestra is heard in a concert hall, and that quest for perfection in performance is central to this aspect of Arups' expertise, but nowadays there is much more to the subject than that.

Tyne & Wear Metro: Byker Viaduct.

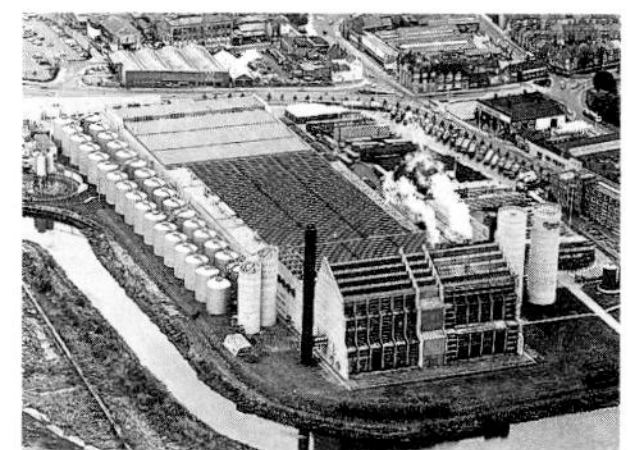

Carlsberg Brewery, Northampton.

Princes Square shopping centre, Glasgow.

Tour de la Liberté, Paris.

Francistown Bridge, Tati River, Botswana.

John Lewis, Kingston-upon-Thames.

Brisbane airport, Australia.

Hopewell Centre, Hong Kong.

Cranfield Institute library.

Mediathèque, Nîmes.

It is just as likely to involve studies of noise and vibration arising from construction and from the location of buildings on crowded and awkward sites.[30]

The most conspicuous common denominator in these developments, apart from their Arup parentage, is the way they have benefited from computers and information systems. Future historians, asked to pinpoint the major developments in late 20th century engineering, are as likely to cite the general impact of computers on analysis and conceptual thinking as they are to refer to specific projects. The boundaries of innovation have been fundamentally altered, allowing shapes, forms and environments to be explored in a manner that was hardly possible before. In the early days of Arups' manual calculators were the most that was available to help an analysis: a Brunswiga calculator was used on the Brynmawr Rubber Factory shells. The first computer, an Elliott 803B, was installed at Arups in 1964, in the era when such machines were hired rather than bought and occupied whole rooms rather than one end of a desk. The reluctance to rely on these machines was overcome by the experience of using a Pegasus computer on the Sydney Opera House: without it the crucial three-dimensional analysis would have been impossible. But what then seemed the height of sophistication was primitive by comparison with what computers can achieve today, in analysis, modelling, and the communication of information. Once looked upon as a labour-saving benefit, they are now the major stimulus to understanding complex geometries and environments. And as well as exploring the use of computers for their own needs, Arups has since 1981 offered communication and information technology as a specialist service.

High speed communications now bind together the Arup offices around the world in a manner that, had he anticipated it, might have allayed Ove Arup's anxieties about the fragmenting effects of global diversification. The international spread that started with offices in Ireland and Africa now extends to countries as diverse as Turkey and America, Germany and the Philippines.

Ove Arup & Partners Hong Kong can stand as an illustration, at the spectacular end of the scale, of how the more recent growth of overseas offices has taken place. Appropriate to Britain's changing status in the world, the starting-point there was not through a London-based architect but via a request from a Hong Kong practice, specifically for assistance in contesting a court case following damage to a block of flats in a major landslide. Unfruitful though that may have seemed it was the harbinger of Arup's reputation in Hong Kong for geotechnics and foundation engineering. The work on structures that followed, including the Hopewell Centre tower and projects relating to the Mass Transit Railway, justified setting up a permanent office in 1976, and the establishment of a local partnership two years later. Within a decade the office was a 300-strong version of the London headquarters (and over 600 by 1996), with its own research and development, geotechnical and acoustic specialists and a sphere of influence extending to Malaysia, Indonesia and Korea.[31] In view of the future of Hong Kong, the most significant development of all in the 1980s was the securing of projects in mainland China. The coal-fired Shajiao B power station in the Guangdong province (1985-7) would have been a challenge enough in terms of size, ground

30. Richard Cowell, 'Arup Acoustics: Ten Years' *The Arup Journal* Vol. 25 No. 2 (Summer 1990), pp. 9-15

31. John Anderson, 'Ove Arup & Partners in Hong Kong – A History' (unpublished typescript, 1993)

conditions, and speed of construction wherever it was built, but to have been completed in a part of the world that few outsiders had ever visited made it yet more extraordinary.

Hong Kong can also serve to illustrate how, 50 years on, the Arup world interconnects. Simultaneously with the extending influence of the Hong Kong office, the building for which the island is architecturally best known, the Hongkong Bank, was being designed by a team in the London office: Arups were in fact appointed prior to the selection of Foster Associates as architects. In a sense the way this project was handled represented a return to the kind of relationship between the centre and the outpost of an earlier era, but the extent to which the globe had shrunk in the meantime dramatically reduced the significance of that distinction. Design work could be done where the right people were available, regardless of distance. A decade after the completion of the Bank the idea of a global office is even more of a reality, with designers in different corners of the world working on the same building at the same time via Internet. Thus a Bangkok transportation project is being shared between the Hong Kong and London offices, with information bouncing back and forth across the time zones. In that kind of interaction the idea of total design is adapted and extended to take in possibilities that were hardly dreamt of half a century ago.

Corby Power Station, Northamptonshire.

Microwave site, Botswana.

Writing the history of Arups is helped by the fact that the practice has generally had a clear view of itself. That is not to say that what has been said and written from within Arups is the only interpretation to be put upon events. There are many grades of meaning between how people describe their intentions and actions and what they actually do, and an outsider may find levels of significance which those directly involved are unaware of. But generally speaking, professional practices, especially in the engineering world, are not given to reflecting publicly upon their progress and aims, so when one of them does so it at least provides a starting-point for understanding what has happened.

The story line at Arups derives originally from the lectures and pronouncements of Ove Arup himself, which collectively presented a view of what should govern the development of the practice and how its success should be measured. In language and outlook what he said remained remarkably consistent, yet with the passage of time he had to adapt his message to take account of what was actually happening to the practice. Thus he was forced to acknowledge that total design or total architecture were as likely to be achieved through diversification within the practice as though collaboration with other firms, even if that meant growing to a size which alarmed him (by the time he died in 1988 there was a total staff, including overseas partnerships, of 3500). Also he saw diversification spread far wider than he had dreamt of earlier on, with far more of an emphasis on civil engineering, building services, and other specialisms, than the practice originally had. The overseas practices were another aspect of growth which was a consequence of Arup's outlook, and of the way he saw the practice developing, though they have multiplied in number and importance far more than was ever first anticipated.

If things had turned out differently there might not be such

Glaxo Wellcome Medicines research centre.

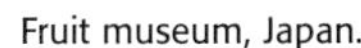

Fruit museum, Japan.

a temptation to return to the sayings of the master. As it is, what is most remarkable about Arups is the extent to which it has followed the pattern of development implicit in his thinking. The creative relationship with architects, for which the practice was best known in its earliest years, is still the core of its reputation, so much so that the roll-call of architects working with Arups reads like a line-up of the best-known names. At the same time the definition of what constitutes total architecture has been allowed to expand, but with the benefits of interaction between the different skills and disciplines still uppermost in mind. Conversely the absence of discontinuities is just as striking. Professional practices do not automatically renew themselves, or maintain their reputation from one generation to the next. At Arups the most painful point came in the early 1970s when the public sector projects which had been the chief means of sustenance began to decrease, and at the same time the original partnership showed signs of becoming an encumbrance. Setting up the new partnership structure may have seemed an abstruse legal exercise but its consequences were profound. It broadened the power structure of the practice and thus allowed talents to find their way to the top who might otherwise have left: through their influence Arups broadened its range of expertise at a critical time. As the guiding presence of Ove Arup began to wane, a new generation was ready to take over. At the same time the independence of the practice was secured at the beginning of an era when the very idea of professional independence began to be called in question.

No professional practice, even one as large and multi-faceted as Arups, can exist as a world unto itself. Ove Arup founded the practice at a particularly auspicious time in the history of the professions. Expertise founded on systematic knowledge was regarded as an essential tool in the remaking of post-war society, and engineering expertise was held in particularly high esteem because of the faith that was put in new technologies. The other main claim of the professional, to be acting disinterestedly for both the client and the community at large, was well in tune with the altruistic sentiments bred by wartime experience. 50 years on, such certainties seem almost unreal. The knowledge base of engineering is far more widely available, with the result that what was once the preserve of the consulting engineer can readily be professed by others in the building world. And those who have gained from that diffusion of knowledge owe a different kind of loyalty to the client and little or none to the world at large. A practice that stands by what it has always believed in is bound to feel beleaguered.[32]

The flexibility and inclusiveness that is built into the Arup system has been an excellent tool for survival so far, but who can say how it will serve the practice in the next half-century? The memory of past successes is a matter for pride and reassurance, but in a fragile world no-one can guarantee that they will always be repeated in the future.

32. John Martin, 'Professionalism…?', *The Arup Journal* Vol. 29 No. 1 (1994), pp.3-4.

Sydney Opera House. The interplay of shells and sky is the most enduring element of Utzon's original concept.

This talk was given on 9 July 1970, at Winchester, during one of the meetings of the Arup Organisation. It has since been known as 'The Key Speech'. This paper was written in response to a collective desire to continue to work together despite the changing pattern whereby the then London Partners were gradually divesting themselves of the ownership and control of the various Partnerships.

Ove Arup

The Key Speech

In its pre-natal stage, this talk has been honoured with the name of 'key speech'. It is doubtful whether it can live up to this name. What is it supposed to be the key to? The future of the firm? The philosophy? The aims? At the moment, sitting in my garden and waiting for inspiration, I would be more inclined to call it: 'Musings of an old gentleman in a garden' – and leave it at that.

I have written before a piece called *Aims and Means* for a conference of Senior and Executive Partners in London on 7 July 1969. It did not manage to deal much with means, however, and it is of course difficult to generalise about means, for they must vary with circumstances. The first part of this paper was published in *Newsletter 37*, November 1969. This you may have read – but I will shortly summarise the aims of the firm as I see them. There are two ways of looking at the work you do to earn a living:

One is the way propounded by the late Henry Ford: Work is a necessary evil, but modern technology will reduce it to a minimum. Your life is your leisure lived in your 'free' time.

The other is: to make your work interesting and rewarding. You enjoy both your work and your leisure.

We opt uncompromisingly for the second way. There are also two ways of looking at the pursuit of happiness:

One is to go straight for the things you fancy without restraints, that is, without considering anybody else besides yourself. The other is to recognise that no man is an island, that our lives are inextricably mixed up with those of our fellow human beings, and that there can be no real happiness in isolation. Which leads to an attitude which would accord to others the rights claimed for oneself, which would accept certain moral or humanitarian restraints. We, again, opt for the second way.

These two general principles are not in dispute. I will elaborate them a little further:

The first means that our work should be interesting and rewarding. Only a job done well, as well as we can do it – and as well as it can be done – is that. We must therefore strive for quality in what we do, and never be satisfied with the second-rate. There are many kinds of quality. In our work as structural engineers we had – and have – to satisfy the criteria for a sound, lasting and economical structure. We add to that the claim that it should be pleasing aesthetically, for without that quality it doesn't really give satisfaction to us or to others. And then we come up against the fact that a structure is generally a part of a larger unit, and we are frustrated because to strive for quality in only a part is almost useless if the whole is undistinguished, unless the structure is large enough to make an impact on its own. We are led to seek overall quality, fitness for purpose, as well as satisfying or significant forms and economy of construction. To this must be added harmony with the surroundings and the overall plan. We are then led to the ideal of 'Total Architecture', in collaboration with other like minded firms or, still better, on our own. This means expanding our field of activity into adjoining fields – architecture, planning, ground engineering, environmental engineering, computer programming, etc. and the planning and organisation of the work on site.

It is not the wish to expand, but the quest for quality which has brought us to this position, for we have realised that only intimate integration of the various parts or the various disciplines will produce the desired result.

The term 'Total Architecture' implies that all relevant design decisions have been considered together and have been integrated into a whole by a well organised team empowered to fix priorities. This is an ideal which can never – or only very rarely – be fully realised in practice, but which is well worth striving for, for artistic wholeness or excellence depends on it, and for our own sake we need the stimulation produced by excellence.

The humanitarian attitude

The other general principle, the humanitarian attitude, leads to the creation of an organisation which is human and friendly in spite of being large and efficient. Where every member is treated not only as a link in a chain of command, not only as a wheel in a bureaucratic machine, but as a human being whose happiness is the concern of all, who is treated not only as a means but as an end.

Of course it is always sound business to keep your collaborators happy – just as any farmer must keep his cattle in good health. But there is – or should be – more in it than that. (We know what happens to cattle.) If we want our work to be interesting and rewarding, then we must try to make it so for all our people and that is obviously much more difficult, not to say impossible. It is again an ideal, unattainable in full, but worth striving for. It leads to the wish to make everybody aware of, and interested in, our aims and to make the environment and working conditions as pleasant as possible within the available means.

This attitude also dictates that we should act honourably in our dealings with our own and other people. We should justify the trust of our clients by giving their interest first priority in the work we do for them. Internally, we should eschew nepotism or discrimination on the basis of nationality, religion, race, colour or sex – basing such discrimination as there must be on ability and character.

Humanitarianism also implies a social conscience, a wish to do socially useful work, and to join hands with others fighting for the same values. Our pursuit of quality should in itself be useful. If we in isolated cases can show how our environment can be improved, this is likely to have a much greater effect than mere propaganda.

There is a third aim besides the search for quality of work and the right human relationships, namely prosperity for all our members. Most people would say that this is our main aim, this is why we are in business. But it would be wrong to look at it as our main aim. We should rather look at it as an essential pre-requisite for even the partial fulfilment of any of our aims. For it is an aim which, if over-emphasised, easily gets out of hand and becomes very dangerous for our harmony, unity and very existence.

It costs money to produce quality, especially when we expand into fields where we have no contractual obligations and can expect no pay for our efforts. We may even antagonise people by poaching on their domain or by upsetting and criticising traditional procedures.

It also costs money to 'coddle' the staff with generosity and welfare, or to lose lucrative commissions by refusing to bribe a minister in a developing country, or to take our duty too seriously if nobody is looking.

Money spent on these 'aims' may be wisely spent in the long term, and may cause the leaders of the firm a certain satisfaction – but if so spent it is not available for immediate distribution among the members, whether partners or staff. So Aim No. 3 conflicts to that extent with Aims 1 and 2. Moreover, if money is made the main aim – if we are more greedy than is reasonable – it will accentuate the natural conflict about how the profit should be distributed between our members – the partners and staff or the different grades of staff.

The trouble with money is that it is a dividing force, not a uniting force, as is the quest for quality or a humanitarian outlook. If we let it divide us, we are sunk as an organisation – at least as a force for good.

So much for our aims. As aims, they are not in dispute. What is debatable, is how vigorously each shall be pursued – which is the most important; how to balance long term against short term aims. Let us first see what these aims imply.

Obviously, to do work of quality, we must have people of quality. We must be experts at what we undertake to do. Again, there are many kinds of quality, and there are many kinds of job to do, so we must have many kinds of people, each of which can do their own job well. And they must be able to work well together. This presupposes that they agree with our aims, and that they are not only technically capable but acceptable to us from a human point of view, so that they fit into our kind of organisation; and that they are effectively organised,so that the responsibility of each is clearly defined and accepted, in short, we must be efficient – individually, in all our sub-divisions, and as a world organisation.

I have tried to summarise the foregoing in a number of points. Like all classification, it is arbitrary and rough – but may nevertheless be useful as a help to understanding and discussion, if its imperfections and its incompleteness are borne in mind.

The main aims of the firm are:
(1) Quality of work
(2) Total architecture
(3) Humane organisation
(4) Straight and honourable dealings
(5) Social usefulness
(6) Reasonable prosperity of members.

If these aims could be realised to a considerable degree, they should result in:
(7) Satisfied members
(8) Satisfied clients
(9) Good reputation and influence.

But this will need:
(10) A membership of quality
(11) Efficient organisation
(12) Solvency
(13) Unity and enthusiasm.

Of course there is not really any strict demarcation between aims (Group A) and means (Group C) and the results (Group B) flowing from the whole or partial fulfilment of the aims in A. And it is not absolutely certain that these results are obtained. For instance, A3 and 4 (a humane organisation and straight dealings) can as well be considered as a means, and in fact all the points are to some extent both aims and means, because they reinforce each other. And there will be members who are dissatisfied no matter how good the firm is, and the same may apply to clients, who may not appreciate quality at all. But on the whole, what I said is true. We should keep the six aims in A in view all the time, and concentrate on the means to bring them about.

But before I do this, I will try to explain why I am going on about aims, ideals and moral principles and all that, and don't get down to brass tacks. I do this simply because I think these aims are very important. I can't see the point in having such a large firm with offices all over the world unless there is something which binds us together. If we were just ordinary consulting engineers carrying on business just as business to make a comfortable living, I can't see why each office couldn't carry on, on its own. The idea of somebody in London 'owning' all these businesses and hiring people to bring in the dough doesn't seem very inspiring. Unless we have a 'mission' – although I don't like the word – but something 'higher' to strive for – and I don't particularly like that expression either – but unless we feel that we have a special contribution to make which our very size and diversity and our whole outlook can help to achieve, I for one am not interested. I suppose that you feel the same, and therefore my words to you may seem superfluous; but it is not enough that you feel it, everybody in the firm should as far as possible be made to feel it, and to believe that we, the leaders of the firm really believe in it and mean to work for it and not just use it as a flag to put out on Sundays. And they won't believe that unless we do.

On the other hand, who am I to tell you and the firm what you should think and feel in the future when I am gone – or before that, for that matter. It wouldn't be any good my trying to lay down the law, and I haven't the slightest inclination to do so. That is my difficulty. I dislike hard principles, ideologies and the like. They can do more harm that good, they can lead to wholesale murder, as we have seen. And yet we cannot live life entirely without principles. But they have in some way to be flexible, to be adaptable to changing circumstances. 'Thou shalt not lie', 'Thou shalt not kill', are all very well, generally, but do not apply if for instance you are tortured by fanatical Nazis or Communists to reveal the whereabouts of their innocent victims. Then it is your duty to mislead. What these commandments should define is an attitude. To be truthful always, wherever it does no harm to other ideals more important in the context, to respect the sanctity of human life and not to destroy life wantonly. But where to draw the line in border cases depends on who you are, what life has taught you, how strong you are.

In the following 13 points, which I must have jotted down some time ago – I found them in an old file – I am grappling with this question, perhaps not very successfully. I give them to you now:

Principles

(1) Some people have moral principles.
(2) The essence of moral principles is that they should be 'lived'.
(3) But only saints and fanatics do follow moral principles always.
(4) Which is fortunate.
(5) Are then moral principles no good?
(6) It appears we can't do without them.
(7) It also appears we can't live up to them.
(8) So what?
(9) A practical solution is what I call the star system.
(10) The star – or ideal – indicates the course. Obstacles in the way are circumnavigated but one gets back on the course after the deviation.
(11) The system is adopted by the Catholic church. Sins can be forgiven if repented – it doesn't affect the definition of good or evil.
(12) That this system can be degenerate into permanent deviation is obvious.
(13) One needs a sense of proportion.

Incidentally, they should not be taken as an encouragement to join the Catholic Church!

I found also another tag:

'The way out is not the way round but the way through.' That's rather more uncompromising, more heroic. It springs from a different temperament. It's equally useful in the right place. But the man that bangs his head against a wall may learn a thing or two from the reed that bends in the wind.

The trouble with the last maxim is that it says something about the way, but not about the goal. The way must be adapted to the circumstances – the goal is much more dependent on what sort of person you are. I admit that the last maxim also says a good deal about the man who propounds it, a man of courage, of action, perhaps not given too much to reflection, perhaps not a very wise man. The wise man will consider whether this way is possible, whether it leads to the desired result. Unless of course his goal is to go through, not to arrive anywhere, like the man in the sports car. But this only shows that it is the goal which is important, whatever it is.

The star system is an attempt to soften the rigidity of moral principles. But it doesn't really solve this dilemma. It is a little more flexible than moral precepts as to the way, but surely the 'stars' must be fixed – for if they can be changed ad lib the whole thing wobbles. And that in a way is what it does – I can't do anything about that. I should have loved to present you with a strictly logical build-up, deducing the aims for the firm from unassailable first principles. Or perhaps this is an exaggeration – for I know very well that this can't be done. All I can do is to try to make the members of the firm like the aims I have mentioned. I would like to persuade them that they are good and reasonable and not too impossible aims, possessing an inner cohesion, reinforcing each other by being not only aims but means to each other's fulfilment.

'Stars' like goodness, beauty, justice have been powerful forces in the history of mankind – but they so often are obscured by a mental fog – or perhaps I should say the opposite – they are created by a mental fog, and when the fog lifts, they are seen to have been illusions. They are man-made. I do not rate them less for that reason – but they are too remote, too indefinable, to be of much practical use as guide-lines. They sustain or are born of the longings of mankind, and belong to the ideal world of Plato – which is fixed for ever. Rigid ideologies feed on them. Not so practical politics.

Our aims on the other hand are not nearly so remote. We will never succeed in fulfilling them in toto, but they can be fulfilled more or less, and the more the better. And they are not grasped arbitrarily out of the sky or wilfully imposed, they are natural and obvious and will, I am sure, be recognised as desirable by all of you: so much so, in fact, that the thing to be explained is not why they are desirable, but why I should waste any words on them.

I do, as I pointed out at the beginning of this argument, because our aims are the only thing which holds us together, and because it is not enough to approve them, we must work for them – and the leaders must be prepared to make sacrifices for them. Temporary diversions there must be, we have to make do with the second best if the best is not within reach, we have to accept expediencies and from a strict point of view all our activities can be considered as expediencies, for in theory they could all be better still – but the important thing is that we always get back on the course, that we never lose sight of the aims. Hence the name star system derived from comparison with old fashioned navigation. But I propose to abandon this expression, partly because its meaning in the film industry may confuse, especially as it is very opposed to our point of view, which is in favour of teamwork rather than stardom: and also because it suggests star-gazing, which I find uncomfortably near the bone because I might with some justification be accused of it. So I am afraid we have to fall back on 'philosophy'. Having dabbled in this subject in my youth I have been averse to seeing the term degraded by talk about the philosophy of pile-driving or hair-dressing, but it is of course useless to fight against the tide. The word has come to stay – and in 'the philosophy of the firm', it is not used quite so badly. So that's what I have been giving you a dose of.

I will now discuss what we have to do in order to live up to our philosophy. And I will do it under the four headings 10 to 13 in my list of aims and means:

(10) Quality staff
(11) Efficiency
(12) Solvency
(13) Unity and enthusiasm

but it will of course be necessary to mix them up to some extent.

Quality of Staff
How do we ensure that our staff is of the right quality, or the best possible quality?

We all realise, of course, that there is a key question. The whole success of our venture depends on our staff. But what can we do about it? We have the staff we have – we must make do with them, of course (and I think we have a larger proportion of really good people than any other firm of our kind). And when we take on new people – the choice is limited. Again we have to take the best we can get. We cannot pay them a much higher salary than our average scale, because that would upset our solvency and sink the boat. Naturally our method of selection is important, and what we can do to educate our staff and give them opportunities to develop is important, but I can't go into details here. All I can say is that staff getting and staff 'treating' must not degenerate into a bureaucratic routine matter, but must be on a personal level. When we come across a really good man, grab him, even if we have no immediate use for him, and then see to it that he stays with us.

The last is the really important point, which in the long run will be decisive. Why should a really good man, a man – or woman – who can get a job anywhere or who could possibly start out on his own, why should he or she choose to stay with us? If there is a convincing and positive answer to that, then we are on the right way.

Presumably a good man comes to us in the first instance because he likes the work we do, and shares or is converted to our philosophy. If he doesn't, he is not much good to us anyhow. He is not mainly attracted by the salary we can offer, although that is of course an important point – but by the opportunity to do interesting and rewarding work, where he can use his creative ability, be fully extended, can grow and be given responsibility. If he finds after a while that he is frustrated by red tape or by having someone breathing down his neck, someone for whom he has scant respect, if he has little influence on decisions which affect his work and which he may not agree with, then he will pack up and go. And so he should. It is up to us, therefore, to create an organisation which will allow gifted individuals to unfold. This is not easy, because there appears to be a fundamental contradiction between organisation and freedom. Strong-willed individuals may not take easily to directions from above. But our work is teamwork and teamwork – except possibly in very small teams – needs to be organised, otherwise we have chaos. And the greater the unit, the more it needs to be organised. Most strong men, if they are also wise, will accept that. Somebody must have authority to take decisions, the responsibility of each member must be clearly defined, understood and accepted by all. The authority should also be spread downwards as far as possible, and the whole pattern should be flexible and open to revision.

We know all this, and we have such an organisation: we have both macro, micro and infra-structure. It has been developed, been improved, and it could undoubtedly be improved still further. We are of course trying to do that all the time. The organisation will naturally be related to some sort of hierarchy, which should as far as possible be based on function, and there must be some way of fixing remuneration, for to share the available profit equally between all from senior partner to office-boy would not be reasonable, nor would it work. And all this is very tricky, as you know, because, as soon as money and status come into the picture, greed and envy and intrigue are not far behind. One difficulty is particularly knotty, the question of ownership, which is connected with 'partnership'. There is dissatisfaction amongst some of those who in fact carry out the functions of a partner – dealing with clients, taking decisions binding on the firm, etc. – because they cannot legally call themselves partners but are 'executive' partners – or have some other title. I have discussed this problem in my paper *Aims and Means*. If some viable way could be found to make 100 partners, I wouldn't mind, but I can't think of any.

In the Ove Arup Partnership we have all but eliminated ownership – the senior partners only act as owners during their tenure of office – because someone has to, according to the laws of the country. And I wish that system could be extended to all our partnerships. It no doubt irks some people that the money invested in the firm may one day (with some contriving) fall into the turban of people who have done nothing to earn it – but what can we do? The money is needed for the stability of the firm, it makes it possible for us to earn our living and to work for a good cause, so why worry?

It may be possible to devise a different and better arrangement than the one we have now, more 'democratic', more fair: it may be possible to build in some defences against the leaders misbehaving and developing boss-complexes and pomposity – and forgetting that they are just as much servants in a good cause as everybody else – only more so. This is partly a legal question depending on the laws of the country. But I have neither the ability nor the time to deal with all that here. What I want to stress is the obvious fact that no matter how wonderful an organisation we can devise, its success depends on the people working in it – and for it. And if all our members really and sincerely believed in the aims which I have enumerated, if they felt some enthusiasm for them, the battle would be nearly won. For they imply a humanitarian attitude, respect and consideration for persons, fair dealings, and the rest, which all tend to smooth human relationships. And anyone having the same attitude who comes into an atmosphere like that, is at least more likely to feel at home in it. And if the right kind of people feel at home with us, they will bring in other people of their kind, and this again will attract a good type of client and this will make our work more interesting and rewarding and we will turn out better work, our reputation and influence will grow, and the enthusiasm of our members will grow – it is this enthusiasm which must start the process in the first place.

And they all lived happily ever after?

Yes, it sounds like a fairy tale, and perhaps it is. But there is something in it. It is a kind of vicious circle – except that it isn't vicious, but benevolent, a lucky circle. And I believe that we have made a beginning in getting onto this lucky circle. I believe that our fantastic growth has something to do with our philosophy. And I believe our philosophy is forward looking, that it is what is needed today, is in tune with the new spirit stirring in our time. But of course there are many other and dangerous spirits about and too much growth may awaken them. Too much growth may also mean too little fruit.

My advice would be:

'Stadig over de klipper',
or if you prefer:
Take it easy!
More haste less speed!
Hâtez-vous lentement!
Eile mit weile!
Hastvaerk er lastvaerk!

It's the fruit that matters. I have a lingering doubt about trying to gain a foothold in various exotic places. Might we not say instead: Thank God that we have not been invited to do a job in Timbuctoo – think of all the trouble we are avoiding. It's different with the work we do in Saudi Arabia, Tehran and Kuwait. There we are invited in at the top, working with good architects, doing exciting work. We are not hammering at the door from outside. But as a rule, grab and run jobs are not so useful for our purpose. I think the Overseas Department agrees with this in principle, if not in practice.

It's also different with civil engineering work, provided we have control – complete control – over the design and are not 'sharing' the job or having a quantity surveyor or 'agent', etc., imposed on it preventing us from doing the job our way. The general rule should be: if we can do a job we will be proud of afterwards, well and good – but we will do it our way. In the long run this attitude pays, as it has already done in the case of Arup Associates. And incidentally, the control of such jobs should be where our expertise resides.

To export Arup Associates' jobs is much more difficult, for whilst we may be able to build a bridge or radio tower in a foreign locality, good architecture presupposes a much more intimate knowledge of the country. Long distance architecture generally fails. But that does not mean that the ideal of Total Architecture is irrelevant to our purely engineering partnerships or divisions. In fact they have been founded on the idea of integrating structure with architecture and construction, and in Scotland for instance they are trying to give architects a service which will unite these domains.

Coming back to my main theme, I realise that when I have been talking about quality, about interesting and rewarding work, about Total Architecture, and attracting people of calibre, you may accuse me of leaving reality behind. 'As you said yourself', you may say, 'our work is teamwork. And most of this work is pretty dull. It is designing endless reinforced concrete floors, taking down tedious letters about the missing bolts, changing some details for the nth time, attending site meetings dealing with trivialities, taking messages, making tea – what is exciting about that? You are discriminating in favour of an elite, it's undemocratic. What about the people who have to do the dull work?'

Equality of opportunity

You have certainly a point there. Of course I am discriminating in favour of quality, and I would do anything to enable our bright people to use their talents. You cannot equate excellence with mediocrity, you cannot pretend they are the same. We would be sunk if we did that. We need to produce works of quality, and we need those who can produce them. One perfect job is more important for the morale of the firm, for our reputation for producing enthusiasm, than 10 ordinary jobs, and enthusiasm is like the fire that keeps the steam-engine going. Likewise one outstanding man is worth 10 men who are only half good. This is a fact of life we cannot change. It is no good pretending that all are equal – they aren't. There should be equality before the law, and as far as possible equality of opportunity, of course. But the fact that you are good at something is something you should be grateful for, not something to be conceited about. It doesn't mean that you are better as a human being. And there are probably many other things you are hopeless at.

No man should be despised or feel ashamed because of the work he does, as long as he does it as well as he can. What we should aim at, naturally, is to put each man on to the work he can do. And, fortunately, there is nearly always something he can do well. We will have square pegs in round holes, we shall have frustrated people, unfortunately – those who are not frustrated one way or another are in the minority. But fortunately people vary, as jobs vary, and few would want to do the job another calls interesting if they are no good at it.

If we can reach a stage where each man or woman is respected for the job they do, and is doing his or her best because the atmosphere is right, because they are proud of what we are and do and share in the general enthusiasm, then we are home. And each job is important. Secretaries, for instance. They could have a tremendously civilising influence on our staff. They could teach them to write English, for instance, a most important and necessary job. But secretaries who can do that are of course at a premium. We must try to find them. It is even more important than that they are good-looking – and nobody could accuse me of being indifferent to that.

Our messengers and cleaners – how important it is that they are reliable and likeable, human, with a sense of humour. A cheerful remark can brighten the day. All our people are part of us, part of our 'image', create the atmosphere we live in.

But it doesn't alter the fact that the services of a messenger are less valuable to the firm than those of a gifted designer or an imaginative mechanical engineer, a fact that even the messenger will understand.

But there are of course people we cannot employ usefully. Masses of them, in fact.

Those we should not take on, obviously, except on a strictly temporary basis. But sometimes they are found inside the firm. They may have been good once, but are on the way down. I am a case in point myself. But their loyal service, their place in the hierarchy, makes it difficult to de-grade them. To deal with them requires much tact, and is embarrassing. But they should not be allowed to pretend to do jobs they are no good at. They must not prevent the good ones from functioning.

It's a problem all firms have, it's one of the cases where humanity and efficiency clash. To resolve it tactfully may be expensive, not to resolve it is fatal.

So far I haven't said much about solvency. Stuart Irons can tell you something about that. I compare it to stability in engineering structures – without it the whole thing collapses but if you have much more money than you need the usefulness of it declines until it becomes distracting and dangerous. That danger need not worry us for the time being.

At the moment the need for solvency is restricting, and is the most frequent cause of having to compromise. That we may have to do – but let's not do it unnecessarily, and let's get back on course.

And Unity and Enthusiasm, the last item, is in a way what my talk has been about. It is a question of giving the firm an identity. What do we mean, when we speak about the firm, about 'we' or 'us'? Is it the whole collection of people in dozens of offices in different places? Are 'we' all of them or some of them, and which?

I think it is unavoidable that 'we' should mean different things in different contexts. Sometimes what is said is only relevant to the upper layers of management, sometimes it is meant to include everybody. What we must aim at is to make 'we' include as many as possible as often as possible. To increase the number of those who have a contribution to make, however small, who agree wholeheartedly with our aims and want to throw in their lot with us. We might think about them as members of our community; the others, who come and go, might be called staff. Of course there can never be any clear line of demarcation – it is not a question of signing a form or bestowing a title – it is a matter of how each feels and what we feel about them. For it is a two-way business.

But what binds our membership together must be loyalty to our aims. And only as long as the leaders of the firm are loyal to these can they expect and demand loyalty from the members. This speech is too long already, and I have not even touched on what you perhaps expected to be the main subject of my talk, the relationship between the Ove Arup Partnership and the Overseas Partnerships. But from the foregoing my point of view should be clear.

The fact that we have these outposts all over the world is of course an enormous source of strength to us and to you, it helps to establish our reputation and power for good, and opens up opportunities for all our members. This is however only because the leaders in these places are our own people, bound to us by common aims and friendships. But as the old leaders retire and growth takes place mainly locally, the ties that bind us together may weaken. We should prevent this by forging more ties, forming new friendships, and always being true to our principles. Improve communications – the universal injunction nowadays. Absence does not make the heart grow fonder, unfortunately. There will always be a need for a strong coordinating body – which is at the moment formed by the senior partners – which has the power to interfere if our principles are seriously betrayed. For should that happen, it would be better to cut off the offending limb, less the poison should spread. Our name must not be allowed to cover practices which conflict with our philosophy. But at the moment there is no danger of that, and we can take comfort from what has been achieved. Perhaps that should have been the gist of my talk? But you are seeing it for yourself. I could also have dwelt on how far we have still to go; it would perhaps have accorded more with my star-gazing habits. But my time is up – my speech should have been condensed to one-third – but it is too late now. I hope at any rate that I haven't deserved the warning which the Duke of Albany addressed to Goneril in King Lear:

> *How far your eyes may pierce I cannot tell.*
> *Striving to better, oft we mar what's well.*
> (Act 1, Scene 4, line 346)

This chronology lists matters concerning the firm, which became Ove Arup & Partners and subsequently was placed under the ownership of a parent trust – Ove Arup Partnership. It is intended to describe the development of the firm's technical business interests, for which purpose we have indicated first main activities even if these were not formally named until a later date. We have excluded individual achievements, apart from appointments as Partners in the UK Partnership (when the firm was still a partnership) and principal officers. The UK Partners were also partners in the partnerships outside the UK. We have shown the formation of various partnerships and subsidiaries of the firm and the opening of offices, and the staff numbers, to indicate the expansion of the firm; however, for clarity, we have not shown all the closures or reductions.

Chronology of the firm

1946 Practice founded in London (UK) by Ove Arup on 1 April (the point at which he gave up his directorships in contracting and commercial activities). Dublin (Ireland) office opened.

1949 Ove Arup formed Ove Arup & Partners, with Ronald Jenkins, Geoffrey Wood and Andrew Young as salaried Partners.

1951 Andrew Young resigned as a Partner in the UK Partnership. Office opened in Ibadan (Nigeria).

1952 Office opened in Salisbury (Rhodesia) – now Harare (Zimbabwe).

1953 Office opened in Lagos (Nigeria).

1954 Offices opened in Lusaka (Zambia), Johannesburg (South Africa), and Bulawayo (Rhodesia, now Zimbabwe).

1955 Ove Arup & Partners West Africa formed. Office opened in Kano (Nigeria).

1956 The number in the firm's employment reached 100 permanent members of staff. A proper Partnership agreement was made, which began the "naked in, naked out" theme that eventually there would not be Partners owning the firm. Peter Dunican taken into partnership in the UK Partnership. Office opened in Accra (Ghana).

1957 Ove Arup & Partners South Africa formed. Offices opened in Nairobi (Kenya) and Sheffield (UK).

1958 Durban (South Africa) and Manchester (UK) offices opened.

1959 Ove Arup & Partners Rhodesia (now Zimbabwe) formed. Library set up in London (UK).

1960 Office opened in Edinburgh (UK). Ove Arup & Partners Scotland formed.

1961 Ronald Hobbs taken into partnership in the UK Partnership.

1962 Geotechnics Group (later Arup Geotechnics) established in London (UK).

1963 Ove Arup & Partners Ireland and Ove Arup & Partners Sierra Leone formed. Offices opened in Freetown (Sierra Leone) and Sydney (Australia). Arup Associates established.

1964 Ove Arup & Partners Australia formed. Offices opened in Kuala Lumpur (Malaysia) and Monrovia (Liberia). Computer Group and Research & Development Group (later to incorporate the Mechanical and Electrical Development group) established in London (UK), later known respectively as Arup Computing and Arup R&D.

1965 The number in the firm's employment reached 500 permanent members of staff. Povl Ahm and Jack Zunz taken into partnership in the UK Partnership. Ove Arup & Partners West Africa dissolved. Ove Arup & Partners Nigeria and Ove Arup & Partners Ghana formed. Cape Town (South Africa) and Windhoek (Namibia) offices opened. Civil Engineering (Highways and Bridges) Group – later Civil Engineering Division – established in London (UK).

1966 Ove Arup & Partners Zambia and Ove Arup & Partners Malaysia formed.

1967 UK Partnership name changed to "Ove Arup & Partners Consulting Engineers". Pretoria (South Africa), Glasgow and Newcastle (UK) offices opened. Housing Division established in London (UK).

1968 The UK Partnership established "Executive Partners". Ove Arup & Partners Jamaica formed. Offices opened in Canberra (Australia) and Birmingham (UK). Transportation Planning Group established in London (UK).

1969 Philip Dowson taken into partnership in the UK Partnership. Perth (Australia) office opened.

1970 The number in the firm's employment reached 1000 permanent members of staff. Meeting at Winchester (UK) of "the Arup Organisations" – later THE ARUP PARTNERSHIPS – at which Sir Ove Arup delivered what came to be known as The Key Speech (see pp.262–273 above). Ove Arup Partnership established, with human and two corporate Partners. The existing human Partners of Ove Arup & Partners became Partners in Ove Arup Partnership, which became the Parent Firm of Ove Arup & Partners (which was then incorporated as a company) and Arup Associates. Ove Arup & Partners Singapore formed. Cardiff, Dundee (UK) and Riyadh (Saudi Arabia) offices opened. Building Engineering established in London (UK).

1971 Singapore office opened.

1972 Ronald Jenkins retired as a Partner in the UK Partnership. Offices opened in Waterford (Ireland), Melbourne (Australia) and Penang (Malaysia).

1973 The number in the firm's employment reached 1500 permanent members of staff.

1974 Offices opened in Teheran (Iran) and Bristol (UK). Fire Engineering Group (later Arup Fire) established in London (UK). Offshore Engineering established in London (UK).

1975 Offices opened in Paris (France), Port Moresby (Papua New Guinea), Kota Kinabalu (Sabah), Doha (Qatar) and Port Louis (Mauritius).

1976 Offices opened in Tripoli, Benghazi (Libya), Cork (Ireland), Kuwait and Hong Kong. Mechanical and Electrical Development group established in London (UK), later part of Arup R&D.

1977 Peter Dunican appointed Chairman of the Ove Arup Partnership, with Jack Zunz as Chairman of Ove Arup & Partners. Geoffrey Wood retired as a Partner in the UK Partnership. Ove Arup dan Rakan Rakan Brunei and Ove Arup dan Rakan Rakan Sabah formed. Offices opened in Lae (Papua New Guinea), Abu Dhabi (United Arab Emirates), Gaborone (Botswana, but subsidiary of the South African practice) and Brunei. Industrial Engineering (later Industrial Engineering Division) established in London (UK).

1979 All the former UK Partners resigned as Partners, leaving the Ove Arup Partnership with two corporate Partners (Ove Arup Incorporated and Arup Partners Incorporated (API)). This was the realisation of the theme of "naked in, naked out" and the conclusion of the firm being owned wholly or partly by individuals. Office opened in Aberdeen (UK). Industrial and Offshore Group established in London (UK).

1980 Office opened in Mmabatho(Bophuthatswana). Arup Acoustics established in London.

1981 Offices opened in Cairo (Egypt), Baghdad (Iraq) and Kuching (Malaysia).

1982 The number in the firm's employment reached 2000 permanent members of staff. Offices opened in Brisbane (Australia), Limerick (Ireland) and Warwick (now Coventry, UK).

1984 Jakarta (Indonesia), Rabaul (PNG) and Cambridge (UK) offices opened. Peter Dunican retired as Chairman. Jack Zunz and Ronald Hobbs succeeded as Co-Chairman of Ove Arup Partnership. Povl Ahm became Chairman of Ove Arup & Partners.

1985 San Francisco (USA) and Auckland (New Zealand) offices opened.

1986 Offices opened in Istanbul (Turkey), Los Angeles (USA), Winchester, Horsham, Nottingham and Wrexham (UK).

1987 The number in the firm's employment reached 2500 permanent members of staff. Environmental Services Group (later Arup Environmental) established in London (UK).

1988 Sir Ove Arup died on 5 February 1988, aged 92. Offices opened in New York (USA) and Leeds (UK).

1989 The number in the firm's employment reached 3000 permanent members of staff. The Ove Arup Foundation, a charitable educational trust, established in memory of Ove Arup. Jack Zunz and Ronald Hobbs retired as Co-Chairmen, and were succeeded by Povl Ahm as Chairman of Ove Arup Partnership with John Martin and Duncan Michael as Deputy Chairmen. Arup Façade Engineering established in Sydney (Australia). Office opened in Gold Coast (Australia). Communications & Information Technology Group (later Arup Communications) established in London (UK).

1990 Arup MMLS (Turkey) formed. Manila (Philippines), East London (South Africa), Tokyo (Japan) and Douglas (Isle of Man) offices opened. Project Management Services Group (later Arup Project Management) established in London (UK).

1991 The number in the firm's employment reached 3500 permanent members of staff. Arup GmbH (Germany) established. Office opened in Leipzig (Germany).

1992 On 1 April, Ove Arup Partnership ceased to be a partnership of two companies and changed its constitution to become an unlimited company, owned by a charitable trust (Ove Arup Partnership Charitable Trust), an employees' trust (Ove Arup Partnership Employees' Trust) and a discretionary trust which has the same role as the employees' trust but for a different class of potential beneficiaries. Office opened in Düsseldorf (Germany). Arup Façade Engineering and Controls & Commissioning Groups established in London (UK). Povl Ahm retired, and was succeeded by John Martin, as Chairman of Ove Arup Partnership, and Duncan Michael as Deputy Chairman.

1993 Offices opened in Moscow (Russia), Berlin (Germany), Madrid (Spain) and Copenhagen (Denmark).

1994 Offices opened in Shanghai and Shenzhen (China).

1995 John Martin retired as Chairman of Ove Arup Partnership and was succeeded by Duncan Michael as Chairman, with Bob Emmerson and Mike Shears as Deputy Chairmen.

Index

Entries in italics indicate illustrations

A

B

C

G

H

I

R

S

T

U

V

WXY

Z

Acknowledgments

All photographs are the copyright of Ove Arup & Partners unless otherwise stated. Photographs are identified according to the key:– t.= top, b.= bottom, l.= left, r.= right, c.= centre.
E.g. (b.l.)=(bottom left); (3.t.)= (third from the top)

20, Renzo Piano Building Workshop; 21, Richard Bryant; 24 (t.l., t.r.), RFR; 24(b.r.), Martin Charles; 25(b.r.) 26,27, Bill Timmerman; 30, 31, 34(l.), Richard Davies, Sir Norman Foster & Partners; 35(l.), Ricardo Bofill; 35(r.), 62(t.), Renzo Piano Building Workshop; 63(t.c.), Jo Reid & John Peck; 65(b.r.), Richard Bryant; 70(t., b.), Katsuaki Furudate; 72(t.l., b.l., b.r.), 73(t.l., t.r., b.), Ian Lambot; 77, BP Petroleum Co. plc.; 80, 81, Morten Kuergaard; 96, John Donat; 111, Royal Institute of British Architects; 116(c.), Patrick Eager; 116(r.c.), Martin Charles; 117(b.l.), Kurt Gahler; 117(b.c.), Paul Rafferty; 117(b.r.), RFR; 117,(b.l.), Richard Bryant; 119, NHPA; 122, 123, Builder Group; 128, 129, Martin Charles; 137(t.l.), Derek Phillips, reproduced with the kind permission of the Dean and Chapter of York; 139, Dennis Gilbert; 142(l.), Deutsche Verlags-Anstalt GmbH; 142(r.), Hunterian Art Gallery, University of Glasgow, Mackintosh Collection; 143(r.), Ben Johnson; 144, 145, Richard Davies, Sir Norman Foster & Partners; 151,152, 155, Science Photo Library; 163(b.), Mike Taylor; 164(t.r.), City Repro: The City of Newcastle-upon-Tyne; 167, NMPFT/Science & Society Picture Library; 169, Martin Charles; 170(r.), Richard Davies, Sir Norman Foster & Partners; 171(l.), George Fessey; 175(l.),178, Nikki Photography; 193(t.r.), Martin Charles; 196, 197, Renzo Piano Building Workshop; 201, Farshad Assassi; 226, Kisho Kurokawa & Assoiates; 230(t.), Adtranz; 237(t.), Royal Institute of British Architects; 237(b.), Avery illustrations; 238(t.), *The Architectural Review*; 239(b.), Sydney W. Newburg; 240(t.), *The Architectural Review*; 240(2.t.), Focus Photography; 240(3.t.), Deegan Photo Ltd.; 240(b.), De Burgh Galwey; 241(t.), H. Tempest Ltd.; 241(4.t.), John Maltby Ltd.; 241(b.), Handford Photography; 242(2.t.), De Burgh Galwey; 242(b.), John Laing; 243(t.), W.J. Toomey, *The Architect's Journal*; 243(2.t.), Aerofilms Ltd.; 243(3.t.), John Laing & Son Ltd; 244(t.), H. Tempest Ltd.; 244 (2.t.), John Maltby; 244(3.t.), De Burgh Galwey, Powell & Moya; 245(2.t.), Behr Photography; 245(b.), John Laing & Son Ltd.; 246(3.t.,4.t.), Colin Westwood; 247(t.,2.t.), John Donat; 248(t.), Henk Snoek; 248(2.t.), Sloman & Pettitt; 248(2.t.), Richard Einzig; 249(t.), Kuo Shang-Wei; 253 (2.t.,b.), O.Baitz; 256(2.t.), Philip Cox; 256(4.t.), Patrick Bingham-Hall; 257(2.t.), Tony Knuff; 257(3.t.), Guthrie Photographs; 257(t.), RFR; 258(4.t.), Sir Norman Foster & Partners.

Production
The following have contributed significantly to the development and production of this book:

Editorial team
Bob Emmerson (Project Director), David Dunster, University of Liverpool (Coordinating Editor), Patrick Morreau, Steven Groák.

Design
Thomas Manss and David Law
Birgit Nitsche
Thomas Manss & Company

Illustrations
Fred English

Index
Annette O'Brien and leading Library staff.

Sub-editing
David Brown

Picture research
Pauline Shirley
Rochelle Hardy
Daniel Imade
Elizabeth Morgan

Typesetting
Thomas Manss & Company

Reproductions
NovaConcept, Berlin

Printed by
DruckConcept, Berlin